Stabilization of Hypersonic Boundary-Layer: Acoustic Metasurfaces

Rui Zhao · Guohua Tu · Jun Liang · Chihyung Wen

Stabilization of Hypersonic Boundary-Layer: Acoustic Metasurfaces

北京理工大学出版社
BEIJING INSTITUTE OF TECHNOLOGY PRESS

Springer

Rui Zhao
School of Aerospace Engineering
Beijing Institute of Technology
Beijing, China

Jun Liang
Institute of Advanced Structure Technology
Beijing Institute of Technology
Beijing, China

Guohua Tu
State Key Laboratory of Aerodynamics
China Aerodynamics Research
and Development Center
Mianyang, Sichuan, China

Chihyung Wen
Department of Aeronautical and Aviation
Engineering
The Hong Kong Polytechnic University
Hong Kong, China

ISBN 978-981-97-8622-0 ISBN 978-981-97-8623-7 (eBook)
https://doi.org/10.1007/978-981-97-8623-7

Jointly published with Beijing Institute of Technology Press
The print edition is not for sale in China (Mainland). Customers from China (Mainland) please order the print book from: Beijing Institute of Technology Press.

This Springer imprint is published by the registered company Springer Nature Singapore Pte Ltd.
The registered company address is: 152 Beach Road, #21-01/04 Gateway East, Singapore 189721, Singapore

Contents

Chapter 1
Introduction and Objective

1.1 Background

Hypersonic vehicles have the benefit of rapid global arrival, high detection difficulty, strong penetration ability, and high combat effectiveness, and therefore become one of the main focuses of international competition. A number of hypersonic flight verification projects have been successfully carried out, but there are still many unknown areas. Boundary layer (BL) transition is one of the most important and inevitable uncertainties [1, 2]. Within the flight altitude, speed, and Reynolds number range of hypersonic vehicles, the BL transition has a high probability of occurrence [3]. As shown in Fig. 1.1, vehicles powered by air-breathing propulsion systems spend most of their atmospheric hypersonic flight in conditions where the flow will be transitional. In a higher attitude with faster speed, the BL transition becomes more complex because of the interaction with real-gas effects. Particularly, it might occure under other flight conditions where the flow around the vehicles is normally thought to be laminar unless roughnesses or defects exist.

Figure 1.1 Typical trajectories of various aircraft and spacecraft [3] Once the transition occurs, the skin friction and heat flux can generally increase 3–5 times after transition, which seriously affects aerodynamic performance and thermal protection system, such as causing serious ablation, flutter aggravation, attitude control difficulties and other problems [4–8]. Therefore, the scientific cognition and effective control of BL transition is a key research topic of great interest to the world's aerospace powers. The US Air Force has identified transition prediction and control as one of the essential issues to be addressed in the development of X-43 and X-51 test aircraft [9, 10]. As shown in Fig. 1.2, the medium- and long-term planning (2020–2030) of the National Hypersonic Basic Research Program (NHFRP) of the United States takes the hypersonic BL transition as one of the main contents [11]. NASA's "CFD 2030 Vision" lists transition as the first urgent physical model problem to be solved [12].

R. Zhao et al., *Stabilization of Hypersonic Boundary-Layer: Acoustic Metasurfaces*,
https://doi.org/10.1007/978-981-97-8623-7_1

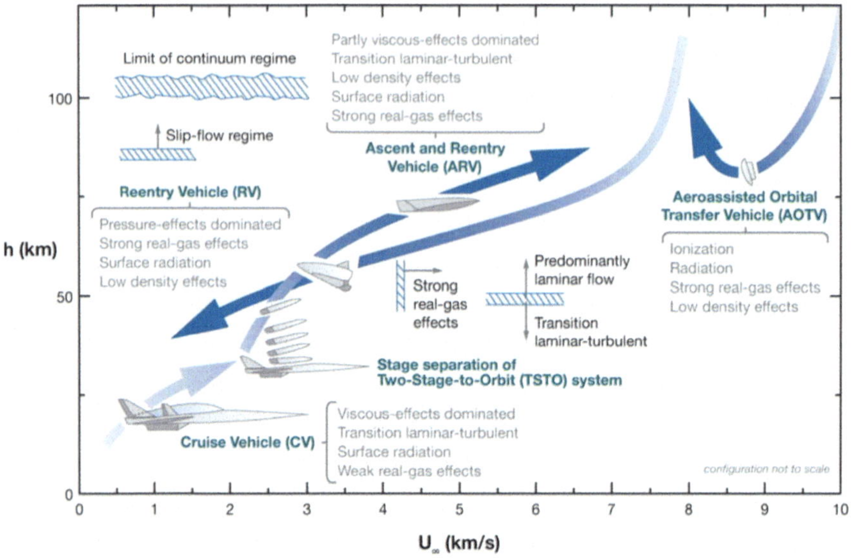

Fig. 1.1 Typical trajectories of various aircraft and spacecraft

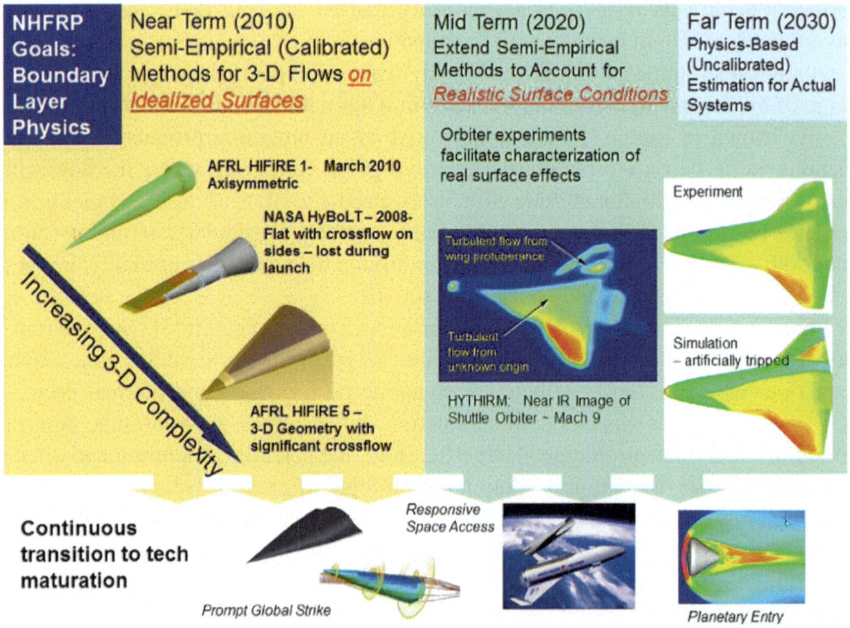

Fig. 1.2 Illustration of the integration of various government research efforts towards achievement of the scientific goals defined in the NHFRP for the Boundary Layer Physics topic of laminar–turbulent transition [11]

Because of the severe heat fluxes and high temperatures around the hypersonic vehicle surface, it renders the many control techniques difficult to implement. Therefore, passive techniques without carrying additional power or mechanical structures are of primary interest. Since the acoustic metasurface concept has been demonstrated with little effect on the mean flow but greatly suppresses the Mack second mode, it could be considered as one of the most promising control technologies. Absorptive porous coating, as a particular type of acoustic metasurface, was issued to Boeing as a patent in 1999 (US patent number 588471). After decades of development, the US Navy advocated a seed fund in 2019, with the aim of designing, fabricating, characterizing, and testing absorptive porous aeroshell materials that successfully damp the Mack second mode instability to delay BL transition on hypersonic boost-glide weapons during pull-up and glide phases [13]. The program includes three phases (Phases I, II, and III). Phase I focuses on determining the desired porosity characteristics that correspond to BL transition delays in flight regimes of interest. Following the success of the Phase I program objectives, the Phase II plan will refine and validate this new technique. The ability of the materials to delay BL transition while remaining thermally and structurally viable for hypersonic flight will be verified using a wind tunnel and arcjet testing. A successful Phase II program will increase the technology readiness level (TRL) from 4 to 6, which will help prepare this technology for a flight test in its operational environment (TRL 7) in Phase III.

1.2 Hypersonic Boundary-Layer Transition

1.2.1 Concept

Laminar-to-turbulent transition generates significant increases in viscous drag and heat flux, leading to severe restrictions on the performance and thermal protection system of hypersonic vehicles. Estimates for the National Aerospace Plane (NASP) show that the payload-to-gross-weight ratio would nearly double if the vehicle boundary layer is fully laminar when compared to the fully turbulent scenario [14]. Moreover, early transition causes higher heating, and this requires an increased performance thermal protection system (TPS), active cooling, or trajectory modification, leading to higher cost and weight of hypersonic vehicles [15]. In order to control boundary-layer transition and maintain the laminar flow as long as possible, extensive work have been focused on the transition mechanisms [16–19]. It has been acknowledged that the transition is a manifold process, which strongly depends on the mean flow and external disturbance. Even for relatively simple two-dimensional or axisymmetric boundary layers on a flat plate or a sharp cone at zero angle of attack, there are several paths for transition [20]. As shown in Fig. 1.3, Path A is the modal growth of BL instabilities, including Mack's first and second modes, the 3D crossflow instability, and the concave-wall Görtler instability. For the high-speed flow over a 2-D flat plate or an axisymmetric body, the dominant instability modes in the boundary layer are Mack's first and second modes. As the external disturbances level increases, laminar to turbulent transition will be triggered through other pathways. Path B is the transient growth of the eigenmodes in the hypersonic BL with higher-level external disturbances. Path C is the transient growth attributed to the non-orthogonal eigenmodes in the HBL. Path D is associated with the internal flows at a considerable turbulence level. Path E is the nonlinear case with such large amplitude disturbances that the linear assumption is invalid. If freestream disturbances are small, transition to turbulence along a smooth vehicle surface occurs due to amplification of the unstable boundary layer mode (path A in [20]) [21, 22]. This path is typical for high-altitude flights in a low disturbance environment. The small environmental disturbances enter the boundary layer and excite the boundary-layer wave modes, through the receptivity process (Fig. 1.4a). Subsequently, the unstable wave modes develop linearly, which could be predicted by stability theory [16] (Fig. 1.4b). Finally, the unstable modes reach certain amplitudes, and non-linear and three-dimensional effects begin to dominate (Fig. 1.4c), leading to the final transition [23]. The current book only considers small amplitude blowing-suction perturbations, which undergo the particular Path A.

In hypersonic boundary layers, modes F and S originate from fast and slow acoustic waves, respectively, in the leading-edge vicinity through the receptivity process. Mode S constitutes both Mack first mode and second mode. The first mode is an inviscid instability wave, associated with the generalized inflection point in the boundary-layer mean profile where $\partial(\rho\partial u/\partial y)/\partial y = 0$. Here ρ and u denote density and streamwise velocity, respectively, and y denotes the axis normal to the wall. This mode is strongly stabilized by wall cooling. On the other side, Mack second mode belongs to the family of trapped acoustic waves (Fig. 1.5). Once the second mode sets in, it becomes the dominant instability with a growth rate that tends to exceed that of the first mode. For insulated surfaces, this occurs for Mach numbers $Ma > 4$. With respect to cooled surfaces, the second mode can dominate at even lower Mach numbers [19, 24]. According to the transition procedure, stabilizations of Mack first mode and second mode are critical to transition control at high Mach number conditions [23].

1.2.2 Control Methods

As documented in Ref. [17], the transition control strategy depends on the flight mission of the hypersonic vehicle. Take X-37 configurations as an example, this kind of reusable launch vehicles usually has a moderate bluntness, and descents through the earths atmosphere at high angles of attack. The windward side endures the harsh heat flux, and the transition control becomes meaningful to the thermal protection system (TPS). Because of the strong bow shock, the local Mack number on the windward boundary layer is relatively low and the second mode might not be found. The radiative cooling provides natural stabilization of the first mode. In this situation, transition is most likely to be forced by roughness such as isolated steps/

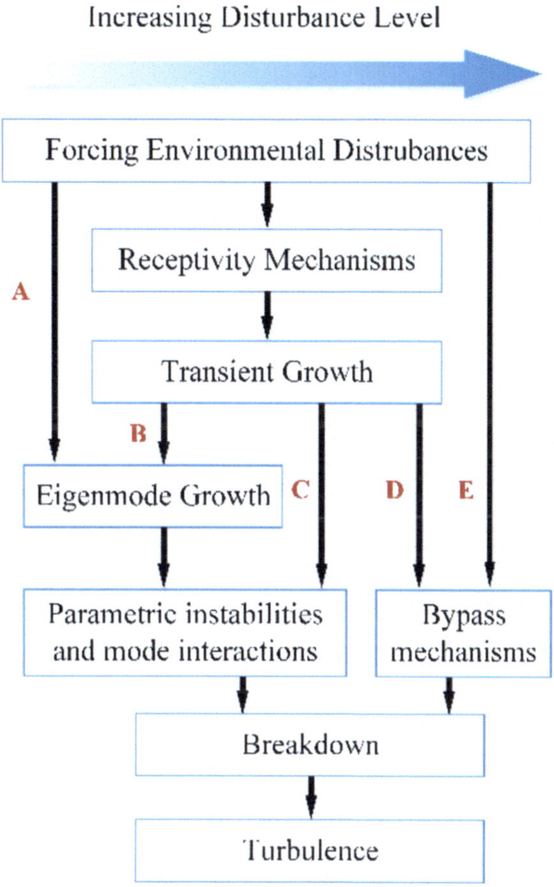

Fig. 1.3 Paths to turbulence in boundary-layer flows. Figure adapted from Morkovin et al. [20]

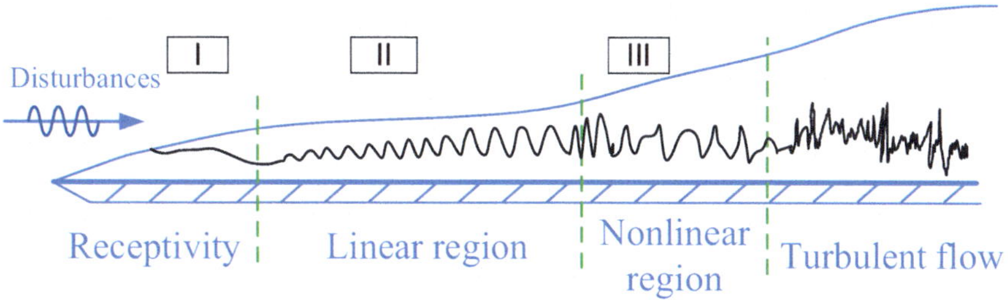

Fig. 1.4 Sketch of the transition process of boundary layer

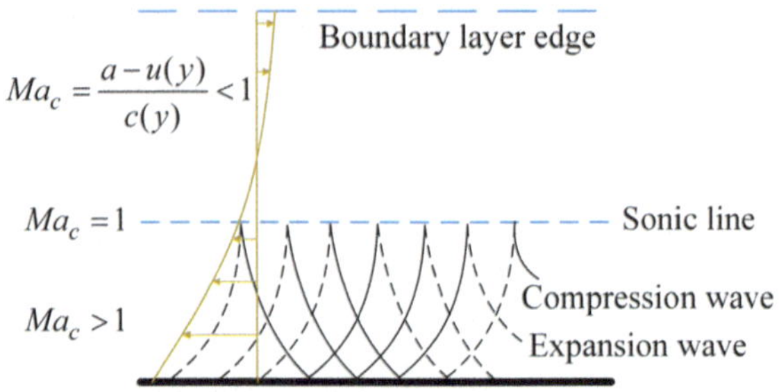

Fig. 1.5 Acoustic modes trapped in the relative supersonic region where the convective Mach number Mac > 1, stationary in this frame of reference, adapted from Morkovin [25]. Here, $u(y)$ and $c(y)$ denote the local speed of flow and sound, respectively, and a denotes the disturbance phase speed

(a) X-37b Configuration (b) Capsule Configuration (c) Cruise Configuration

Fig. 1.6 Typical hypersonic vehicles

gaps between tiles of TPS (Fig. 1.6a). For blunt capsules configurations (Fig. 1.6b), the surface ablation causes the degradation, roughness and pyrolysis gas injection, and that would greatly promote transition. The transition control strategy for these two types of hypersonic vehicles should focus on the shaping and development of new TPS materials to diminish the body-generated forcing and suppress the bypass transition.

Hypersonic cruise vehicles flying at low angles of attack and high lift-to-drag ratios, such as the X-43A (Fig. 1.6c), are normally designed with predominantly two-dimensional (2D) shapes and relatively sharp leading edges. Because of small oblique shock angles, the post-shock Mach numbers could maintain high enough to excite the Mack second mode. This suggests the transition delay could be realized by stablization of the second mode.

Kimmel [26] identified the following categories of transition control techniques that can suppress or attenuate the Mack second mode: (1) passives, such as global or local shaping [27, 28], acoustic metasurfaces, such as **porous coatings** [15], and (2) actives, such as retuned blow-suction [29], CO_2 injection [30, 31], and plasma actuators [32]. A comprehensive survey of different types of transition control techniques has been reported in a recent review by Fedorov [33]. Because of the severe heat fluxes and high temperatures around the hypersonic vehicle surface, it renders the active control techniques difficult to implement. Therefore, passive techniques are of primary interest. Since the acoustic metasurface concept has been demonstrated with little effect on the mean flow but greatly suppresses the Mack second mode, it could be considered as one of the most promising control technologies.

1.3 Acoustic Metasurfaces

1.3.1 Concept

Wave manipulation using artificial materials is a central topic in materials physics. The design of a material has an important role in determining the characteristics of waves inside and outside the material and in achieving such manipulation. Since 2011, a family of thin, 2D artificial materials, namely, metasurfaces, has emerged. The concept of metasurfaces was introduced to the materials science and physics communities for optical waves via the generalized Snell's law, which opened up opportunities for optical wave manipulation and led to new applications [34, 35]. Inspired by this pioneering work, and because acoustic waves also follow Snell's law, acoustic metasurfaces have emerged to control sound waves. Acoustic metasurfaces are planar metamaterial structures constructed with monolayer or multilayer stacks of subwavelength building blocks, which have significantly broadened the horizon of acoustic wave manipulation from wave-front modulation [36–41] (Fig. 1.7), to sound insulation, and absorption [42–44] (Fig. 1.8). The recent emergence of the acoustic metasurface concept has provided more possibilities for controlling the propagation of the Mack second mode.

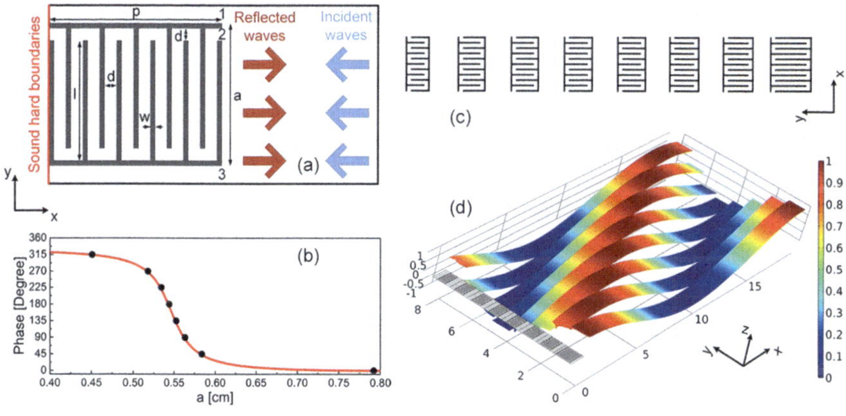

Fig. 1.7 An acoustic metasurface for generalized Snell's law [36]

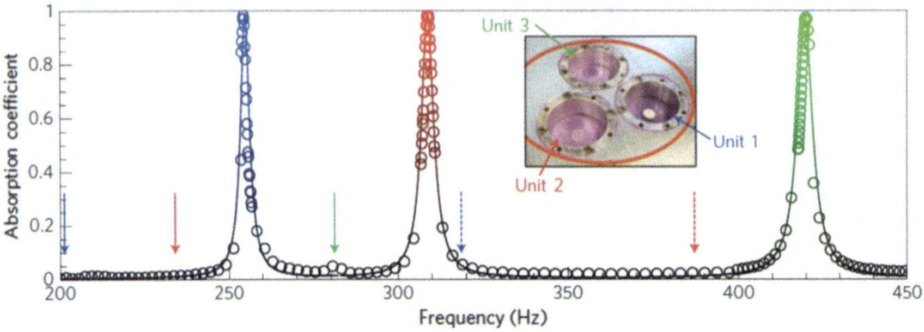

Fig. 1.8 Unity (>0.99) absorption attained at tunable multiple frequencies [42]

1.3.2 Classification

(1) *Absorptive acoustic metasurface*

Compared to the definition of the acoustic metasurface, the aforementioned porous coating (of thickness ~ 0.1 of the wavelength of the Mack second mode) belongs to the category of an absorptive metasurface. It was first focused on a coating with regular micropores [45]. Fedorov et al. [15] deduced the surface impedance boundary condition to model the acoustic effect of realistic microstructures and applied the LST to investigate the stabilization effect of various inflow conditions and pore shape parameters (Fig. 1.9). Further, their theoretical finding was verified in the GALCIT T-5 shock tunnel [46] with a 5° half-angle sharp cone at a free-stream Mach number that ranges from 4.59 to 6.4. The BL on the half cone covered by the porous surface was mostly laminar, whereas a transition was observed halfway along the other half cone with an untreated solid surface. Given the motivation for a practical coating to be symbiotic with the actual materials of TPS, Fedorov et al. [47] experimentally and theoretically studied the effect of a felt-metal coating with random microstructures on hypersonic BL instabilities (Fig. 1.10); they showed that this type of porous coating strongly stabilized the second mode and marginally destabilized the first mode. Based on the promising results on porous surfaces with regular and random microstructures and the fact that a majority of TPS materials have random microstructures, the German Aerospace Center (DLR) pioneered the use of carbon–carbon (C/C) (Fig. 1.11), an intermediate state of C/C-SiC already employed in hypersonic vehicles [48], to control the Mack second mode waves. Wind tunnel experiments at Mach 7.5 were conducted; the stabilization of the second mode and an increase in the laminar portion of the BL over the porous surface were observed [49–51]. Wagner et al. [51] optimized the manufacturing process of C/C-SiC to increase its porosity and enhance the stabilization effect on the hypersonic BL.

Based on the abundant investigations of various absorptive acoustic metasurfaces, it is reasonable to suspect that most of porous materials with characteristic pore diameter much less than 1 mm could stabilize the Mack second mode and delay the hypersonic BL transition. The first mode should be considered when designing the

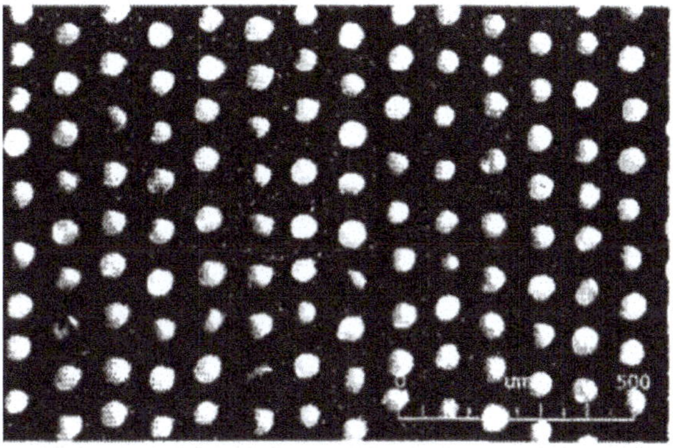

Fig. 1.9 Microphotograph of porous surface (Porosity $\phi = 0.09$, pore diameter $d = 0.06$ mm, and pore depth $H = 0.5$ mm) [15]

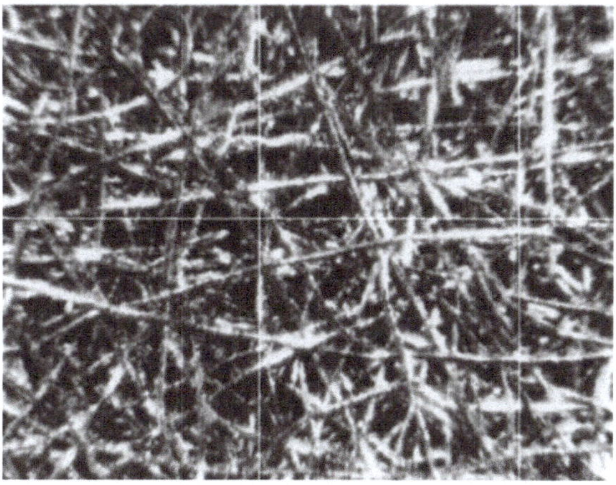

Fig. 1.10 Microphotograph of felt-metal coating ($\phi = 0.75$, averaged $d = 0.1$ mm, and averaged $H = 0.75$ mm) [47]

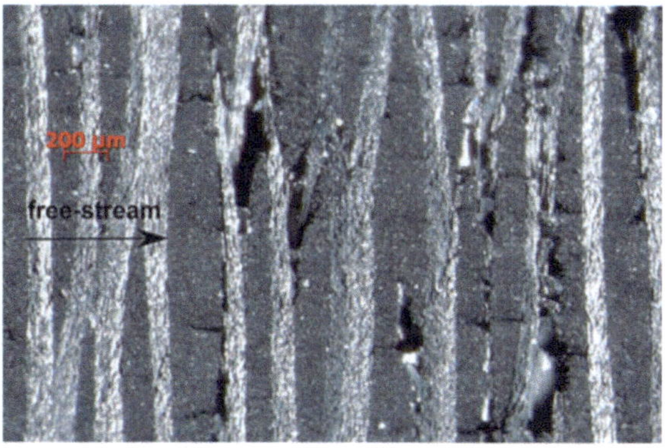

Fig. 1.11 Microphotograph of the C/C surface ($\phi \approx 0.15$, dominant $d = 0.028$–0.03 mm, and dominant $H = 5$–17 mm) [49]

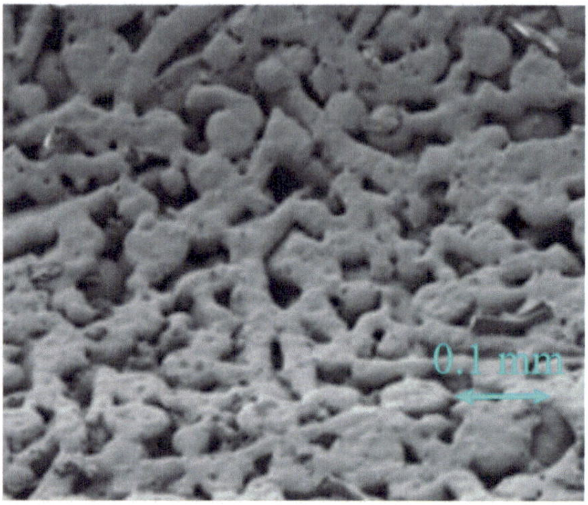

Fig. 1.12 Microphotograph of porous steel (Air volume ratio = 0.2–0.3 and average aggregate radius ≈ 0.14 mm) [55]

optimal diameter, because the influence on the first mode can grow if the diameter is selected too large. Recently, Zhu et al. [52–54] investigated the stabilization effect of the permeable steel (Fig. 1.12) on the second mode in a Mach 6 BL flow. It showed this permeable material could effectively weaken the nonlinear interaction and delay the transition, although the second mode grew faster on the material surface. They also found the permeable surface could reduce the pressure-dilatation-induced heat flux by approximately 28%, through the modification of surface sound admittance [55].

(2) *Impedance-near-zero acoustic metasurface*

Matasurfaces could depict specific acoustic characteristics by carefully designing the microstructures. Zhao et al [56] designed an impedance-near-zero metasurface and showed that even with little damping, the Mack second mode can be greatly suppressed by introducing the near-zero surface acoustic impedance. Figure 1.13 shows the acoustic characteristics of the designed impedance-near-zero metasurface without considering the fluid field. The impedance-near-zero metasurface was designed as a rigid surface periodically corrugated with subwavelength slits (lower map in Fig. 1.13), which resemble the traditional absorptive porous coating [57, 58] (middle map in Fig. 1.13 but had comparatively shallower cavities. For the impedance-near-zero metasurface, the absolute acoustic pressure field generated by a near-surface point source is extremely weak at the interface $y = 0$, as a result of the out-of-phase behavior between the incident and reflected waves. Sound waves are localized within the central cavity and hardly transmit to neighboring ones, corresponding to a band gap of the surface acoustic wave. This is in stark contrast to the sound field of the absorptive metasurface, wherein most incident waves are absorbed compared with the case of a rigid surface (upper map in Fig. 1.13), but the acoustic pressure oscillation at the interface remains strong. When the hypersonic BL flow passes the metasurfaces (Fig. 1.14), the typical two-cell Mack second mode structures

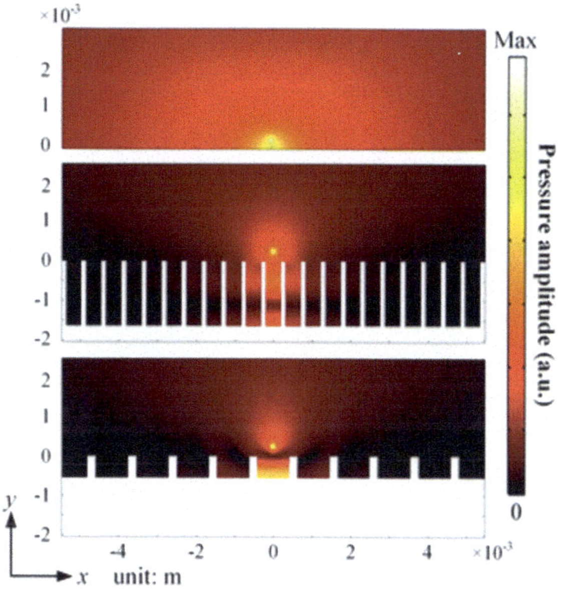

Fig. 1.13 Simulated absolute acoustic pressure fields for different types of surfaces in the absence of hypersonic fluid flow. Upper: rigid surface; middle: absorptive metasurface with matched impedance; and lower: impedance-near-zero metasurface [56]

at the rigid surface change differently for the two types of metasurfaces. For both types of metasurfaces, the upper cells join the lower part when moving through the metasurface region. The intensity is considerably weakened in the cavities because of the dissipation effect of the absorptive one (middle map in Fig. 1.14), whereas the minimum intensity is observed at the surface because of the phase opposition effect for the impedance-near-zero one (lower map in Fig. 1.14). The impedance-near-zero metasurface divides the mode structures in the cavity and above the surface (lower map in Fig. 1.14); this inhibits the amplification process of the Mack second mode. In addition, the shallow-cavity design is advantageous for manufacturing and maintenance.

(3) *Reflection-controlled acoustic metasurface*

In addition to the impedance-near-zero metasurface, Zhao et al [59] proposed a reflection-controlled metasurface for suppressing the Mack second mode by controlling the reflection directions of disturbance waves. The idea of reflection direction control is based on the generalized Snell's law [60, 61], i.e., the relationship between the reflection and incident angles in terms of the gradient of the reflection phase along the wall.

$$\sin\theta_r - \sin\theta_i = \frac{\lambda}{2\pi}\frac{\mathrm{d}\Phi(x)}{\mathrm{d}x} \tag{1.1}$$

where θ_r is the reflection angle, θ_i is the incident angle, λ is the wavelength of the incident wave, and $\Phi(x)$ is the reflection phase, which is equal to the reflection coefficient phase when the phase of the incident wave is zero. For second-mode

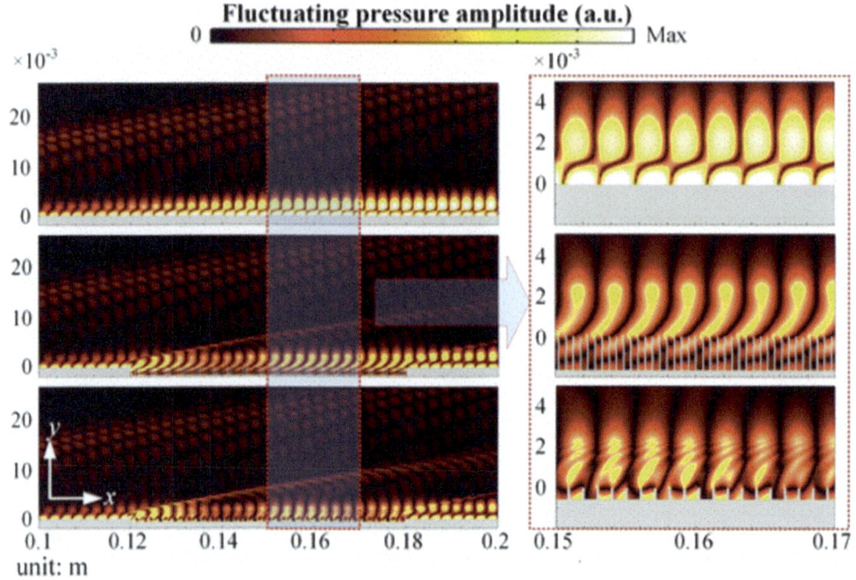

Fig. 1.14 Fluctuating pressure amplitude contours when a Mach 6 flow passes the metasurface. Upper: rigid surface, middle: absorptive metasurface, lower: impedance-near-zero metasurface (lower), with the corresponding zoom-in views shown in the right column [59]

instability waves, a normal incidence can be hypothesized (i.e., $\theta_i = 0$) [60], and therefore, the required θ_{r} can be achieved by designing the metasurface to produce a constant gradient $\Phi(x)$ along the streamwise direction. Thus, the reflection-controlled metasurfaces comprises periodic slit groups (Fig. 1.15); each group has N slits with the same half-width b and unit-cell period s but different depths H_i (i = 1–N). The reflection phase $\Phi(x)$ can be designed to change linearly from 0 to 2π in each group because of the varied H_i values.

Zhao et al [59] subsequently designed three metasurfaces with different reflection angles, namely $\theta_r = \pm 90°$ and $\theta_r = -45°$ at the same disturbance frequency 138.74 kHz. As can be seen in Fig. 1.16, the DNS results reveal that the typical "two-cell" structure is divided into several segments when moving through the designed metasurface, and this change is thought to break down the Mack second mode propagation pattern in the BL. It also found that the most effective way to suppress disturbance amplification is to tune reflection wave propagation downstream in the direction parallel to the surface ($\theta_r = 90°$) in Fig. 1.17. However, there are too few reflection angles tested in their study to conclude $\theta_r = 90°$ is the optimum angle to stabilize the Mack second mode. Similar to the impedance-near-zero acoustic metasurface, the reflection-controlled ones are designed at a specific disturbance frequency, and that means the stabilization effect would be weakened at other frequencies (Fig. 1.17).

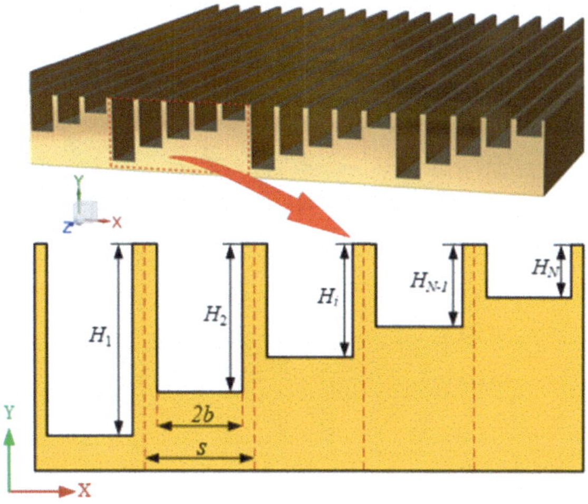

Fig. 1.15 Schematic of reflection-controlled metasurfaces with subwavelength slits [59]

Besides the long-proposed absorptive acoustic metasurface, the recent artificial impedance-near-zero and reflection-controlled metasurfaces provide two new alternatives for suppressing the Mack second mode. Further, it indicates that the type of porous surfaces with random microstructures such as felt-metal and C/C composite present not only the absorptive but also other acoustic characteristics when applied to stabilize the hypersonic BL. The manufacturing process can be further optimized to achieve a better stabilization effect. Indeed, the inherent abundant wave manipulation choices through interdisciplinary investigations of metasurfaces and BL flow can benefit the full control of hypersonic BL transition in the future.

1.4 Objective

Hypersonic boundary layer (BL) transition generates a significant increase in viscous drag and heat flux, which leads to severe restrictions on the performance and thermal protection systems of hypersonic vehicles. Among various passive/active transition control strategies, acoustic metasurfaces demonstrate minimal effects on the mean flow but significantly suppress the Mack second mode. Therefore, it can be considered one of the most promising transition control technologies.

Acoustic metasurfaces are planar metamaterial structures that comprise monolayer or multilayer stacks of subwavelength microstructures, which affect unstable modes via acoustic wave manipulations. This book first presents a review of the research progress made on acoustic metasurfaces for hypersonic BL stabilization over the past two decades, and then outlines the mathematical modeling of acoustic metasurfaces. The design strategies of the acoustic metasurface are proposed in detail, and the possible stabilization mechanisms are explored. In the end, other stabilization methods are also included for reference.

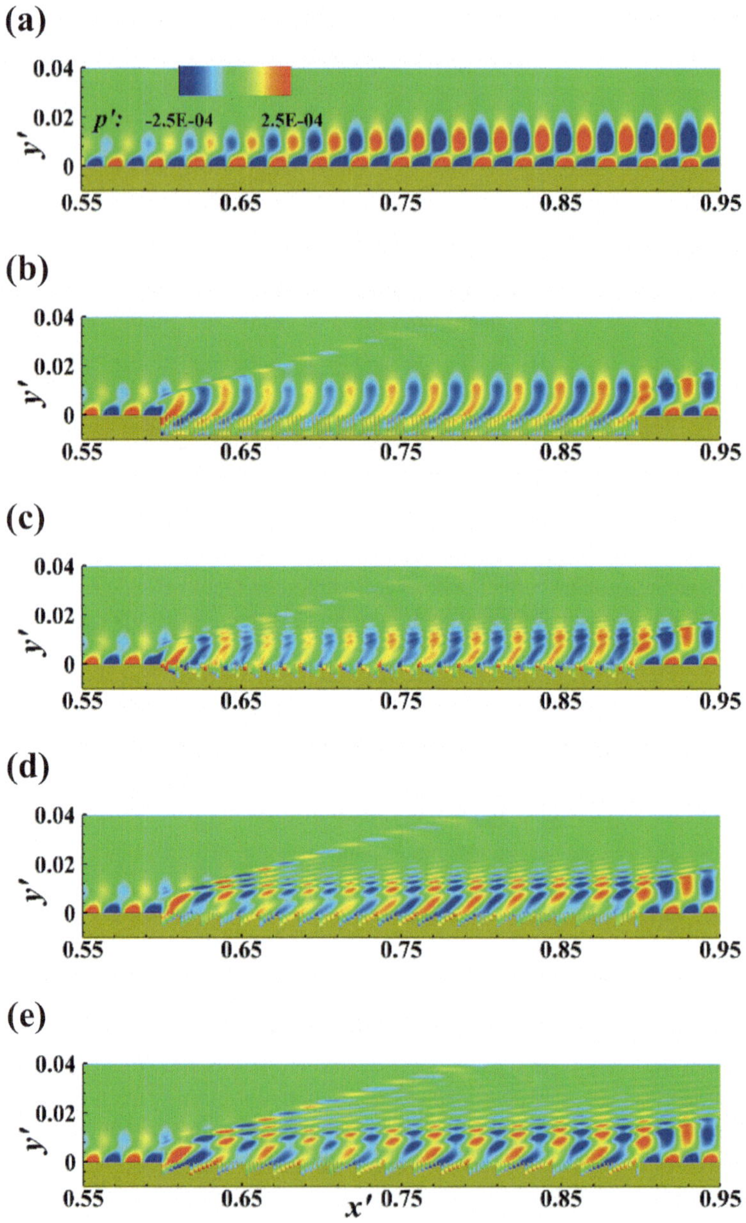

Fig. 1.16 Fluctuating pressure contours for: **a** rigid surface, **b** optimized absorptive metasurface, and three reflection-controlled surfaces: **c** Case 1: $\theta_r = 90^\circ$, **d** Case 2: $\theta_r = -90^\circ$, and **e** Case 3: $\theta_r = -45^\circ$

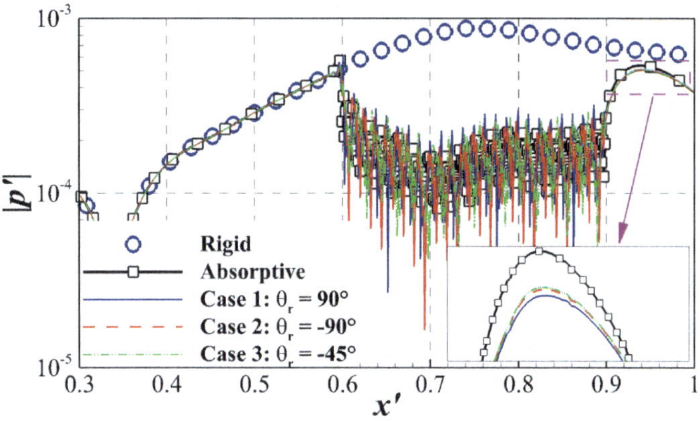

Fig. 1.17 Comparison of fluctuating pressure amplitudes along the wall

References

1. Bertin JJ, Cummings RM (2003) Fifty years of hypersonics: where we've been, where we're going. Prog Aerosp Sci 39:511–536
2. Fuxing A, Lei L, Wei S, Wenling L, Chao D (2021) Key issues in hypersonic vehicle aerodynamic design. SCIENTIA SINICA Physica, Mechanica Astronomica 51(10):6–25 (in Chinese)
3. Bertin JJ, Cummings RM (2006) Critical Hypersonic Aerothermodynamic Phenomena. Annu Rev Fluid Mech 38:129–157. https://doi.org/10.1146/annurev.fluid.38.050304.092041
4. Yu P, Duan Y, Chen J (2015) Some Aerodynamic issues in hypersonic flight. Acta Aeronautica Sinica 36(1):7–23 (in Chinese)
5. Jianqiang C, Guohua T, Yifneg Z, Guoliang XU, Xianxu YU, Cheng CH (2017) Hypersonic Boundary Layer Transition: What We Know, Where shall we go. Acta Aerodynamica Sinica 35(03):311–337 (in Chinese)
6. Wubing Y, Qing S, Dehuaet Z, Min YU, Zhiyong LI (2018) Tendency and current status of hypersonic boundary layer transition. Acta Aerodynamica Sinica 36(02):183–195 (in Chinese)
7. Zhiwen Li, Haitao Y, Bin H et al (2021) The Hypersonic boundary layer transition: a perspective from the view of system design. Acta Aerodynamica Sinica 39(04):26–38 (in Chinese)
8. Qiang L, Guohua T, Zhenbin L (2022) Progress in hypersonic boundary layer transition delay contral. Acta Aerodynamica Sinica 2021:1–15 (in Chinese)
9. Berry S, Daryabeigi K, Wurster K, Bittner R (2010) Boundary-layer transition on X-43A. J Spacecr Rocket 47(6):922–934
10. Joseph MH, James SM, Richard CM (2008) The X-51A scramjet engine flight demonstration program. In: AIAA Paper 2008, p 2540
11. Schmisseur JD (2015) Hypersonics into the 21st century: a perspective on AFOSR-sponsored research in aerothermodynamics. Prog Aerosp Sci 72:3–16
12. Slotnick J, Khodadoust A, Alonso J, Darmofal D, Gropp W, Lurie E, Mavriplis D (2014) CFD vision 2030 study: a path to revolutionary computational aerosciences. NASA/CR-2014, p 218178
13. America's Seed Fund (2022) Development of ultrasonically absorptive aeroshell materials for hypersonic boundary layer transition (BLT) delay. Accessed 24 Jan 2022. https://www.sbir.gov/node/1532091
14. Whitehead A (1989) NASP aerodynamics. In: AIAA paper. AIAA-89, p 5013
15. Fedorov AV, Malmuth ND, Rasheed AR, Hornung HG Stabilization of hypersonic boundary layers by porous coatings. In: AIAA paper. AIAA-2001, p 0891

16. Mack LM (1969) Boundary-layer stability theory. Jet Propulsion Laboratory, California Institute of Technology, Pasadena, California
17. Reshotko E (1976) Boundary layer stability and transition. Annu Rev Fluid Mech 8:311–349
18. Reshotko E (1991) Hypersonic stability and transition. In: Desideri JA, Glowinski R, Periaux J (eds) Hypersonic flows for reentry problems, vol 1. Springer, Berlin, pp 18–34
19. Fedorov AV (2011) Transition and stability of high-speed boundary layers. Annu Rev Fluid Mech 43:79–95
20. Morkovin MV, Reshotko E, Herbert T (1994) Transition in open flow systems: a reassessment. Bull APS 39(9):1–31
21. Malik MR, Zang TA, Bushnell DM (1990) Boundary layer transition in hypersonic flows. In: AIAA Paper. AIAA-90, p 5232
22. Fedorov AV (2015) Prediction and control of laminar-turbulent transition in high-speed boundary-layer flows. Procedia IUTAM 14:3–14
23. Wang XW, Zhong XL (2012) The stabilization of a hypersonic boundary layer using local sections of porous coating. Phys Fluids 24:034105
24. Alba CR, Casper KM, Beresh SJ, Schneider S (2010) Comparison of experimentally measured and computed second-mode disturbances in hypersonic boundary layers. In: AIAA paper. AIAA-2010, p 897
25. Morkovin MV (1987) Transition at hypersonic speeds. NASA Contract. Report No 178315
26. Kimmel RL (2003) Aspects of hypersonic boundary layer transition control. In: AIAA Paper 2003, p 772. https://doi.org/10.2514/6.2003-772
27. Malik MR (1989) Prediction and control of transition in supersonic and hypersonic boundary layers. AIAA J 27(11):1487–1493. https://doi.org/10.2514/3.10292
28. Zhong X, Wang X (2012) Direct numerical simulation on the receptivity, instability, and transition of hypersonic boundary layers. Annu Rev Fluid Mech 44:527–561. https://doi.org/10.1146/annurev-fluid-120710-101208
29. Thomas ASW (1983) The control of boundary-layer transition using a wave-superposition principle. J Fluid Mech 137:233–250. https://doi.org/10.1017/s0022112083002384
30. Leyva IA, Jewell JS, Laurence S, Shepherd J (2009) On the impact of injection schemes on transition in hypersonic boundary layers. In: AIAA Paper 2009, p 7204. https://doi.org/10.2514/6.2009-7204
31. Fedorov AV, Soudakov VG, Leyva IA (2014) Stability analysis of high-speed boundary-layer flow with gas injection. In: AIAA Paper 2014, p 2498. https://doi.org/10.2514/6.2014-2498
32. Riherd M, Roy S, Balachandar S (2014) Local stability effects of plasma actuation on a zero pressure gradient boundary layer. Theor. Comput. Fluid Dyn. 28:65–87. https://doi.org/10.1007/s00162-013-0302-5
33. Fedorov AV (2015) Prediction and control of laminar-turbulent transition in high-speed boundary-layer flows. Procedia IUTAM. 14:3–14. https://doi.org/10.1016/j.piutam.2015.03.017
34. Assouar B, Liang B, Wu Y, Li Y, Cheng JC, Jing Y (2018) Acoustic metasurfaces. Nat Rev Mater 3(12):460–472. https://doi.org/10.1038/s41578-018-0061-4
35. Yu N et al (2011) Light propagation with phase discontinuities: generalized laws of reflection and refraction. Science 334:333–337
36. Li Y, Liang B, Gu ZM, Zou XY, Cheng JC (2013) Reflected wavefront manipulation based on ultrathin planar acoustic metasurfaces. Sci Rep 3:2546. https://doi.org/10.1038/srep02546
37. Zhao J, Li B, Chen Z, Qiu CW (2013) Manipulating acoustic wavefront by inhomogeneous impedance and steerable extraordinary reflection. Sci Rep 3:2537. https://doi.org/10.1038/srep02537
38. Xie Y, Wang W, Chen H, Konneker A, Popa BI, Cummer SA (2014) Wavefront modulation and subwavelength diffractive acoustics with an acoustic metasurface. Nat Commun 5:6553. https://doi.org/10.1038/ncomms6553
39. Tang K, Qiu C, Ke M, Lu J, Ye Y, Liu Z (2014) Anomalous refraction of airborne sound through ultrathin metasurfaces. Sci Rep 4:06517. https://doi.org/10.1038/srep06517

40. Li Y, Jiang X, Li RQ, Liang B, Zou XY, Yin LL, Cheng JC (2014) Experimental realization of full control of reflected waves with subwavelength acoustic metasurfaces. Phys Rev Appl 2:064002. https://doi.org/10.1103/physrevapplied.2.064002
41. Li Y, Jiang X, Liang B, Cheng JC, Zhang LK (2015) Metascreen-based acoustic passive phased array. Phys Rev Appl 4:024003. https://doi.org/10.1103/physrevapplied.4.024003
42. Ma G, Yang M, Xiao S, Yang Z, Sheng P (2014) Acoustic metasurface with hybrid resonances. Nat Mater 13:873–878. https://doi.org/10.1038/nmat3994
43. Cheng Y, Zhou C, Yuan BG, Wu DJ, Wei Q, Liu XJ (2015) Ultra-sparse metasurface for high reflection of low- frequency sound based on artificial Mie resonances. Nat Mater 14:1013–1019. https://doi.org/10.1038/nmat4393
44. Li Y, Assouar BM (2016) Acoustic metasurface-based perfect absorber with deep subwavelength thickness. Appl Phys Lett 108:063502. https://doi.org/10.1063/1.4941338
45. Malmuth ND, Fedorov AV, Shalaev V, Cole J, Khokhlov A (1998) Problems in high speed flow prediction relevant to control. In: AIAA Paper 98, p 2695. https://doi.org/10.2514/6.1998-2695
46. Rasheed A, Hornung HG, Fedorov AV, Malmuth ND (2002) Experiments on passive hypervelocity boundary-layer control using an ultrasonically absorptive surface. AIAA J 40:481–489. https://doi.org/10.2514/2.1671
47. Fedorov AV, Shiplyuk AN, Maslov AA, Burov E, Malmuth ND (2003) Stabilization of a hypersonic boundary layer using an ultrasonically absorptive coating. J Fluid Mech 479:99–124. https://doi.org/10.1017/S0022112002003440
48. Turner J, Horschgen M, Jung W, Stamminger A, Turner P (2016) SHEFEX—hypersonic re-entry flight experiment—vehicle and subsystem design, flight performance and prospects. In: AIAA Paper 2006, p 8115. https://doi.org/10.2514/6.2006-8115
49. Wagner A, Kuhn M, Martinez Schramm J, Hannemann K (2013) Experiments on passive hypersonic boundary layer control using ultrasonically absorptive carbon-carbon material with random microstructure. Exp Fluids 54:1606. https://doi.org/10.1007/s00348-013-1606-3
50. Wagner A (2014) Passive hypersonic boundary layer transition control using ultrasonically absorptive carbon–carbon ceramic with random microstructure. Ph.D. thesis, Katholieke Universiteit Leuven, Belgium, vol 31, pp 1–146. https://doi.org/10.13140/RG.2.1.4246.8968
51. Wagner A, Wartemann V, Hannemann K, Kuhn M, Dittert C (2015) The potential of ultrasonically absorptive TPS materials for hypersonic vehicles. In: AIAA Paper 2015, p 3576. https://doi.org/10.2514/6.2015-3576
52. Zhu W, Shi M, Zhu Y, Lee C (2020) Experimental study of hypersonic boundary layer transition on a permeable wall of a flared cone. Phys Fluids 32:011701. https://doi.org/10.1063/1.5139546
53. Zhu W, Shi M, Lee C (2020) Experimental and theoretical study of the hypersonic boundary layer transition on the permeable wall of a flared cone. In: AIAA Paper 2020, p 2964. https://doi.org/10.2514/6.2020-2964
54. Zhu W, Chen X, Zhu Y, Lee C (2020) Nonlinear interactions in the hypersonic boundary layer on the permeable wall. Phys Fluids 32:104110. https://doi.org/10.1063/5.0028698
55. Zhu Y, Gu D, Zhu W, Chen S, Lee C, Oran ES (2021) Dilatational-wave-induced aerodynamic cooling in transitional hypersonic boundary layers. J Fluid Mech 911:1–21. https://doi.org/10.1017/jfm.2020.1044
56. Zhao R, Liu T, Wen CY, Zhu J, Cheng L (2019) Impedance-near-zero acoustic metasurface for hypersonic boundary-layer flow stabilization. Phys Rev Appl 11:044015. https://doi.org/10.1103/PhysRevApplied.11.044015
57. Brès GA, Colonius T, Fedorov AV (2010) Acoustic properties of porous coatings for hypersonic boundary-layer control. AIAA J 48:267–274. https://doi.org/10.2514/1.40811
58. Brès GA, Inkman M, Colonius T, Fedorov AV (2013) Second-mode attenuation and cancellation by porous coatings in a high-speed boundary layer. J Fluid Mech 726:312–337. https://doi.org/10.1017/jfm.2013.206
59. Zhao R, Dong Y, Zhang X, Wen C, Long T, Yuan W (2021) Control of reflected waves with acoustic metasurfaces for hypersonic boundary-layer stabilization. AIAA J 59:1533. https://doi.org/10.2514/1.J060282

60. Yu N, Genevet P, Kats MA, Aieta F, Tetienne J-P, Capasso F, Gaburro Z (2011) Light propagation with phase discontinuities: generalized laws of reflection and refraction. Science 334:333–337. https://doi.org/10.1126/science.1210713
61. Asadchy VS, Albooyeh M, Tcvetkova SN, Diaz-Rubio A, Ra'di Y, Tretyakov SA (2016) Perfect control of reflection and refraction using spatially dispersive metasurfaces. Phys Rev B 94:075142. https://doi.org/10.1103/PhysRevB.94.075142

Chapter 2
Hypersonic Boundary-Layer Stabilization

2.1 Introduction

Various instability modes can exist in the hypersonic boundary layer depending on the flow condition. Mack [1] first categorized the flow disturbances into the first mode and the second mode. In hypersonic boundary layers, the second mode plays an important role, and it behaves like a trapped acoustic wave propagating in a waveguide between the wall and the sonic line, but has a phase velocity larger than the speed of sound [2]. Gushchin and Fedorov [3] showed that the second mode instability occurs in a region where the two modes of the continuous spectrum are synchronized leading to the branching of the discrete spectrum, and these discrete modes are categorized as the fast mode (Mode F) and the slow mode (Mode S) by Fedorov [4] based on their asymptotic behavior near the leading edge. The branching pattern of the discrete spectrum depends on the Mach number at a given Reynolds number [5], and the first mode and second mode are related to Mode F and Mode S. When the phase speed of Mode F and Mode S are synchronized, the resonance between them triggers the unstable second mode [4, 6]. Since the wall temperature of the hypersonic vehicles is generally lower than the total temperature under flight conditions, the first mode disturbance is suppressed, while the second mode is promoted instead. Therefore, under hypersonic conditions, how to effectively suppress the growth of the second-mode disturbance has remained one hot issue in transition control.

In three-dimensional boundary layers, crossflow instabilities emerge due to the inflectional flow profile, in addition to the first and second mode instabilities. Earlier research on the crossflow instability has been thoroughly reviewed by Bippes [7], Saric et al. [8], Reed and Saric[9], Kocian et al. [10]. It is known that surface roughnesses could effectively excite stationary crossflow vortices, [11] while free-stream disturbances tend to trigger traveling crossflow waves in boundary layers [7, 12]. As the stationary crossflow mode grows to a sufficiently large amplitude, it will modify the primary instabilities (second mode and traveling crossflow mode) of the boundary layer [13, 14] and is also conducive to multiple rapidly amplifying

R. Zhao et al., *Stabilization of Hypersonic Boundary-Layer: Acoustic Metasurfaces*,
https://doi.org/10.1007/978-981-97-8623-7_2

secondary instabilities that quickly lead to transition [15–19]. Traveling crossflow waves with frequencies generally below 70 kHz are more unstable than the stationary counterpart [20] and have been detected both in ground tests with either quiet or noisy conditions [15, 18, 19, 21, 22] and in a flight test [20]. Traveling and stationary crossflow waves can coexist in a boundary layer, and their interaction leads to multiple traveling crossflow modes [23] Moyes and Reed [24] emphasized that if the influence of traveling crossflow is ignored in the calculation of heat flux, results will be inconsistent with the data obtained in the quiet wind tunnel experiment.

In the natural transition of the boundary layer, two-dimensional waves generally undergo an initial instability (primary instability) and grow slowly as predicted by linear theory. Subsequently, due to nonlinear self-interaction, the growth rate of the two-dimensional waves gradually slows down until it reaches a saturated state. The saturated two-dimensional waves superimpose with the original boundary layer flow, forming a new flow field with spatiotemporal periodic variations (if observed in a reference frame moving with the disturbance wave velocity, a stable flow field with periodic structures in the flow direction will be seen). When the amplitude of the two-dimensional waves exceeds a certain threshold (for incompressible flow, when the perturbation velocity reaches 1% of the main flow velocity [25]), this new flow field becomes highly unstable to specific three-dimensional disturbances, resulting in a secondary instability. The physical cause of the secondary instability may be due to inflection points in the profile of the new flow field (non-viscous instability), while mathematically, it may be related to parametric resonance. Benney and Lin [26] first studied the influence of saturated streamwise vortices on the stability of boundary layer flow. Bayly [27] suggested that the secondary instability is caused by inviscid elliptic instability, but his theory lacks comparison with experimental results [28]. Herbert [25, 29] used Floquet theory to attempt to explain the cause of the secondary instability. Herbert's calculations showed that three-dimensional waves with frequencies half that of the two-dimensional primary wave (subharmonic) or the same as the two-dimensional primary wave (fundamental wave) are unstable over a large range of wave angles, with maximum growth rates even one order of magnitude higher than predicted by linear theory. Herbert proposed that the instability of subharmonics is an extension of the C-type resonance, while the instability of the fundamental wave corresponds to the K-type resonance. Compared to the C-type resonance condition proposed by Craik (i.e., three-dimensional wave angles around 60°) [30], Herbert's subharmonic resonance condition is much broader, so it is also referred to as H-type resonance. The early stage of K-type transition may be related to K-type resonance, but the later stage is much more complex, and there is currently no general explanation available.

Having delineated various boundary layer instability modes, the forthcoming subsection aims to elucidate contemporary methodologies employed in investigating the stability of boundary layers. In this regard, analytical techniques, such as linear stability analysis (LST) and parabolic stability equations (PSE), and direct numerical simulation (DNS) are widely used to predict and resolve the disturbance growth in BL.

2.2 Analysis Methods

2.2.1 Linear Stability Theory

For the three-dimensional boundary layer, assuming that the base flow does not vary in the streamwise direction x and neglecting the variation of the base flow along normal direction y, while setting the normal velocity $v = 0$ to satisfy the continuity equation, the base flow becomes a function of only y. In this case, the disturbance can be expressed in the form of traveling waves as follows:

$$\phi'(x, y, z, t) = \hat{\phi}(y)e^{i(\alpha x+\beta z-\omega t)} + c.c \tag{2.1}$$

$$\hat{\phi}(y) = \left[\hat{\rho}(y), \hat{u}(y), \hat{v}(y), \hat{w}(y), \hat{T}(y)\right]^T \tag{2.2}$$

where α and β are the dimensionless wavenumbers in the x- and z-directions (streamwise and spanwise directions), and ω denotes the angular frequency. Taking spatial modes as an example, the disturbance equation can be written as follows:

$$\begin{aligned}
&L_0 + \alpha L_1\hat{\phi} + \alpha^2 L_2\hat{\phi} = 0\\
&L_0 = D + i\beta C - i\omega\Gamma + \beta^2 V_{zz} + \left(B - i\beta V_{yz}\right)\frac{\partial}{\partial y} - V_{yy}\frac{\partial^2}{\partial y^2}\\
&L_1 = iA + \beta V_{xz} - iV_{xy}\frac{\partial}{\partial y}\\
&L_2 = V_{xx}
\end{aligned} \tag{2.3}$$

For discrete modes, the disturbance amplitude function satisfies the Dirichlet boundary condition at the boundaries:

$$\begin{aligned}
\hat{u} = \hat{v} = \hat{w} = \hat{T} = 0, \quad y = 0\\
\hat{u} = \hat{v} = \hat{w} = \hat{T} = 0, \quad y \to \infty
\end{aligned} \tag{2.4}$$

Hence, the disturbance solution becomes a generalized eigenvalue problem for solving ordinary differential equations. In the case of spatial modes, this nonlinear eigenvalue problem can be transformed into a linear eigenvalue problem.

$$H_0\Phi = \alpha H_1\Phi \tag{2.5}$$

$$H_0 = \begin{bmatrix} L_0 & L_1 \\ 0 & -I \end{bmatrix}, \ H_1 = \begin{bmatrix} 0 & -L_2 \\ -I & 0 \end{bmatrix}, \ \Phi = \begin{bmatrix} \hat{\phi} \\ \alpha\hat{\phi} \end{bmatrix} \tag{2.6}$$

2.2.2 Parabolized Stability Equations

For small disturbances, neglecting the nonlinear interactions between disturbances, the linear PSE method can be used to solve the linear development process of a specific disturbance. For a two-dimensional boundary layer, we can still assume that the disturbance is a periodic function of time and the spanwise direction. The disturbance can be written as:

$$\phi'(x, y, z, t) = \hat{\phi}(x, y)e^{i\left(\int_{x_0}^{x} \alpha(x)dx+\beta z-\omega t\right)} + c.c \tag{2.7}$$

$$\hat{\phi}(x, y) = \left[\hat{\rho}(x, y), \hat{u}(x, y), \hat{v}(x, y), \hat{w}(x, y), \hat{T}(x, y)\right]^T \tag{2.8}$$

the government equation is:

$$\begin{gathered} V_{xx}\frac{\partial^2\hat{\phi}}{\partial x^2} + V_{yy}\frac{\partial^2\hat{\phi}}{\partial y^2} + V_{xy}\frac{\partial^2\hat{\phi}}{\partial x\partial y} + \hat{A}\frac{\partial\hat{\phi}}{\partial x} + \hat{B}\frac{\partial\hat{\phi}}{\partial y} + \hat{D}\hat{\phi} = 0 \\ \begin{cases} \hat{A} = A^l + 2i\alpha V_{xx}^l + i\beta V_{xz}^l \\ \hat{B} = B^l + i\alpha V_{xy}^l + i\beta V_{yz}^l \\ \hat{D} = -i\omega\Gamma^l + i\alpha A^l + i\beta C^l + D^l + \left(i\frac{d\alpha}{dx} - \alpha^2\right)V_{xx}^l - \alpha\beta V_{xz}^l - \beta^2 V_{zz}^l \end{cases} \end{gathered} \tag{2.9}$$

Ignoring the second-order derivative term in the equation, the LPSE control equation is obtained as:

$$V_{yy}\frac{\partial^2\hat{\phi}}{\partial y^2} + \hat{A}\frac{\partial\hat{\phi}}{\partial x} + \hat{B}\frac{\partial\hat{\phi}}{\partial y} + \hat{D}\hat{\phi} = 0 \tag{2.10}$$

In the LPSE equation, there are two unknown physical quantity: the shape function $\hat{\phi}(x, y)$ and the streamwise wavenumber $\alpha(x)$. In order to satisfy the slowly varying characteristics of the shape function along the streamwise direction, the following normalization condition is adopted in this study:

$$\begin{gathered} \alpha^{new} = \alpha^{old} - i\frac{1}{E}\int_0^\infty \overline{\rho}\left(\hat{u}^c\frac{\partial\hat{u}}{\partial x} + \hat{v}^c\frac{\partial\hat{v}}{\partial x} + \hat{w}^c\frac{\partial\hat{w}}{\partial x}\right)dy \\ E = \int_0^\infty \overline{\rho}\left(|\hat{u}|^2 + |\hat{v}|^2 + |\hat{w}|^2\right)dy \end{gathered} \tag{2.11}$$

2.2.3 Direct Numerical Simulation

The basic governing equations of this study are the compressible Navier–Stokes (N-S) equations, which can be nondimensionalized by selecting appropriate reference variables and scales for the physical and coordinate variables. Generally, the boundary layer thickness is chosen at a certain streamwise location as reference length. In a Cartesian coordinate system, the dimensionless form of the conservative compressible Navier–Stokes (N-S) equations is given by:

$$\frac{\partial U}{\partial t} + \frac{\partial E}{\partial x} + \frac{\partial F}{\partial y} + \frac{\partial G}{\partial z} = \frac{\partial E_v}{\partial x} + \frac{\partial F_v}{\partial y} + \frac{\partial G_v}{\partial z} \tag{2.12}$$

where U represents the conservative flux and E, F, G are the convective terms, and E_v, F_v, G_v are the viscous terms,

$$\begin{aligned}
U &= [\rho, \rho u, \rho v, \rho w, \rho e]^T \\
E &= \left[\rho u, \rho u^2 + p, \rho uv, \rho uw, u(\rho e + p)\right]^T \\
F &= \left[\rho v, \rho uv, \rho v^2 + p, \rho vw, v(\rho e + p)\right]^T \\
G &= \left[\rho w, \rho uw, \rho vw + p, \rho w^2 + p, w(\rho e + p)\right]^T \\
E_v &= \left[0, \tau_{xx}, \tau_{yx}, \tau_{zx}, u\tau_{xx} + v\tau_{yx} + w\tau_{zx} + \kappa\frac{\partial T}{\partial x}\right]^T \\
F_v &= \left[0, \tau_{xy}, \tau_{yy}, \tau_{yz}, u\tau_{xy} + v\tau_{yy} + w\tau_{zy} + \kappa\frac{\partial T}{\partial y}\right]^T \\
G_v &= \left[0, \tau_{xz}, \tau_{yz}, \tau_{zz}, u\tau_{xz} + v\tau_{yz} + w\tau_{zz} + \kappa\frac{\partial T}{\partial z}\right]^T
\end{aligned} \tag{2.13}$$

where $e = C_v T + 0.5(u^2 + v^2 + w^2)$, $C_v = C_p/\gamma$, $C_p = 1/Ma^2(\gamma - 1)$. The state equation is

$$p = \frac{\rho T}{\gamma Ma^2} \tag{2.14}$$

And the viscosity coefficient μ is obtained from Sutherland's law:

$$\mu = \frac{1}{\mathrm{Re}_0} T^{1.5} \frac{1 + T_s^*/T_\infty^*}{T + T_s^*/T_\infty^*} \tag{2.15}$$

where $T_s^* = 110.4K$.

2.3 Research Progress of Acoustic Metasurface

In the past two decades, many studies have been performed to study the stabilization effect of acoustic metasurfaces on hypersonic BL flows using LST, DNS, or wind tunnel experiments. With the acoustic impedance model implemented as the boundary condition, the LST can predict the growth ratio of any single mode affected by metasurfaces in the linear stage of the BL transition. In addition to the use of the impedance model to save computational resources, direct numerical simulations of the flow field within the microstructures can provide more information on the interaction mechanisms. The wind tunnel experiments can directly prove the stabilization effect on the unstable modes and help evaluate the delay performance by measuring the transition onset locations. Table 2.1 reviews and summarizes the major achievements that have been conducted theoretically, numerically, and experimentally.

2.3.1 Theoretical Analysis

Fedorov et al. [31] conducted parametric studies on the stabilization effect of an acoustic metasurface with equally spaced cylindrical blind micropores. It was found that deep pores (pore depth is approximately five diameters) with a relatively small radius could stabilize the Mack second mode in a wide frequency band at various Reynolds numbers. The maximum growth rate gradually diminished as the porosity increased. As the depth increased, the growth rate decreased to a limit. The roughness effect was also estimated and that indicated that the coating could be treated as aerodynamically smooth in the unit Reynolds number range required for transition experiments [33]. Further, Brès et al. [34] also indicated that the optimal pore depth of the minima of the reflection coefficient could be obtained at which the reflection waves from the solid wall and the pore bottom were in opposition of phase at the coating surface. Wartemann et al. [32] confirmed that the maximum growth rate of the second Mack mode decreased to a limit as the depth became larger than approximately five pore diameters in the LST analysis of the Mach 6 BL flow. Further, an optimal depth was found, and it was considerably smaller than the limit pore depth (Fig. 2.1).

Brès et al. [34–36] considered metasurfaces operating in attenuative regimes where pores were relatively deep and cancellation/reinforcement regimes with alternating regions of local minima and maxima of the metasurface acoustic absorption depending on the disturbance frequency. As shown in Fig. 2.2, the stabilization effect is considerable when the metasurface operates in the cancelation regime. Further, in this best-case scenario, the pores are shallower with the added convenience of manufacturing and maintenance. With the hypothesis that the maximum absorption occurs when the reflection from the pore bottom and metasurface is in opposition to the phase, they presented a simple model to predict the frequencies of the minimum

Table 2.1 Summary of theoretical, numerical, and experimental progresses on the metasurface transition control technology

Primary theoretical studies				
Ref.	Metasurface type	Model	Method	Main points
Fedorov et al. [31] Wartemann et al. [32]	Regular: equally spaced cylindrical micropores	Mach 6, flat-plate	LST	Parametric study on the effect of pore diameter, depth, and porosity
Fedorov and Malmuth [33]	Regular: equally spaced cylindrical micropores	Mach 7, 10, sharp cone	LST, e^N	Broadband parametric studies demonstrating robustness and estimating the roughness effect
Brès et al. [34–36]	Regular: equally spaced microslits	Mach 6, flat-plate	LST	Report the cancellation/reinforcement working regime of metasurfaces
Zhao et al. [37]	Regular: equally spaced micro-slits	Mach 6, flat-plate	LST	Report the impedance-near-zero metasurface
Zhao et al. [38]	Regular: equally spaced micro-slits	Mach 6, flat-plate		Propose the numerical optimization method for absorptive metasurfaces
Fedorov et al. [39] Tritarelli et al. [40]	Random: felt-metal	Mach 6, sharp cone	LST	Report the stabilization of Mack second mode and destabilization of first mode
Wartemann et al. [41]	Random: C/C material	Mach 7.5, sharp cone	LST, e^N	Suggest the impedance model for the LST analysis of the C/C material
Wartemann et al. [42]	Random: C/C material	Mach 7, HEXAFLY wind tunnel model	LST, e^N	Valid the transition delay performance for HEXAFLY-INT hypersonic glider
Stephen and Michael [43]	Regular: equally spaced cylindrical micropores; Random: mesh layers; Random: felt-metal	Mach 5.3, sharp cone	LST	Demonstrate the destabilization of the nonaxisymmetric first mode
Tian et al. [44]	Regular: gradient micro-slits	Mach 4, flat-plate	LST	Propose a metasurface to suppress the second mode without amplifying the first mode distinctly

(continued)

Table 2.1 (continued)

Primary theoretical studies				
Ref.	Metasurface type	Model	Method	Main points
Zhao et al. [45]	Regular: gradient micro-slits	Mach 4, flat-plate	LST	Propose a metasurface to suppress the first mode
Xu et al. [46]	Regular: equally spaced cylindrical micropores	Mach 4.5, flat-plate	SIT	Investigate the stabilization on secondary instability
Primary numerical studies				
Ref.	Metasurface type	Model	Method	Main points
Egorov et al. [47]	Regular: equally spaced cylindrical blind micropores	Mach 6, flat-plate, sharp cone; Mach 5, compression corner	DNS_BC	Confirm robustness of the metasurface stabilization concept
Wang and Zhong [48–50]	Random: felt-metal	Mach 6, flat-plate	DNS_BC	Emphasize on the effect of metasurface location and the first-mode destabilization
Wang and Zhong [51]	Regular: equally spaced microslits; Random: felt-metal	Mach 6, flat-plate	DNS_BC	Compare the stabilization efficiencies between the regular one and the felt-metal one
Sousa et al. [52, 53]	Random: C/C material	Mach 7.5, sharp cone	DNS_TDIBC	Attempt to simulate the full transition along the metasurface
Chen and Scalo [54]	A single-pole Helmholtz oscillator	Mach 1.5, 3.5, and 6: channel	DNS_TDIBC	Investigate the effect of a wall with complex impedance on the supersonic/hypersonic flows
Brès et al. [35, 55, 56]	Regular: equally spaced microslits	Mach 6, flat-plate	TDNS	Demonstrate the cancellation/ reinforcement working regime of metasurfaces

(continued)

Table 2.1 (continued)

Primary theoretical studies				
Ref.	Metasurface type	Model	Method	Main points
Sandham and Lüdeke [57]	Regular: equally spaced microslits, equally spaced rectangular micropores	Mach 6, flat-plate	TDNS	Confirm the consistence with LST predictions
Tulio and Sandham [58]	Regular: equally spaced square micropores	Mach 6, flat-plate	TDNS	Report the stabilization of primary and secondary instabilities, and destabilization of first mode
Hader et al. [59, 60]	Regular: equally spaced square micropores	Mach 6, flat-plate	TDNS	Focus on the effect of metasurfaces on the nonlinear regime of transition
Zhu et al. [61]	Regular: equally spaced square micropores	Mach 6, flat-plate	SDNS	Compare the stabilization effect of different pore distributions
Zhao et al. [62]	Regular: equally spaced microslits	Mach 6, flat-plate	SDNS	Systematically investigate the metasurface effect on the outflow fields, accuracy of the theoretical model, and the location effect
Zhao et al. [37, 63]	Regular: microslits	Mach 6, flat-plate	SDNS	Demonstrate the stabilization effect of the impedance-near-zero and reflection-controlled metasurfaces
Guo et al. [64]	Regular: equally spaced slits	Mach 6, flat-plate	SDNS	Investigate the stabilization effect of relatively large cavities
Primary experimental studies				
Ref.	Metasurface type	Model	Tunnel	Main points
Rasheed et al. [65]	Regular: equally spaced cylindrical micropores	Mach 5, sharp cone	GALCIT T-5 shock tunnel	Verify the metasurface transition control concept

(continued)

Table 2.1 (continued)

Primary theoretical studies				
Ref.	Metasurface type	Model	Method	Main points
Fedorov et al. [66]	Regular: equally spaced cylindrical micropores	Mach 6, sharp cone	ITAM T-326 blowdown wind tunnel	Compare the metasurface stabilization effect with LST predictions
Fedorov et al. [39]	Random: felt-metal	Mach 6, sharp cone	ITAM T-326 blowdown wind tunnel	Compare the metasurface stabilization effect with LST predictions
Maslov et al. [67]	Random: felt-metal	Mach 12, sharp cone	ITAM AT-303 high-enthalpy wind tunnel	Investigate the transition onset under varied inflow Reynolds numbers, and make comparisons with e^{N} predictions
Lukashevich et al. [68]	Random: several layers of a stainless-steel wire mesh	Mach 6, sharp cone	Transit-M hypersonic short-duration wind tunnel	Valid an optimal coating approximately five times thinner than the traditional deep one
Lukashevich et al. [69, 70]	Random: three layers of a nylon mesh	Mach 5.8, sharp cone	Transit-M hypersonic short-duration wind tunnel	Systematically investigate the effect of the metasurface length and location
Wagner et al. [71–73]	Random: C/C material	Mach 7.5, sharp cone	High enthalpy shock tunnel Göttingen	Report the stabilization effect and transition delay performance
Chokani et al. [74], Bountin et al. [75]	Regular: equally spaced cylindrical micropores	Mach 6, sharp cone	ITAM T-326 blowdown wind tunnel	Bispectral analysis of the metasurface effect on the nonlinear regime of transition
Zhu et al. [76–78]	Random: permeable steel materials	Mach 6, sharp cone	Peking University hypersonic wind tunnel	Clarify the stabilization mechanisms

DNS_BC: DNS with the metasurface modeled as an impedance boundary condition for a single-frequency disturbance (Eq. (2.5)); DNS_TDIBC: DNS with the metasurface modeled as a time-domain impedance boundary condition for broadband disturbances (Eq. (2.6)), TDNS: temporal DNS, SDNS: spatial DNS

reflection coefficient [34, 36]:

$$\frac{fH}{a_{\mathrm{w}}} = (2n - 1)/4 \tag{2.16}$$

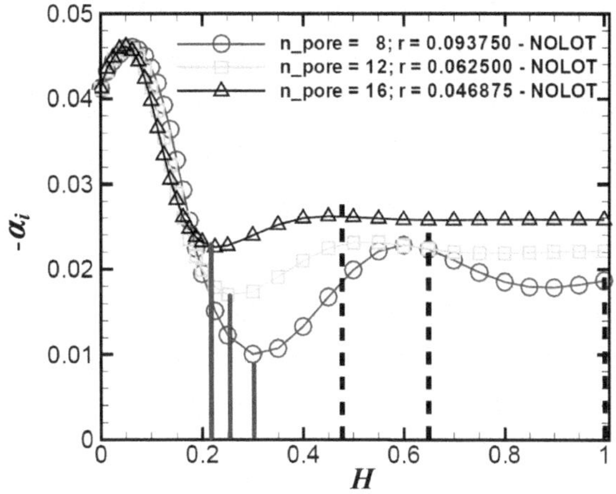

Fig. 2.1 Growth rate − αi against pore depth H at various pore radius r in a Mach 6 BL flow. The dashed line shows the limit-pore depth that is about five times of the pore diameter, and the solid line denotes the optimal depth [32]

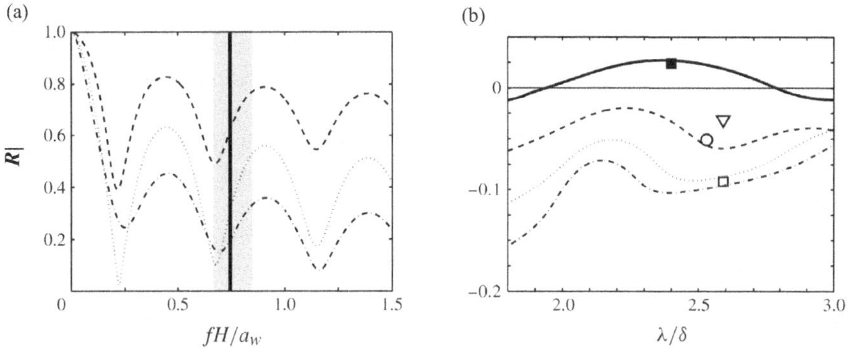

Fig. 2.2 Performance improvement for the metasurface in the cancellation regime: $Ma_\infty = 6$, $Re_\delta = 31{,}300$, cavity aspect ratio $Ar = 0.12$, and porosity $\phi = 0.2$ (dashed line), $\phi = 0.48$ (dotted line), and $\phi = 0.8$ (dot-dashed line). (a) Reflection coefficient amplitude from theory. The grey shaded area corresponds to the frequencies of the unstable modes and the vertical thick line represents the frequency of the most unstable mode. (b) A linear growth rate from LST for the metasurfaces and the rigid flat plate (thick line) [35]

where f denotes the disturbance frequency, H represents the pore depth, a_w denotes the sound velocity at the wall, and n denotes an integer with $n = 1, 2, 3$, which leads to specific frequencies of 0.25,0.75, and 1.25 that correspond to the local minima of the reflection coefficient (Fig. 2.2a).

According to the impedance model, we proposed a numerical solution to obtain the minimum reflection coefficient by carefully selecting the geometrical parameters of the microslits to achieve a reflection coefficient equal to 0 directly. The optimization was achieved by solving Eq. (2.4) with program loops in the following order: loop 1: $0.2 \le \phi \le 0.8$, loop 2: $0.06 \le Ar \le 1.5$, and loop 3: $0 < f_{acs} < \min(\phi/Ar, 2.0)$, where $Ar = 2b/H$ was the aspect ratio and $f_{acs} = fH/c$ was the normalized frequency.

These parameters span the range of interest for practical applications [35] and ensure accuracy. With this method, the obtained $|R|$ at a given frequency f can be optimized close to zero, and the second mode can be suppressed to the maximum extent using the sound absorptive mechanisms of the metasurfaces. The optimized "cancellation" design can only prominently suppress the Mack second mode at around the designed frequency; however, its performance quickly deteriorates compared to that of the conventional "deep-pores" design (Fig. 2.3). Under this condition, the metasurface completely dissipates acoustic disturbances; however, it is not the only choice to stabilize the acoustic Mack second mode. Zhao et al. [37] showed that the growth rate continued to decrease as the impedance became smaller than the background impedance, i.e., the metasurface tended to become nondissipative, and yet, the Mack second mode was effectively suppressed in an even better manner. They attributed this phenomenon to the minimized acoustic pressure perturbation at the antinode of the Mack second mode and designed the impedance-near-zero metasurface as a practical realization.

In the LST analysis of the stabilization effect of the felt-metal microstructures [39, 40], the strong stabilization of second-mode disturbances and the minor destabilization of the first-mode disturbances were observed. The second mode was shifted to lower frequencies, which is in accordance with the results for metasurfaces with regular microstructures [31]. Wartemann et al. [41] compared the effectiveness of different impedance boundary conditions to model the stabilization effect of C/C materials, and they found that the LST based on the felt-metal model [39] delivered the best agreement with the experiments. Stephen and Michael [43] theoretically considered the effect of various metasurfaces on the first mode of a hypersonic BL on a sharp slender cone; they found that the metasurface significantly destabilized

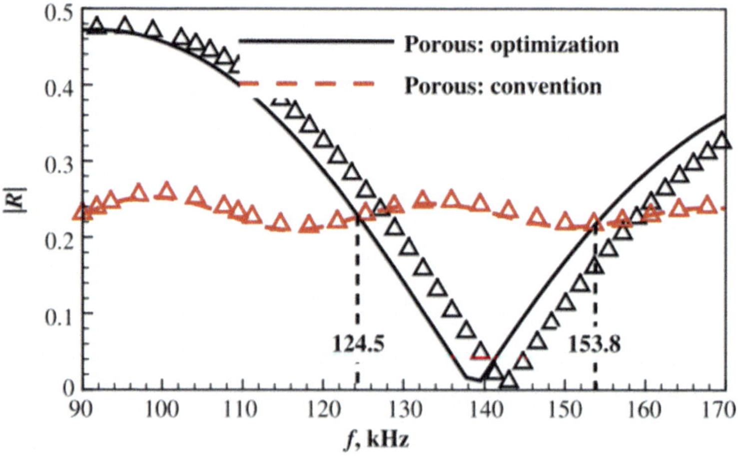

Fig. 2.3 Comparison of reflection coefficient distributions in broadband frequencies ($Ma\infty = 6$; $Re\infty = 10.5 \times 106\ m^{-1}$; and the designed frequency for the optimized metasurface is 138.74 kHz) [38]

nonaxisymmetric modes. Tian et al. [44] used the LST to investigate the stabilization effect of the metasurface admittance phase θ and magnitude $|A|$ systematically for broadening the stabilization range of unstable modes; they found that the stabilization of the Mack modes was closely dependent on phase θ. Further, the corresponding stabilization or destabilization effect was intensified by an increase in the admittance magnitude $|A|$. The first mode is marginally stabilized if θ approaches 0.5π and is greatly destabilized near $\theta = \pi$. In contrast, the second mode is restrained when $\theta \geq 0.75\pi$ and amplified as θ approaches 0.5π. Following the theoretical analysis, they designed a metasurface with gradient microslits to suppress the second mode across a wide frequency band without amplifying the first mode distinctly in a Mach 4 flat-plate BL flow. Considering that the first mode is the dominant instability that induces a transition in the supersonic condition, Zhao et al. [45] further theoretically proposed a metasurface to suppress the first mode in a wide frequency band. The designed metasurface was also corrugated with gradient microslits, but it focused to achieve the phase approaching 0.5π under the most amplified first mode frequency at different streamwise locations. Although the LST results prove that the acoustic metasurface can stabilize the first mode, the roughness effect is ignored in the LST analysis, and it can otherwise excite the first mode. Recently, Xu et al. [46] used the floquet-based secondary instability theory (SIT) [79] to investigate the stabilization effect of acoustic metasurfaces on the secondary instability. They found that there are no strongly preferred interaction modes dominating the secondary instability that corresponds to different amplitudes of the primary Mack mode disturbances affected by the porosity parameters. This finding is in contrast with that of the smooth wall. Further, they found that the suppression of the fundamental mode became more efficient with a larger pore size or porosity.

2.3.2 Numerical Simulations

Generally, there are two numerical strategies to directly simulate the BL flow that interacts with the acoustic metasurfaces. One strategy adopts the impedance boundary condition, similar to the LST ($v'_w = p'_w/Z$). The impedance $Z(\omega)$ is originally defined in the frequency domain and should be solved in the time domain to study the disturbance propagation over the acoustic metasurface. In the coupling with the (N–S) equations for a single disturbance frequency ω, this relationship is written as [47]

$$v_w = p'_w \mathrm{Real}(1/Z) - \frac{1}{\omega}\frac{\partial p'_w}{\partial t}\mathrm{Imag}(1/Z) \tag{2.17}$$

Egorov et al. [47] performed the DNS of the flows over a flat plate, sharp cone, and compression corner at freestream Mach numbers 5–6, which use the impedance model of the metasurface that comprises equally spaced cylindrical blind micropores [31]. In the flat-plate BL flow, they compared the fluctuating wall pressure with the

LST and showed a consistent result. The coating end effects that are associated with the discontinuity of the boundary conditions at the juncture between the solid and the metasurface were found to be localized over 2–3 disturbance wavelengths and can be neglected. Further, the effect of the metasurface on receptivity to acoustic disturbances was studied. The metasurface weakly affected the acoustic disturbances and initial amplitudes of the BL modes, whereas it strongly suppressed the second-mode amplification. The compression corner flow showed that the metasurface weakly affected high-frequency disturbances in the separation region and strongly stabilized them in the reattached flow. Their numerical studies confirmed the robustness of the acoustic metasurface stabilization concept. Using the felt-metal impedance model [39], Wang and Zhong [48–50] studied the stabilization of a Mach 5.92 flat-plate BL using local sections of the metasurface with an emphasis on the effect of metasurface location and first-mode destabilization. They found that the disturbances were destabilized or stabilized when the metasurface was located upstream or downstream of the synchronization point. The synchronization point is the streamwise location where the phase speed of mode F synchronizes with that of mode S at a particular frequency of disturbance. In addition, they compared the stabilization efficiencies of two types of metasurface impedance models at approximately the same porosity, i.e., the regular model comprising the microslits and felt-metal one [51], and they concluded that the regular metasurface was less efficient in stabilizing the second mode and tended to destabilize the first mode to a smaller extent.

As mentioned previously, Eq. (2.5) is applicable only to a single disturbance frequency. It can deal with the metasurface stabilization effect on the most unstable mode, but fails to simulate the response of the broadband disturbances in the real flight or most wind tunnel conditions. Mathematically, the time-domain equivalent of $Z(\omega)$ can be derived by taking its inverse Fourier transform. However, the inverse Fourier transform of the frequency-domain impedance boundary condition leads to a convolution integral. The computation of the convolution integral is time-consuming, especially when coupled with DNS studies [80]. If the impedance can be represented by a sum of certain special functions, the evaluation of the convolution can be performed in a simple systematic manner. Therefore, the metasurface stabilization effect on the broadband instabilities, even in the full transition process and turbulence, can be evaluated. In 2019, Sousa et al. [52] conducted the DNS of a spatially developing hypersonic BL flow over a sharp cone at Mach 7.5. They developed the time-domain impedance boundary conditions (TDIBC) to model the acoustic response of a C/C metasurface to a broadband disturbance. The TDIBC application requires that the acoustic impedance be specified as a set of complex poles and residues representing a superposition of causal second-order oscillators [81–83]. These poles and residues were obtained by fitting the wall softness coefficient that corresponds to the evaluated broadband acoustic impedance from the IHS (Sect. 3.2).

$$\hat{S}(\omega) = \hat{R}(\omega) + 1 = \frac{2}{1 + Z_*(\omega)} \approx \sum_{k=1}^{n_0} \left[\frac{\mu_k}{s - p_k} + \frac{\tilde{\mu}_k}{s - \tilde{p}_k} \right] \tag{2.18}$$

where $\widehat{S}$ and $\widehat{R}$ denote the respective softness and reflection coefficients, respectively; Z_* denotes the nondimensional impedance; n_0 denotes the number of oscillators; μ_k and p_k represent the k-th residue and k-th pole of $\widehat{S}$, respectively; $s = j\omega$; and the superscript $\sim$ denotes the complex conjugate. The fitting residues and poles are provided to predict the next time ($t + \Delta t$) reflected wave from the metasurface, and the incident wave is evaluated by extrapolating the value of the incoming wave in time based on the truncated Taylor series expansion in space [83]. The fluctuating wall-normal velocity at $t + \Delta t$ is given by

$$v'(t + \Delta t) = \frac{1}{2}\left[v^+(t + \Delta t) + v^-(t + \Delta t)\right] \tag{2.19}$$

where v^+ denotes the incident wave and v^- denotes the reflected wave. Using the TDIBC method, Sousa et al. [84] conducted a 3D DNS of transition to turbulence in a hypersonic sharp cone flow with a C/C metasurface. The simulation was started at the onset of transition; however, it collapsed after the appearance of shocklets inside turbulent spots during the transition owing to the low robustness of their numerical scheme. However, the effectiveness of the C/C metasurface in attenuating second-mode waves was demonstrated again. Further, the impedance boundary was reactive to the 3D overlying transitional turbulent flow, which resulted in an effective decrease in the wall heat flux (Fig. 2.4). In addition, they indicated that the impedance boundary may not be applicable for turbulence flow because turbulent structures exhibit smaller scales in the order of the metasurface microstructure. Recently, Chen and Scalo [54] investigated the effect of an isothermal wall with complex impedance on the compressible turbulent channel flow up to the bulk Mach number 6 via the TDIBC. The results indicated that the wall-shear stress was enhanced because of the new emerging streamwise-travelling waves which entail high–low alternating wall pressure patterns. The resonating frequency and resistance values applied in their research are not necessarily the case in real applications, and the revealed wave structures need to be confirmed via experiments.

The other strategy involves directly resolving the detailed BL flow fields that include microstructures. This strategy includes both, a temporal direct numerical simulation (TDNS) and a spatial direct numerical simulation (SDNS). In the TDNS studies (Fig. 2.5a), the initial value problem was solved using the periodic boundary conditions in the x-and z-directions. The forcing terms are included on the right-hand side of the N–S equations to maintain the base flow independent of time. The DNS is initialized with an artificial disturbance of a small amplitude at $t = 0$; the eigenmode is allowed to develop from the initial field. After the transient process, the disturbance exponentially increases with time. The temporal growth rate is calculated as $\alpha_i = d\,(\ln a)/dt$, where $a(t)$ denotes the disturbance amplitude at a fixed point (x, y, z) [85], and it is consistent with parallel flow approximations typically made in the LST. The TDNS can be considered a tractable DNS, i.e., its computational region only needs one wavelength of the Mack second mode. Although TDNS largely saves computational resources, the first-mode instability will not be captured (even if unstable) because its wavelength is longer than that permitted in the

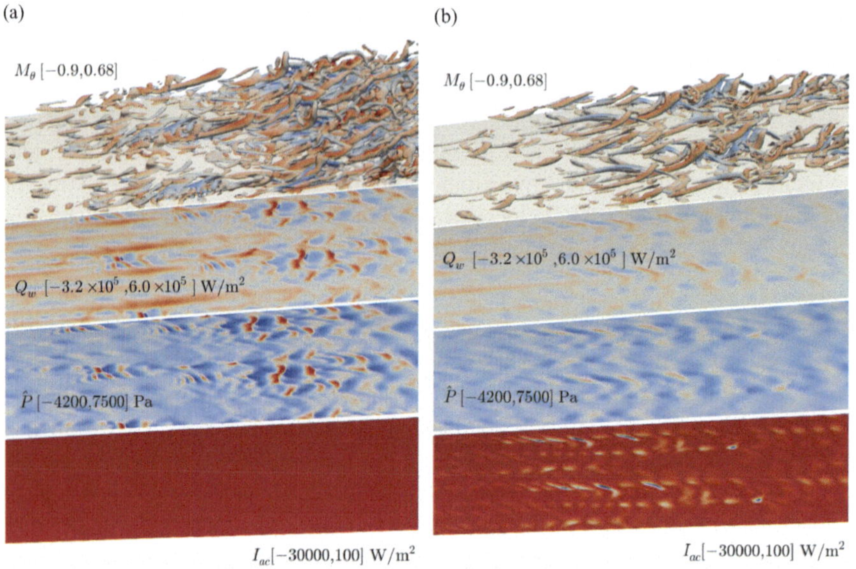

Fig. 2.4 Comparison between flow quantities in the transitional region between **a** the solid wall and **b** the C/C metasurface. From top to bottom: *Q*-criterion iso-surfaces colored by the Mach number in the azimuthal direction, instantaneous heat flux to the wall, pressure oscillations, and acoustic flux [84]

computational domain [35, 55]. In addition, the nonparallel velocity effect, believed to stabilize the Mack second mode and destabilize the first mode, was neglected [62]. The SDNS resolves instability waves that interact with metasurfaces along the streamwise direction; however, this requires abundant computational grids based on the metasurface length.

Brès et al. [35, 55, 56] performed a TDNS of BL flows over a metasurface corrugated with subwavelength microslits. They indicated that metasurface stabilization was directly related to the acoustic absorptive capability of the surface and showed the

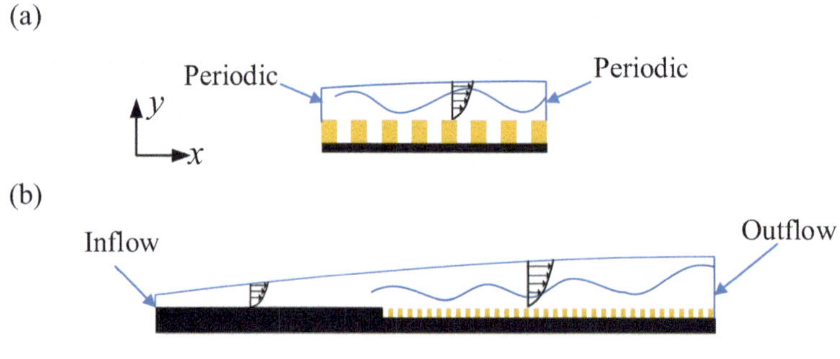

Fig. 2.5 Schematic of **a** the TDNS and **b** SDNS of the stabilization effect of acoustic metasurfaces

presence of an unstable mode that appeared in the presence of high porosity with cavities of a moderate aspect ratio that can potentially decrease the overall performance of the metasurface. The stabilization performances of metasurfaces with relatively deep cavities operating in attenuative regimes and of shallow cavities operating in cancellation/reinforcement regimes with alternating regions of local minima and maxima of the acoustic absorption were compared. The numerical simulations confirmed the results of the linear instability theory, which employed impedance boundary conditions. Sandham and Lüdeke [57] conducted TDNS over metasurfaces comprising equally spaced microslits (2D case) and pores with a rectangular cross-section (3D case). They found that the stabilizing effect was stronger than predictions from the LST, with growth rates that are 10%–30% smaller, depending on porosity and pore depth. Fedorov [85] further demonstrated that the discrepancy was caused by the incorrect impedance boundary condition of LST, wherein 2D slits were modeled as cylindrical pores of the equivalent hydraulic radius [57]. For all cases considered in [85], including the 2D microslits and 3D pores of rectangular cross-sections, the LST results agreed well with the numerical solutions if the cross-sectional pore shapes were considered in the calculations of the metasurface impedance. They also found in the 3D case of one cavity per disturbance wavelength, the discrepancy between the theoretical and DNS growth rates was less than 1%. It is somewhat questionable and needs further validation because the impedance model used in LST actually degrades precision when the incident wavelength approaching the cavity period [86]. Tulio and Sandham [58] investigated turbulence breakdown and reported a reduction in the growth rate of primary and secondary instabilities caused by the presence of regular pores. Further, they found that the oblique first mode wave was destabilized by the metasurfaces, and once the oblique first mode was excited, the flow became turbulent because of the nonlinear interactions without the need for secondary instabilities. Hader et al. [59, 60] applied TDNS to a Mach 6 BL flow and focused on the effect of metasurfaces on the nonlinear regime of transition. A comparison of the amplitude evolution and phase speed locking indicated that the metasurface changed the nature of the nonlinear interaction between the modes and delayed the transition onset (Fig. 2.6). However, a higher resolution is required to simulate the breakdown regime.

Zhu et al. [61] used SDNS to investigate the metasurface stabilization effect in a Mach 6 flat-plate BL flow. Their metasurfaces were corrugated with square pores of sequential and interlaced distributions. The DNS results were compared to those of the LST. They found that the metasurface could effectively inhibit the growth rate of the Mack second mode, and the sequence-distributed pores stabilized the BL flow more effectively than the interlace-distributed pores. However, the streamwise computational region was only approximately 2.7δ, where δ denotes the inflow BL thickness. The narrow range neglected the nonparallel and edge effects. Zhao et al. [62] systematically simulated the spatial evolution of disturbances that interacted with the metasurface microstructure, and they emphasized the metasurface effect on the outflow fields, accuracy of the theoretical model, and effect of metasurface locations. In the steady calculation, cavity edges induced alternating expansion and

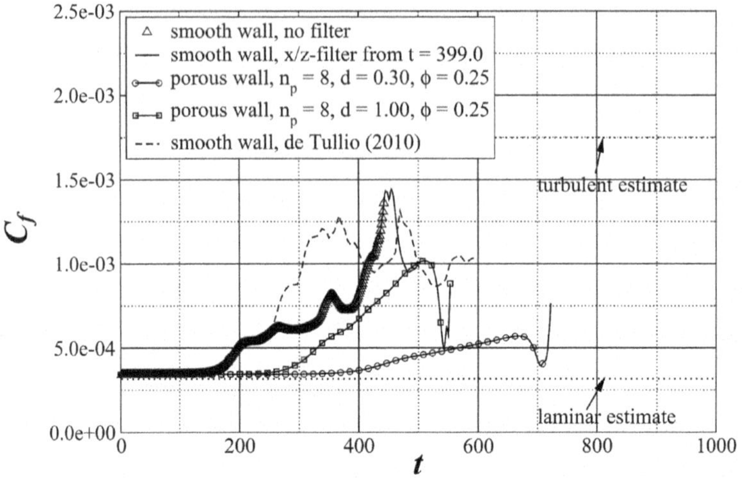

Fig. 2.6 Comparison of the evolution of averaged skin friction coefficient for smooth wall and metasurface [59]

compression waves, and recirculation zones were observed inside cavities. Streamlines were essentially straight near the metasurface (Fig. 2.7). The BL thickness δ decreased slightly at the metasurface region; however, it quickly recovered to its undisturbed condition on the rigid surface after a short relaxation distance of approximately 5–7δ from the junction. The effect of the metasurface on the mean flow field was sufficiently small to satisfy the "aerodynamically smooth" hypothesis. The impedance boundary could accurately model the metasurface stabilization effect by providing an average treatment of the acoustic characteristics of microcavities when comparing the fluctuating pressure in the metasurface region (Fig. 2.8). Note that this situation may be different at which the nonlinear interaction is appreciable. The Mack second mode was destabilized or stabilized when a metasurface was located upstream (Case 1 in Fig. 2.8) or downstream (Cases 2–3 in Fig. 2.8) of the synchronization point; this trend was slightly overpredicted by the theoretical mode. Further, they applied SDNS to investigate the BL stabilization effect of the impedance-near-zero metasurface and reflection-controlled one in a Mach 6 flat-plate BL flow [37, 63]. Guo et al. [64] investigated the effect of large cavities in the second mode. The cavity width was approximately 1/3 of the BL thickness δ; however, no new instability mode appeared in the spectrum of the sampling point. In addition, they found that large-sized cavities could help reduce the frictional drag; a reduction of over 40% in the total drag could be achieved when the porosity was 60%. Because the large cavities weaken the surface acoustic characteristics, the recirculation zones among the cavities would produce a more stable free shear layer than the BL, and that may reduce the disturbance growth [86, 87].

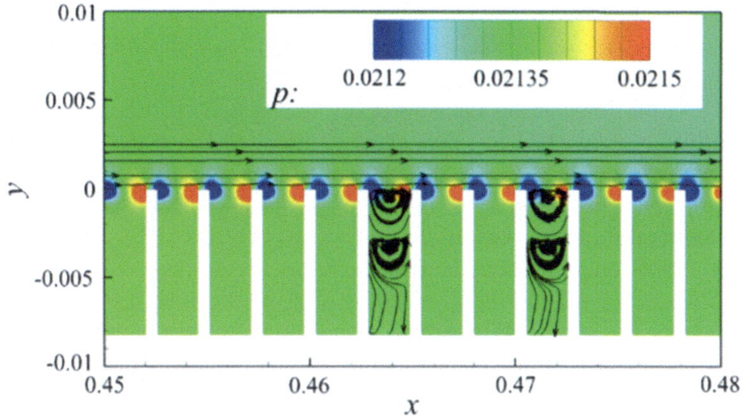

Fig. 2.7 Local pressure contour around cavities [62]

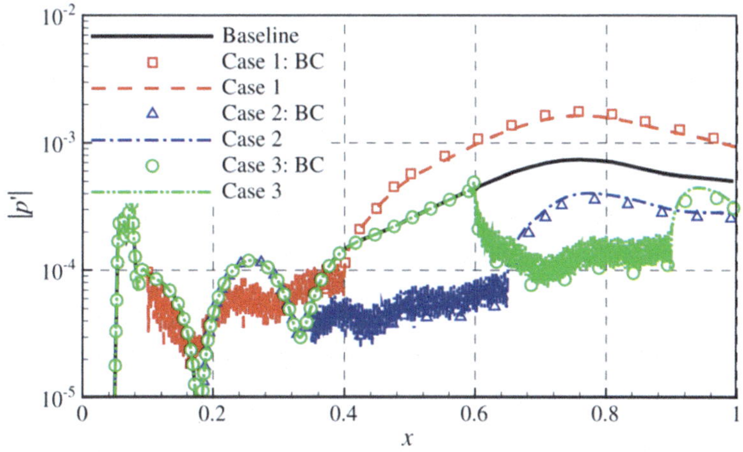

Fig. 2.8 Comparison of pressure perturbation amplitude distributions predicated by DNS with impedance boundary condition and SDNS along the wall. Baseline is the smooth wall and cases 1–3 denote different metasurface locations [62]

2.3.3 *Wind Tunnel Experiments*

Wind tunnel experiments are the most effective approach for evaluating the BL transition control performance of acoustic metasurfaces. The stabilization effect can be checked by measuring the fluctuating flow quantities, and the transition onset point can be determined from the streamwise Stanton number or skin friction distributions. In addition to the general considerations for wind tunnel experiments of hypersonic BL transition, the acoustic metasurface should be carefully designed, manufactured, and installed. Specially, the metasurface acoustic characteristics [38, 88], parametric influences [33], and roughness effect [33, 65] are suggested to be evaluated first before their manufactures for the wind tunnel or flight experiments. Rasheed et al. [65] verified the metasurface transition control concept in the GALCIT T-5 shock tunnel, California Institute of Technology, by testing a 5° half-angle sharp cone. The cone in these tests had one half of its surface solid, and the other half was covered by a porous sheet perforated with equally spaced blind cylindrical holes. Through the laser drilling process, the slightly conical holes were approximately 10–20 holes per disturbance wavelength and with a porosity of 28%. The two halves of the flat sheets (one perforated and one smooth) were rolled to form two longitudinal halves of a cone and then welded by laser fusion along the seams. It took advantage of the mismatch in the thermal expansion coefficients of the stainless-steel cone sheet and aluminum base cone by cooling both parts to 190 K in special freezers. This assembly technique resulted in the cone sheet being stretched tightly over the base cone, which provides blind micropores for the metasurface and eliminates any surface imperfections. Further, it eliminated the need for any mechanical fasteners that would disturb the BL and allow the cone sheet to be nondestructively removable (and thus reusable) by simply reversing the thermal interference fit process. Experiments were performed for ranges of the free-stream total enthalpy, $4.18 \leq H_0 \leq 13.34$ MJ kg^{-1}, and the free-stream Mach number, $4.59 \leq M_\infty \leq 6.4$. For most of the runs, the laminar BL sustained up to the model base on the porous surface half, whereas the transition on the rigid surface was observed halfway along the cone (Fig. 2.9). Further, they observed that the metasurface could delay the transition under a broad range of inflow Reynolds numbers, whereas the inhibition efficiency decreased as the Reynolds number increased. One possible explanation is that the roughness effect of the micropores will become prominent to trip the BL transition as the Reynolds number increased [33, 65]. In their work, the roughness was characterized by the Reynolds number based on the pore diameters, Re^*_{d}. The stabilization becomes qualitatively less effective at Re^*_{d} of about 130 and prematurely trips the BL transition at $Re^*_{\mathrm{d}} > 300$. The LST comparison was not feasible because the stability characteristics of the BL disturbances were not measured in Ref. [65]; this motivated Fedorov et al. [66] to perform a series of stability experiments (Mach number 5.95) in the open-jet test section of the Institute of Theoretical and Applied Mechanics (ITAM) T-326 hypersonic blowdown wind tunnel. The model was a 7° half-angle sharp with one half of the base part covered by a stainless-steel perforated sheet with equally spaced cylindrical pores of depth $H = 450$ μm with an approximate porosity of 20%. The perforated sheet was tightly stretched on the cone surface using a custom-built tension mechanism. In addition to the inflow natural disturbances, a 3D actuator was mounted near the cone tip to provide high-frequency artificial wave packets. Using the hot-wire measurement, the metasurface was found to stabilize the second mode and weakly affect the first mode in the natural disturbance case; the LST predictions agreed well with the measured amplitudes of the second-mode disturbances in the artificially excited wave packets. Willems et al. [89] tested the damping effect of regular and C/C metasurfaces using the same 3° half-angle sharp cone and the same data acquisition system in the quiet wind tunnel BAM6QT and the conventional wind tunnel H2K under similar conditions at Mach 6. Their results indicate an amplification of the second mode and an earlier transition. The reason could be the lower frequencies (~100 kHz) of the second mode and the lower porosity (~12%) of metasurfaces. At these frequencies, the absorptive capacity of the tested metasurfaces is relatively weak. In addition to the roughness effect [33, 65], the acoustic metasurfaces failed to suppress the BL transition.

In the same wind tunnel (ITAM, T-326), experiments were performed to evaluate the stabilization effect of the felt-metal materials [39]. The stainless-steel fibers of diameter $d = 30$ μm were hard-sintered randomly on a solid stainless-steel sheet and rolled to a porosity of 75%. This resulted in a total thickness of the felt metal sheet of 1 mm with a random pore depth of 0.75 mm. The average pore size was approximately 100 μm, which led to 20 pores per BL disturbance wavelength. After attachment to the basic cone, the felt-metal sheet was ground until the surface irregularity was reduced to 0.05 mm. Like their regular metasurface experiments, the experimental results indicated that the felt-metal strongly stabilized the second mode and marginally destabilized the first mode. Theoretical predictions by LST were in good quantitative agreement with the stability measurements for artificially excited wave packets associated with the second mode. Maslov et al. [67] performed transition experiments on a sharp cone with a felt-metal coating in the Mach 12 hypersonic high-enthalpy wind tunnel AT-303 of ITAM. The porous parameters were the same as those in Ref. [39]. Further, as shown in Fig. 2.10, the laminarization effect increases with the unit Reynolds number. This trend is unlike that reported in the literature [65], and it may be attributed to the large inflow Mach number under which the Mack second mode is more unstable. The laminar range on the metasurface was doubled that on the solid wall at $Re \approx 2.4 \times 10^7$/m. Further, they applied the e^N method to

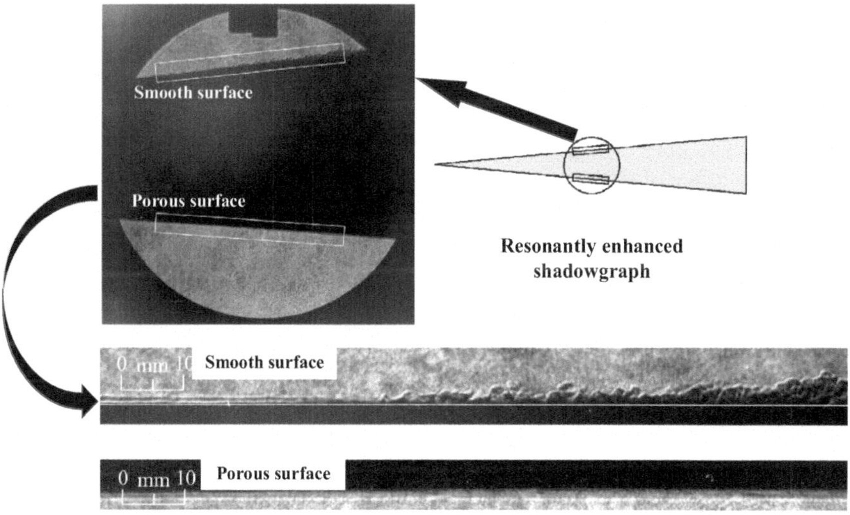

Fig. 2.9 Resonantly enhanced shadowgraph (inflow total pressure P0 = 48.2 MPa and total enthalpy H0 = 9.8 MJ/kg) [65]

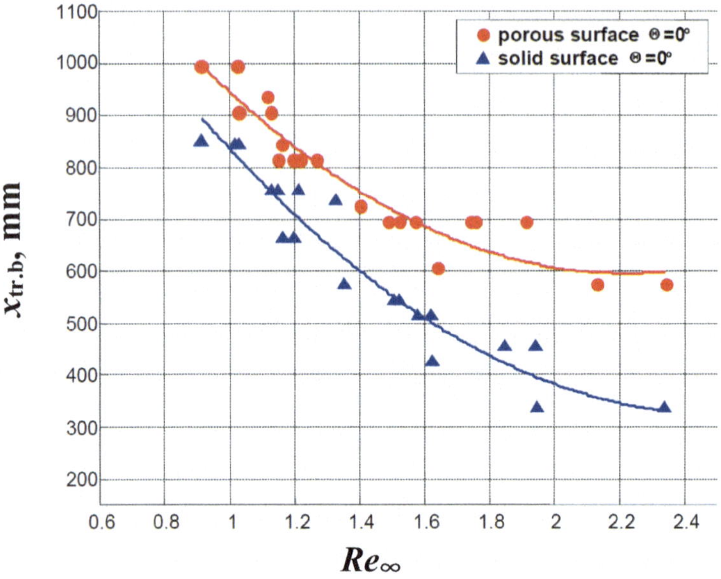

Fig. 2.10 Points of transition beginning versus the freestream unit Reynolds number [67]

predict the transition onset point and found that the location could agree satisfactorily with the experiment when the critical amplification factor $N \approx 5$. Lukashevich et al. [69, 70] systematically investigated the effect of the metasurface length and location on disturbances in a Transit-M hypersonic short-duration wind tunnel under Mach number 5.8. The metasurface was composed of three layers of nylon mesh with a porosity of 44%. They concluded that the effective method to delay transition is to place the acoustic metasurface in a region of second-mode instability, which confirmed the numerical results of Wang and Zhong [49, 50].

Wagner et al. [71–73] conducted a series of wind tunnel experiments to study the stabilization effect of C/C materials in a high-enthalpy shock tunnel in Göttingen of the German Aerospace Center. The C/C material represents an intermediate state of C/C-SiC that has been successfully used as TPS in hypersonic vehicles [90]. The final C/C-SiC can achieve high porosity without losing the advantages of siliconized ceramics (i.e., oxidation resistance and increased material strength) by optimizing the porous parameters of the C/C ceramic [73]. The free-stream Mach number was 7.5 with the unit Reynolds number varying over a range of $Re = 1.5 \times 10^6\ \text{m}^{-1}$ to $6.4 \times 10^6\ \text{m}^{-1}$. The model is a 7° half-angle sharp cone with one-third of its surface covered by C/C. The results indicated that the second-mode instability significantly damped on the C/C surface; the transition onset location was obviously delayed, especially under a high inflow Reynolds number. On the C/C surface, the heat flux scatter around the predicted laminar and turbulent heat flux levels was higher than that of the smooth surface, and this was assumed to be caused by small transducer misalignments on the model surface (Fig. 2.11).

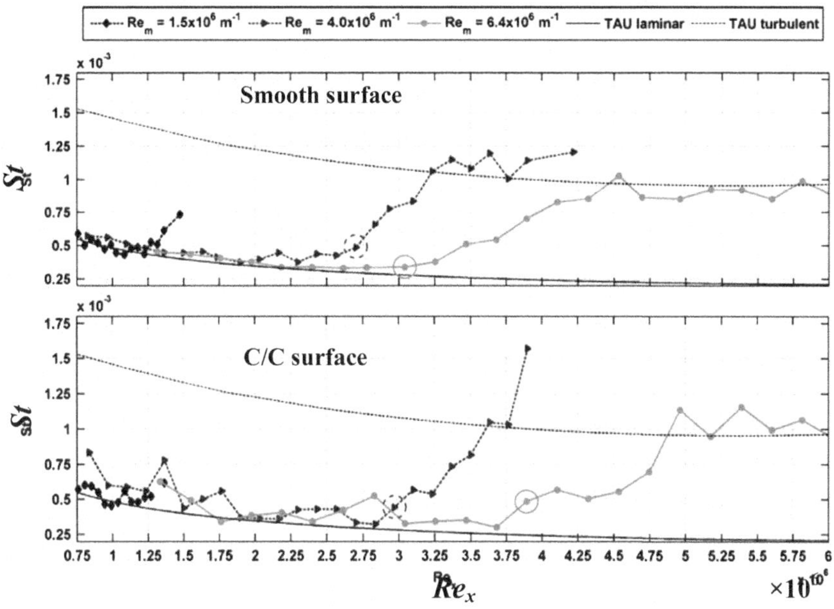

Fig. 2.11 Normalized heat flux distribution on smooth (upper map) and C/C surface (lower map) for different unit Reynolds numbers (transition Reynolds numbers marked by circles) [71]

Chokani et al. [74] and Bountin et al. [75] used bispectral analysis to study the effect of metasurfaces on the nonlinearity of the second mode. The wind tunnel, test model, and type of metasurface were the same as those in Ref. [66]. The bispectral measurements suggested that the subharmonic and harmonic resonances of the second mode were significantly modified on the metasurface. The harmonic resonance, which is known to be the primary nonlinear mechanism for the breakdown to turbulence in the hypersonic BL on solid surfaces, was completely absent on the metasurface. Further, they observed the subharmonic resonance of the first mode, which was not present on the solid surface. Zhu et al. [76–78] investigated the effect of permeable steel materials on the BL transition in a Mach 6 hypersonic wind tunnel at Peking University. They found that the second mode grew faster in the linear stage and lasted over a longer distance along the flow direction on the permeable wall. However, the bispectral analysis showed that the permeable material disrupts the phase-locked relationship and prevents the growth of fundamental oblique waves, which results in a delayed transition.

2.4 Conclusion

This chapter presents a review of the analysis methods, and also the research progress made on acoustic metasurfaces for hypersonic BL stabilization over the past two decades. Although there have been many fruitful achievements, the following aspects of the acoustic metasurface transition control technology could be further developed.

(1) Mathematical impedance model of acoustic metasurfaces can be used to optimize the metasurface microstructure to obtain specific acoustic characteristics; it can be used as the boundary condition of LST and DNS. However, the current models do not consider flow effects. Influences of recirculation zone inside pores and shear layers at pore edges on the disturbance waves require further study. Further, when the disturbance wavelength is close to or less than the microstructure width, the impedance model is partially or no longer valid. Under this condition, the influence of the metasurface on the unstable mode is prospected.
(2) There have been a few experiments that failed to reproduce the metasurface stabilization effect. It implies a competitive mechanism between the metasurfaces acoustic stabilization and roughness destabilization effects on the unstable modes in the BL. The acoustic characteristics, e.g. the absorption, the length and the position are suggested to be optimized by parametric studies before the metasurface manufactures. Also, the roughness effect on the unstable modes should be further clarified to propose a physical-based roughness criterion to aid in the applications.
(3) There is a lack of sufficient studies on the full hypersonic transition process on acoustic metasurfaces. Interaction mechanisms between nonlinear/turbulent structures and metasurface microstructures should be clarified. The skin friction and heat flux affected by the metasurface in the turbulent region should be evaluated before implementing this technology in applications. Similar to the prospects of impedance models, the range of applications of the TDIBC method should be qualified.
(4) Currently, acoustic metasurfaces are mainly applied to suppress the second mode. However, the dominant mode of transition may change as the flight Mach number and attitude change. The suppression of transition over a wide frequency range by acoustic metasurfaces is an important research direction. Further, most hypersonic vehicles have 3D configurations wherein the crossflow mode may dominate the transition, and thus, the wave control ability of the metasurface needs to be further investigated. Joint active control technologies may be required; for example, the development of micro-blowing/metasurface joint control technology, to achieve a more effective and robust transition control capability.

References

1. Mack LM (1984) Special course on stability and transition of laminar flow. AGARD report, No. 709
2. Morkovin, MV (1987) Transition at hypersonic speeds. NASA Contract, Report No. 178315
3. Gushchin VR, Fedorov AV (1990) Excitation and development of unstable disturbances in a supersonic boundary layer. Fluid Dyn 25(3):344–352
4. Fedorov A (2011) Transition and stability of high-speed boundary layers. Annu Rev Fluid Mech 43(1):79–95. https://doi.org/10.1146/annurev-fluid-122109-160750
5. Fedorov A, Tumin A (2010) Branching of discrete modes in high-speed boundary layers and terminology issues. In: 40th fluid dynamics conference and exhibit. https://doi.org/10.2514/6.2010-5003
6. Ma Y, Zhong X (2003) Receptivity of a supersonic boundary layer over a flat plate. Part 1. Wave structures and interactions. J Fluid Mech 488:31–78. https://doi.org/10.1017/S0022112003004786
7. Bippes H (1999) Basic experiments on transition in three-dimensional boundary layers dominated by crossflow instability. Prog Aerosp Sci 35:363–412
8. Saric WS, Reed HL, White EB (2003) Stability and transition of threedimensional boundary layers. Annu Rev Fluid Mech 35:413–440
9. Reed HL, Saric WS (1989) Stability of three-dimensional boundary layers. Annu Rev Fluid Mech 21:235–284
10. Kocian TS, Moyes AJ, Reed HL, Craig SA, Edelman JB (2018) Hypersonic crossflow instability. J Spacecr Rockets 56:432–446
11. Balakumar P, Owens LR (2010) Stability of hypersonic boundary layers on a cone at an angle of attack. AIAA Paper No. 2010-4718
12. Adriano C, Neil S (2020) Transition mechanisms in cross-flow-dominated hypersonic flows with freestream acoustic noise. J Fluid Mech 896:A21
13. Dong S, Chen J, Yuan X, Chen X, Xu G (2020) Wall pressure beneath a transitional hypersonic boundary layer over an inclined straight circular cone. Adv Aerodyn 2:20
14. Li X, Fu D, Ma Y (2010) Direct numerical simulation of hypersonic boundary layer transition over a blunt cone with a small angle of attack. Phys Fluids 22:025105
15. Ward C, Henderson R, Schneider SP (2015) Possible secondary instability of stationary crossflow vortices on an inclined cone at Mach 6. AIAA Paper No. 2015-2773
16. Fei L, Choudhari M, Paredes P, Lian D (2016) High-frequency instabilities of stationary crossflow vortices in a hypersonic boundary layer. Phys Rev Fluids 1:053603
17. Choudhari M, Li F, Paredes P (2018) Nonlinear evolution and breakdown of azimuthally compact crossflow vortex pattern over a yawed cone. AIAA Paper No. 2018-1823
18. Craig SA, Saric WS (2016) Crossflow instability in a hypersonic boundary layer. J Fluid Mech 808:224–244
19. Craig SA, Saric WS (2015) Crossflow instability on a yawed cone at Mach 6. Procedia IUTAM 14:15–25
20. Wan B, Tu G, Yuan X, Chen J, Zhang Y (2021) Identificatification of traveling crossflow waves under real hypersonic flight conditions. Phys Fluids 33:044110
21. Henderson RO (2014) Crossflow transition at Mach 6 on a cone at low angles of attack. Ph.D. thesis (Purdue University, 2014)
22. Ward CAC (2014) Crossflow instability and transition on a circular cone at angle of attack in a Mach-6 quiet tunnel. Ph.D. thesis (Purdue University, 2014)
23. Moyes AJ, Paredes P, Kocian TS, Reed HL (2017) Secondary instability analysis of crossflow on a hypersonic yawed straight circular cone. J Fluid Mech 812:370–397
24. Moyes A, Reed HL (2019) Nonlinear boundary-layer stability analysis of BOLT and HIFiRE-5. AIAA Paper No. 2019-2972
25. Herbert T (1988) Secondary instability of boundary layers. Ann Rev Fluid Mech 20:487–526
26. Benney DJ, Lin CC (1960) On the secondary motion induced by oscillations in a shear flow. Phys Fluids 3(4):656–657

27. Bayly BJ (1986) Three-dimensional instability of elliptical flow. Phys Rev Lett 57(17):2160
28. Bayly BJ, Orszag SA, Herbert T (1988) Instability mechanisms in shear-flow transition. Ann Rev Fluid Mech 20:359–391
29. Herbert T (1983) Secondary instability of plane channel flow to subharmonic three-dimensional disturbances. Phys Fluids 26
30. Craik AD (1971) Non-linear resonant instability in boundary layers. J Fluid Mech 50:393–413
31. Fedorov AV, Malmuth ND, Rasheed A, Hornung HG (2001) Stabilization of hypersonic boundary layers by porous coatings. AIAA J 39:605–610. https://doi.org/10.2514/3.14776
32. Wartemann V, Lüdeke H, Sandham ND (2009) Stability analysis of hypersonic boundary layer flow over microporous surfaces. AIAA Paper 2009-7202. https://doi.org/10.2514/6.2009-7202
33. Fedorov AV, Malmuth ND (2008) Parametric studies of hypersonic laminar flow control using a porous coating of regular microstructure. AIAA Paper 2008-588. https://doi.org/10.2514/6.2008-588
34. Brès GA, Colonius T, Fedorov AV (2010) Acoustic properties of porous coatings for hypersonic boundary-layer control. AIAA J 48:267–274. https://doi.org/10.2514/1.40811
35. Brès GA, Inkman M, Colonius T, Fedorov AV (2013) Second-mode attenuation and cancellation by porous coatings in a high-speed boundary layer. J Fluid Mech 726:312–337. https://doi.org/10.1017/jfm.2013.206
36. Brès G, Colonius T, Fedorov A (2008) Interaction of acoustic disturbances with micro-cavities for ultrasonic absorptive coatings. AIAA Paper 2008–3903. https://doi.org/10.2514/6.2008-3903
37. Zhao R, Liu T, Wen CY, Zhu J, Cheng L (2019) Impedance-near-zero acoustic metasurface for hypersonic boundary-layer flow stabilization. Phys Rev Appl 11:044015. https://doi.org/10.1103/PhysRevApplied.11.044015
38. Zhao R, Liu T, Wen CY, Zhu J, Cheng L (2018) Theoretical modeling and optimization of porous coating for hypersonic laminar flow control. AIAA J 56:2942–2946. https://doi.org/10.2514/1.J057272
39. Fedorov AV, Shiplyuk AN, Maslov AA, Burov E, Malmuth ND (2003) Stabilization of a hypersonic boundary layer using an ultrasonically absorptive coating. J Fluid Mech 479:99–124. https://doi.org/10.1017/S0022112002003440
40. Tritarelli RC, Lele SK, Fedorov AV (2015) Stabilization of a hypersonic boundary layer using a felt-metal porous coating. J Fluid Mech 769:729–739. https://doi.org/10.1017/jfm.2015.156
41. Wartemann V, Wagner A, Eggers T, Hannemann K (2015) Passive hypersonic boundary layer control by means of ultrasonically absorptive carbon-carbon ceramics: investigations of different boundary conditions. AIAA Paper 2015-3577. https://doi.org/10.2514/6.2015-3577
42. Wartemann V, Wagner A, Eggers T (2016) Passive hypersonic boundary layer control: the potential of an ultrasonically absorptive ceramic for HEXAFLY-INT. AIAA Paper 2016-4250. https://doi.org/10.2514/6.2016-4250
43. Stephen SO, Michael V (2013) Effects of porous walls on hypersonic boundary layers over a sharp cone. AIAA J 51:1234–1244. https://doi.org/10.2514/1.J052082
44. Tian X, Zhao R, Long T, Wen CY (2019) Reverse design of ultrasonic absorptive coating for the stabilization of Mack modes. AIAA J 57:2264–2269. https://doi.org/10.2514/1.J058105
45. Zhao R, Yan H, Xi K, Wen CY (2020) Research on acoustic metasurfaces for the suppression of the first mode. Aeronau Sci Tech 31:104–112. https://doi.org/10.19452/j.issn1007-5453.2020.11.013 (in Chinese)
46. Xu J, Liu J, Mughal S, Yu P, Bai J (2020) Secondary instability of Mack mode disturbances in hypersonic boundary layers over micro-porous surface. Phys Fluids 32:044105. https://doi.org/10.1063/5.0001914
47. Egorov IV, Fedorov AV, Novikov AV, Soudakov VG (2007) Direct numerical simulation of supersonic boundary-layer stabilization by porous coatings. AIAA Paper 2007-948. https://doi.org/10.2514/6.2007-948
48. Wang X, Zhong X (2008) Role of the synchronization point on boundary layer stabilization using porous coating. AIAA Paper 2008-4382. https://doi.org/10.2514/6.2008-4382

49. Wang X, Zhong X (2010) Effect of porous coating on boundary-layer instability. AIAA Paper 2010-1243. https://doi.org/10.2514/6.2010-1243
50. Wang X, Zhong X (2012) The stabilization of a hypersonic boundary layer using local sections of porous coating. Phys Fluids 24:034105. https://doi.org/10.1063/1.3694808
51. Wang X, Zhong X (2011) Numerical simulations on mode S growth over feltmetal and regular porous coatings of a Mach 5. 92 flow. AIAA Paper 2011-375. https://doi.org/10.2514/6.2011-375
52. Sousa V, Patel D, Chapelier J-BB, Wartemann V, Wagner A, Scalo C (2019) Numericalinvestigation of second-mode attenuation over carbon/carbon porous surfaces. J Spacecr Rockets 56:319–332. https://doi.org/10.2514/1.A34294
53. Sousa VCBB, Patel D, Chapelier J-BB, Wartemann V, Wagner A, Scalo C (2019) Numerical investigation of second-mode attenuation over Carbon/Carbon porous surfaces. J Spacecr Rockets 56:319–332. https://doi.org/10.2514/1.A34294
54. Chen Y, Scalo C (2021) Trapped waves in supersonic and hypersonic turbulent channel flow over porous walls. J Fluid Mech 920:1–37. https://doi.org/10.1017/jfm.2021.428
55. Bres GA, Colonius T, Fedorov AV (2008) Stability of temporally evolving supersonic boundary layers over micro-cavities for ultrasonic absorptive coatings. AIAA Paper 2008-4337. https://doi.org/10.2514/6.2008-4337
56. Bres GA, Inkman M, Colonius T, Fedorov AV (2009) Alternate designs of ultrasonic absorptive coatings for hypersonic boundary layer control. AIAA Paper 2009-4217. https://doi.org/10.2514/6.2009-4217
57. Sandham ND, Lüdeke H (2009) Numerical study of Mach 6 boundary-layer stabilization by means of a porous surface. AIAA J 47:2243–2252. https://doi.org/10.2514/1.43388
58. De Tullio N, Sandham ND (2010) Direct numerical simulation of breakdown to turbulence in a Mach 6 boundary layer over a porous surface. Phys Fluids 22:094105. https://doi.org/10.1063/1.3481147
59. Hader C, Brehm C, Fasel HF (2013) Numerical investigation of transition delay using porous walls. AIAA Paper 2013-2740. https://doi.org/10.2514/6.2013-2740
60. Hader C, Brehm C, Fasel HF (2014) Numerical investigation of transition delay for various controlled breakdown scenarios in a Mach 6 boundary layer using porous walls. AIAA Paper 2014-2500. https://doi.org/10.2514/6.2014-2500
61. Zhu DH, Liu ZY, Yuan XJ (2016) Mechanism of transition delay by porous surface in hypersonic boundary layers. Chin J Comput Phys 33:163–169 (in Chinese). https://doi.org/10.3969/j.issn.1001-246X.2016.02.005
62. Zhao R, Wen CY, Long TH, Tian XD, Zhou L, Wu Y (2019) Spatial direct numerical simulation of the hypersonic boundary-layer stabilization using porous coatings. AIAA J 57:5061–5065. https://doi.org/10.2514/1.J058467
63. Zhao R, Dong Y, Zhang X, Wen C, Long T, Yuan W (2021) Control of reflected waves with acoustic metasurfaces for hypersonic boundary-layer stabilization. AIAA J 59:1533. https://doi.org/10.2514/1.J060282
64. Guo QL, Tu GH, Chen JQ, Yuan XX, Wan BB (2020) Control of hypersonic boundary layer instability by transverse rectangular micro-cavities. J Aerosp Power 35:135–143 (in Chinese). https://doi.org/10.13224/j.cnki.jasp.2020.01.016
65. Rasheed A, Hornung HG, Fedorov AV, Malmuth ND (2002) Experiments on passive hypervelocity boundary-layer control using an ultrasonically absorptive surface. AIAA J 40:481–489. https://doi.org/10.2514/2.1671
66. Fedorov AV, Kozlov VF, Shiplyuk AN, Maslov AA, Malmuth ND (2006) Stability of hypersonic boundary layer on porous wall with regular microstructure. AIAA J 44:1866–1871. https://doi.org/10.2514/1.21013
67. Maslov A, Shiplyuk A, Sidorenko A, Polivanov P, Fedorov A, Kozlov V, Malmuth N (2006) Hypersonic laminar flow control using a porous coating of random microstructure. AIAA Paper 2006-1112. https://doi.org/10.2514/6.2006-1112.
68. Lukashevich SV, Maslov AA, Shiplyuk AN, Fedorov AV, Soudakov VG (2012) Stabilization of high-speed boundary layer using porous coatings of various thicknesses. AIAA J 50:1897–1904. https://doi.org/10.2514/1.J051377

69. Lukashevich SV, Morozov SO, Shiplyuk AN (2013) Experimental study of the effect of a passive porous coating on disturbances in a hypersonic boundary layer. 1. Effect of the porous coating length. J Appl Mech Tech Phys 54:572–577. https://doi.org/10.1134/S002189441304007X
70. Lukashevich SV, Morozov SO, Shiplyuk AN (2016) Experimental study of the effect of a passive porous coating on disturbances in a hypersonic boundary layer 2. Effect of the porous coating location. J Appl Mech Tech Phys 57:873–878. https://doi.org/10.1134/S002189441605014X
71. Wagner A, Kuhn M, Martinez Schramm J, Hannemann K (2013) Experiments on passive hypersonic boundary layer control using ultrasonically absorptive carbon-carbon material with random microstructure. Exp Fluids 54:1606. https://doi.org/10.1007/s00348-013-1606-3
72. Wagner A (2014) Passive hypersonic boundary layer transition control using ultrasonically absorptive carbon–carbon ceramic with random microstructure. Ph.D. thesis, Belgium: Katholieke Universiteit Leuven 31:1–146. https://doi.org/10.13140/RG.2.1.4246.8968
73. Wagner A, Wartemann V, Hannemann K, Kuhn M, Dittert C (2015) The potential of ultrasonically absorptive TPS materials for hypersonic vehicles. AIAA Paper 2015-3576. https://doi.org/10.2514/6.2015-3576
74. Chokani N, Bountin DA, Shiplyuk AN, Maslov AA (2005) Nonlinear aspects of hypersonic boundary-layer stability on a porous surface. AIAA J 43:149–155. https://doi.org/10.2514/1.9547
75. Buntin DA, Maslov AA, Chimytov TA, Shiplyuk AN (2010) Bispectral analysis of nonlinear processes in the hypersonic boundary layer on a porous cone surface. Fluid Dyn 45:415–421. https://doi.org/10.1134/S0015462810030087
76. Zhu W, Shi M, Zhu Y, Lee C (2020) Experimental study of hypersonic boundary layer transition on a permeable wall of a flared cone. Phys Fluids 32:011701. https://doi.org/10.1063/1.5139546
77. Zhu W, Shi M, Lee C (2020) Experimental and theoretical study of the hypersonic boundary layer transition on the permeable wall of a flared cone. AIAA Paper 2020-2964. https://doi.org/10.2514/6.2020-2964
78. Zhu W, Chen X, Zhu Y, Lee C (2020) Nonlinear interactions in the hypersonic boundary layer on the permeable wall. Phys Fluids 32:104110. https://doi.org/10.1063/5.0028698
79. Herbert T (1988) Secondary instability of boundary layers. Annu Rev Fluid Mech 20:487–526. https://doi.org/10.1146/annurev.fl.20.010188.002415
80. Li XY, Li XD, Tam CKW (2012) Improved multipole broadband time-domain impedance boundary condition. AIAA J 50:980–984. https://doi.org/10.2514/1.J051361
81. Fung KY, Ju H (2004) Time-domain impedance boundary conditions for computational acoustics and aeroacoustics. Int J Comut Fluid Dyn 18:503–511. https://doi.org/10.1080/10618560410001673515
82. Poinsot TJ, Lele SK (1992) Boundary conditions for direct simulations of compressible viscous flows. J Comput Phys 101:104–129. https://doi.org/10.1016/0021-9991(92)90046-2
83. Scalo C, Bodart J, Lele SK (2015) Compressible turbulent channel flow with impedance boundary conditions. Phys Fluids 27:035107. https://doi.org/10.1063/1.4914099
84. Sousa V, Wartemann V, Wagner A, Scalo C (2019) Towards direct numerical simulation of hypersonic transition delay via distributed wall porosity. AIAA Paper 2019-2151. https://doi.org/10.2514/6.2019-2151
85. Fedorov AV (2010) Temporal stability of hypersonic boundary layer on porous wall: comparison of theory with DNS. AIAA Paper 2010-1242. https://doi.org/10.2514/6.2010-1242
86. Egorov IV, Novikov AV, Federov AVF (2010) Direct numerical simulation of supersonic boundary layer stabilization using grooved wavy surface. AIAA Paper 2010-1245. https://doi.org/10.2514/6.2010-1245
87. Bountin D, Chimitov T, Maslov A, Novikov A, Egorov I, Fedorov A, Utyuzhnikov S (2013) Stabilization of a hypersonic boundary layer using a wavy surface. AIAA J 51:1203–1210. https://doi.org/10.2514/1.J052044
88. Zhao R, Zhang XX, Wen CY (2020) Theoretical modeling of porous coatings with simple microstructures for hypersonic boundary-layer stabilization. AIAA J 58:981–986. https://doi.org/10.2514/1.J058403

89. Willems S, Gülhan A, Ward CAC, Schneider SP (2017) Free transition on a slender cone in a quiet and a conventional wind tunnel and the effect of ultrasonically absorptive materials. Prog Flight Phys 9:497–516. https://doi.org/10.1051/eucass/2016090497
90. Turner J, Horschgen M, Jung W, Stamminger A, Turner P (2006) SHEFEX—hypersonic re-entry flight experiment—vehicle and subsystem design, flight performance and prospects. AIAA Paper 2006-8115. https://doi.org/10.2514/6.2006-8115

Chapter 3
Acoustic Metasurface Modelling

3.1 Introduction

The stabilization effect of acoustic metasurfaces on hypersonic BL flow is caused by the interaction between their acoustic characteristics and instability waves. In addition to the use of direct numerical simulations of the flow field within microstructures, an acoustic impedance boundary condition of the vertical velocity at the wall ($v'_{\mathrm{w}} = p'_{\mathrm{w}}/Z$) is adopted to model the metasurface. Here, v'_{w} and p'_{w} denote the vertical velocity and pressure perturbations, respectively. Z denotes the surface impedance, which is a complex quantity that depends on the properties of the wall material, porosity parameters, mean flow characteristics on the wall surface, and flow perturbation parameters such as wave frequencies and wavelengths. This treatment saves computational resources for numerical simulations and assists in the LST analysis. The influence of the microstructure geometry parameters such as depth, diameter, and porosity can be systematically investigated using the impedance model. It should be noted the applications of the impedance model are independent of the nature of the BL disturbances, e.g., inviscid second mode or viscous TS waves [1]. Gaponov applied the impedance model to investigate the porosity effect on TS waves for subsonic [2, 3] and moderate supersonic speeds [4].

The equation for admittance was first derived by applying the theory of sound wave propagation in a thin long tube [1]. It considered the pore radius and spacing to be much less than the disturbance wavelength, and in this case, the porosity was thought fine enough to avoid disturbing the laminar BL by other mechanisms associated with effective surface roughness. Boundary conditions representing the effect of acoustic metasurfaces on 2D disturbance of a fixed frequency are expressed as:

$$u'_{\mathrm{w}} = 0, \quad v'_{\mathrm{w}} = Ap'_{\mathrm{w}}, \quad w'_{\mathrm{w}} = 0, \tag{3.1}$$

$$T'_{\mathrm{w}} = Bp'_{\mathrm{w}}$$

R. Zhao et al., *Stabilization of Hypersonic Boundary-Layer: Acoustic Metasurfaces*,
https://doi.org/10.1007/978-981-97-8623-7_3

where A ($\equiv 1/Z$) and B represent the sound and thermal admittances, respectively. In Equation,

$$A = -(\phi/Z_0)tanh(\Lambda h)$$
$$B = -\phi(\gamma - 1)Ma^2 T_w J_2(k_t)/J_0(k_t) \tag{3.2}$$

where ϕ is porosity, h is the porous-layer thickness, Z_0 is characteristic impedance, Λ is the propagation constant, J_0 and J_2 are Bessel functions, and k_t measures the ratio of the tube radius to the thermal BL thickness on the tube surface [1]. Z_0 and Λ are calculated using the transmission line formalism [5, 6], and B can be assumed to be zero because the temperature perturbations on the metasurface weakly affect the disturbance growth rate after investigations [1]. The same analytical solution of Z_0 and Λ can be obtained in terms of the complex dynamic density $\tilde{\rho}$ and compressibility $\tilde{C}$ [7], which is more convenient for the analysis of acoustic disturbances in randomly porous materials and is widely applied in subsequent research. As indicated in Eq., A and B are originally deduced for a single cylindrical pore and they neglect the interactions among neighboring pores. The integral acoustic characteristics of the metasurfaces are evaluated by multiplying ϕ.

Under wind-tunnel conditions, the fluid density is relatively small, and the rarefaction effects inside the pores may be important. Maslov [8] stated that the rarefaction effect would make metasurfaces more transparent with respect to the disturbance, and this would lead to the stronger stabilization of the second mode. As the Knudsen number increased to $Kn = 0.588$, the growth ratio of the Mach second mode calculated by the LST decreased by 15–20% considering the rarefaction effect. Relatively comprehensive rarefaction models for regular metasurfaces of various cross-sectional microstructures (circle, rectangular, triangular, and slit) were proposed by Kozlov et al. [9]. All these models ignore the end effects associated with the scattering of incoming acoustic waves at the cavity/pore mouth. Further, there is no coupling between disturbances from neighboring cavities or pores, and this may not be true for closely spaced cavities. To consider the coupling between disturbances from neighboring cavities, Brès et al. [10] hypothesized a slip velocity at the wall with expressions such as $u'_w = f(v'_w, p'_w)$ for the LST analysis. Nevertheless, they indicated that this boundary condition should be further studied and refined because of the large number of fitting parameters.

Although this theoretical model, hereinafter called Fedorov's model, has been widely applied, it does not take into account the high-order diffracted waves, and thus underestimates the coupling among adjacent cavities as well as its contribution to the overall admittance [11–13]. This simplification is considered to be responsible for the low-frequency shift of the reflection curves [10, 14]. The design of this metasurface device and its integration with thermal protection systems require the development of accurate models of the effect of the acoustic metasurface. In particular, further optimization requires frequency matching between the most amplified Mack second-mode instability wave and the minimum reflection property of microstructures. Our team considered the high-order diffracted modes of plane ultrasonic acoustic waves impinging on metasurfaces with 2D microslits and 3D micropores, which have improved the accuracy of the predicted acoustic characteristics [15, 16].

3.2 Mathematical Modelling and Verification

3.2.1 Metasurface Corrugated with Microgrooves

As shown in Fig. 3.1, the metasurface is a rigid surface periodically corrugated with subwavelength grooves (two-dimensional cavities) being infinitely extended in the x direction. The background medium is assumed to have uniform and constant density ρ_w and sound speed c_w. The subscript w denotes the local physical quantity at the wall. The half-width and depth of the cavities are b and H, respectively, with the unit-cell period being s. The porosity and aspect ratio are $\phi = 2b/s$ and $Ar = 2b/H$, respectively. These definitions are in accordance with previous research [10, 14].

The acoustic field and the surface admittance of such textured surface can be analytically obtained on the plane wave expansion method [11–13]. An obliquely incident plane acoustic wave can be written as (the time dependence $e^{-j\omega t}$ is omitted for simplicity)

$$\begin{aligned} p_i &= e^{jk_x x} e^{-jk_y y}, \\ v_{y,i} &= \frac{1}{j\omega\rho_w}\frac{\partial p_i}{\partial y} = -\frac{k_y}{\rho_w \omega} e^{jk_x x} e^{-jk_y y}, \end{aligned} \tag{3.3}$$

where p_i is the incident pressure, $v_{y,i}$ is the y component of the particle velocity, and $j = \sqrt{-1}$. k_x is the parallel momentum and $k_y = (k_0^2 - k_x^2)^{1/2}$ is the perpendicular momentum, in which $k_0 = \omega/c_w$ is the wavenumber with ω being the angular frequency. The reflected pressure field p_r^n and y-component particle velocity $v_{y,r}^n$ of

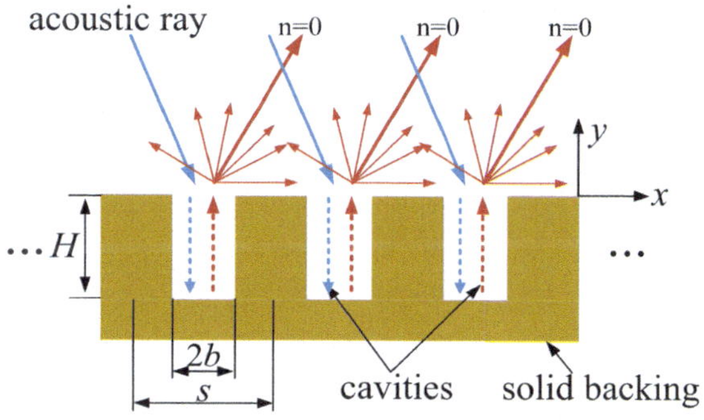

Fig. 3.1 Schematic drawing of reflection of acoustic waves from equally spaced two-dimensional cavities

the nth-order diffracted wave are expressed as

$$
\begin{aligned}
p_r^{(n)} &= R_n e^{jk_x^{(n)}x} e^{jk_y^{(n)}y}, \\
v_{y,r}^{(n)} &= \frac{k_y^{(n)}}{\rho_\mathrm{w}\omega} R_n e^{jk_x^{(n)}x} e^{jk_y^{(n)}y},
\end{aligned}
\tag{3.4}
$$

where $k_x^{(n)} = k_x + 2\pi n/s$, $k_y^{(n)} = [k_0^2 - (k_x^{(n)})^2]^{1/2}$, $n \in \mathbf{Z}$ and R_n is the reflection coefficient of the nth-order diffraction. Inside the cavities, the fundamental wave mode dominates for long wavelengths limit ($2b \ll \lambda_\mathrm{acs}$, where λ_acs is the wavelength of the incident acoustic wave), and the sound pressure and particle velocity within the cavity are denoted as

$$
\begin{aligned}
p_\mathrm{c} &= C_1 e^{jk_\mathrm{c}y} + C_2 e^{-jk_\mathrm{c}y}, \\
v_{y,\mathrm{c}} &= \frac{k_\mathrm{c}}{\tilde{\rho}\omega}\left(C_1 e^{jk_\mathrm{c}y} - C_2 e^{-jk_\mathrm{c}y}\right),
\end{aligned}
\tag{3.5}
$$

where the dynamic density $\tilde{\rho}$, compressibility $\tilde{C}$, and wavenumber k_c are complex and frequency-dependent quantities owing to the thermal and viscous boundary layers inside the narrow cavity,

$$
\begin{aligned}
\tilde{\rho} &= \rho_\mathrm{w}/\Psi_v,\ \tilde{C} = \frac{\gamma - (\gamma - 1)\Psi_t}{\rho_\mathrm{w} c_\mathrm{w}^2}, \\
k_\mathrm{c}^2 &= \omega^2 \tilde{\rho}\tilde{C} = k_0^2 \frac{\gamma - (\gamma - 1)\Psi_t}{\Psi_v}.
\end{aligned}
\tag{3.6}
$$

Here, $\Psi_i = 1 - \tan(k_i b)/k_i b$ with

$$
k_i^2 = \begin{cases} k_v^2 = j\omega \dfrac{\rho_\mathrm{w}}{\mu}, & \text{viscous wave number} \\ k_t^2 = j\omega \dfrac{\rho_\mathrm{w} C_p}{\kappa}, & \text{thermal wave number} \end{cases}.
\tag{3.7}
$$

The subscript i is either v or t to denote the effects of viscous or thermal boundary layers, respectively. In the above equations, κ is thermal conductivity, μ is viscosity, $\gamma = C_p/C_v$ is the ratio of the specific heat at constant pressure, C_p, to specific heat at constant volume, C_v.

The bottom of the cavity is rigid ($v_{y,\mathrm{c}}|_{y=-H} = 0$), thus $C_1 = C_2 e^{2jk_\mathrm{c}H} \equiv Ce^{2jk_\mathrm{c}H}$. At the interface between the upper half space and the cavity opening mouth, the sound pressure should be continuous:

$$
\frac{1}{2b}\int_{x=-b}^{x=b}\left(e^{jk_x x} + \sum_{n=-\infty}^{+\infty} R_n e^{jk_x^{(n)}x}\right)dx = C\left(1 + e^{2jk_\mathrm{c}H}\right).
\tag{3.8}
$$

Equation (3.9) is then deduced as

$$\sum_{n=-\infty}^{+\infty} \left(\delta_{n,0} + R_n\right) S_n = C\left(1 + e^{2jk_c H}\right), \tag{3.9}$$

where $S_n = (2b)^{-1} \int_{-b}^{b} e^{jk_x^{(n)}x} dx = \mathrm{sinc}(k_x^{(n)} b)$ is the overlap integral between the nth-order diffracted mode and the fundamental mode inside the cavity, and $\delta_{n,0}$ is the Kronecker delta function defined as $\delta_{n,0} = 1$ for $n = 0$ and $\delta_{n,0} = 0$ otherwise.

As for the requirement of continuous particle velocity v_y at the interface, one may further obtain

$$-\frac{k_y}{\rho_w \omega} e^{jk_x x} + \sum_{n=-\infty}^{+\infty} \frac{k_y^{(n)}}{\rho_w \omega} R_n e^{jk_x^{(n)} x} = \begin{cases} \frac{k_c}{\tilde{\rho}\omega} C\left(e^{2jk_c H} - 1\right), & x \in (-b, b) \\ 0, \quad x \notin (-b, b). \end{cases} \tag{3.10}$$

Multiplying Eq. (3.7) by $e^{-jk_x^{(r)}x}$ ($r \in \mathbf{Z}$), and averaging over the unit-cell area, we have

$$R_r = \delta_{r,0} - C\left(1 - e^{2jk_c H}\right)\phi \frac{\rho_w k_c}{\tilde{\rho} k_y^{(r)}} S_r. \tag{3.11}$$

Substituting Eq. (3.8) into the pressure continuity condition (6) yields

$$2S_0 - C\left(1 - e^{2jk_c H}\right)\phi^2 \frac{\rho_w}{\tilde{\rho}} \sum_{r=-\infty}^{+\infty} \frac{k_c}{k_y^{(r)}} S_r^2 = C\left(1 + e^{2jk_c H}\right). \tag{3.12}$$

The coefficient C can be determined from Eq. (3.9) and is then substituted into Eq. (3.8). The reflection coefficients of Eq. (3.9) are determined to be

$$R_n = \delta_{n,0} + \frac{2j \tan(k_c H)\phi \frac{\rho_w}{\tilde{\rho}} S_n \frac{k_c}{k_y^{(n)}}}{1 - j\tan(k_c H)\phi \frac{\rho_w}{\tilde{\rho}} \sum_{r=-\infty}^{+\infty} \frac{k_c}{k_y^{(r)}} S_r^2}. \tag{3.13}$$

For second-mode instability waves, normal incidence may be hypothesized, i.e., $k_x = 0$ and $k_y = k_0$ [14]. As a result, the formulations of v_i, p_i and $v_r^{(n)}$, $p_r^{(n)}$ are greatly simplified. Assuming that the periodicity $s \ll \lambda_{acs}$, the effective admittance A can be derived as

$$A = \frac{v}{p}\bigg|_{y=0} = \frac{s^{-1}\int_{-s/2}^{s/2}\left(v_i + \sum_{n=-\infty}^{+\infty} v_r^{(n)}\right)\bigg|_{y=0} dx}{s^{-1}\int_{-s/2}^{s/2}\left(p_i + \sum_{n=-\infty}^{+\infty} p_r^{(n)}\right)\bigg|_{y=0} dx}$$
$$= \frac{\int_{-s/2}^{s/2}\left(-\frac{k_0}{\rho_W\omega} + \sum_{n=-\infty}^{+\infty}\frac{\sqrt{k_0^2-\left(\frac{2\pi n}{d}\right)^2}}{\rho_W\omega}R_n e^{j\frac{2\pi n}{s}x}\right)dx}{\int_{-s/2}^{s/2}\left(1+\sum_{n=-\infty}^{+\infty} R_n e^{j\frac{2\pi n}{s}x}\right)dx},$$
$$= \frac{\frac{k_0}{\rho_W\omega}\int_{-s/2}^{s/2}(R_0-1)dx}{\int_{-s/2}^{s/2}(R_0+1)dx} = \frac{1}{\rho_W c_W}\frac{R_0-1}{R_0+1} \tag{3.14}$$

where the reflection coefficient of the zero-order diffraction (specular refraction) is

$$R_0 = 1 + \frac{2j\tan(k_c H)\phi\frac{\rho_w}{\tilde{\rho}}\frac{k_c}{k_0}}{1 - j\tan(k_c H)\phi\frac{\rho_w}{\tilde{\rho}}\sum_{r=-\infty}^{+\infty}\frac{k_c}{\sqrt{k_0^2-\left(\frac{2\pi r}{s}\right)^2}}S_r^2}. \tag{3.15}$$

Note that all higher-order diffracted modes are evanescent and cannot radiate to far field for normal incidence at low frequencies ($s/\lambda_{acs} \le 1$).

Fedorov's model is first calculated for an isolated deep cavity and then multiplied by porosity ϕ to treat the structured surface as a homogenous surface of uniform effective acoustic admittance [1]. Therefore, the diffracted waves coming from the opening mouth are not considered. To evaluate the overall effect when a plane wave penetrates a porous surface, the present model considers an infinite array of grooves. Because higher-order diffracted modes are included in the derivation, the mutual coupling between disturbances from neighboring cavities is taken into account. In particular, if we neglect all higher-order modes and let the porosity ϕ approach 0 (the local oscillation inside each cavity is independent), Eq. (3.13) reduces to Fedorov's model [9, 10],

$$A = \frac{\phi}{Z_c} j\tan(k_c H), \tag{3.16}$$

where $Z_c = \tilde{\rho}\omega/k_c$ is the characteristic impedance inside the cavity.

Figure 3.2 shows the reflection coefficients of plane monochromatic ultrasonic waves penetrating a porous surface in a quiescent and uniform-temperature atmosphere, with the two above-mentioned theoretical models, numerical results of Brès et al. [10], and by a finite-element solver (COMSOL Multiphysics®). The porous parameters are taken from Brès et al.[10], and listed as a combination of $Ar = 0.12$, 0.3 and $\phi = 0.2, 0.48, 0.8$. It should be noted that some of the high-order diffracted modes become propagative and thus radiate into the propagating mode in free space, when $s/\lambda_{acs} > 1$ [11]. The specular reflection is obfuscated with high-order diffractions in the upper half-space and cannot be separated from the superposed sound field in our full-wave simulation by COMSOL. Therefore, we did not provide the data when $f_{acs} > 0.67$, i.e. $s/\lambda_{acs} > 1$ for the case in Fig. 3.2a. Here, $f_{acs} = fH/c_w = H/\lambda_{acs}$ is the normalized incidence-wave frequency [10], and $H/s = \phi/Ar = 0.67$ for the case in Fig. 3.2a. Also, we would like to emphasize that $s/\lambda_{acs} = 1$ is a reasonable upper limit of the theoretical model. Actually, the numerical results of Brès et al. [10] were also cut off before $f_{acs} = 0.67$ for the case in Fig. 3.2a, but they declared the calculation was strongly affected by the presence of a resonant acoustic mode of frequency $f_{res} = \phi/Ar$ [17]. In addition, they only showed the results before $f_{res}/2$ for the case in Fig. 3.2b, and attributed the reason to the first subharmonic mode [17]. Nevertheless, the influences of the first subharmonic mode are not observed in other cases. For all cases, the predictions of the proposed model basically coincide with the numerical results from COMSOL and Brès et al. [10], even at high frequency when the wavelength λ_{acs} decreases to the same order of magnitude as s and the interaction of the scattered waves at the porous surface becomes strong; for instance, at $f_{acs} = 0.67$ in Fig. 3.2a, when the ratio s/λ_{acs} approaches unity. Comparatively, Fedorov's model tends to shift the predicted frequency because it neglects the higher-order diffracted modes. Despite the homogenous hypothesis of $s \ll \lambda_{acs}$ for both theoretical models, the present model shows the advantage of predicting results that are consistent with numerical results up to the limit $s/\lambda_{acs} = 1$. In particular, when $s/\lambda_{acs} = 1$, a coupling acoustic mode is reportedly induced by the interaction of waves scattered from adjacent cavities [10, 17] and greatly decreases the absorption of the acoustic metasurface. This coupling mode is well reproduced by the proposed model, with the predicted $|R|$ approaching unity. For further illustration, we calculate the distributions of $|R|$ as a function of porosity ϕ at constant $Ar = 0.3$. As shown in Fig. 3.3, the maximum $|R|$ appears along the diagonal (dashed line) of the contour plot, where the incident-wave frequency f_{acs} approaches ϕ/Ar. This observation is exactly the same expression of coupling frequency f_{res} advocated by Brès et al. based on direct numerical simulation [10, 14, 17]. Additionally, when the incident-wave frequency is much lower than f_{res}, both theoretical models above give results that are consistent with numerical results Fig. 3.2d. Moreover, Fig. 3.3 shows that the minimum $|R|$ can be achieved by optimizing cavity-shape parameters.

3.2.2 Metasurface Corrugated with Micropores

The section extends the plane wave expansion method to derive impedance models for acoustic metasurfaces with three-dimensional (3D) pores of circular and square cross-sections, which are considered more practical for engineering applications.

As shown in Fig. 3.4, the metasurface is taken to be a rigid surface that is periodically corrugated with subwavelength pores of circular or square cross-section, and is of infinite extension in the x and z directions. In accordance with the definitions of the geometrical parameters in previous research [15], H and s are the pore depth and unit-cell period, respectively, and b is the radius in the case of a circular pore and the half-width in the case of a square pore. The aspect ratio is defined as $Ar = 2b/H$ for both kinds of pores, and the porosity is defined as $\phi = \pi b^2/s^2$ for a circular pore and $\phi = 4b^2/s^2$ for a square pore. The coating is assumed to be located in a quiescent atmosphere with uniform and constant density ρ and sound speed c.

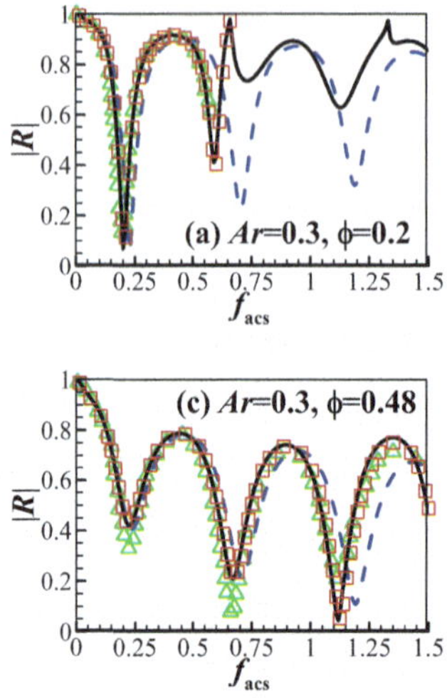

Fig. 3.2 Reflection coefficient amplitudes at normal incidence from numerical simulation by COMSOL (squares), numerical results of Brès et al. [10] (triangle), Fedorov's model (dashed line), and proposed model (solid line). $f_{\mathrm{acs}} = fH/c_{\mathrm{w}}$ is the normalized incidence-wave frequency

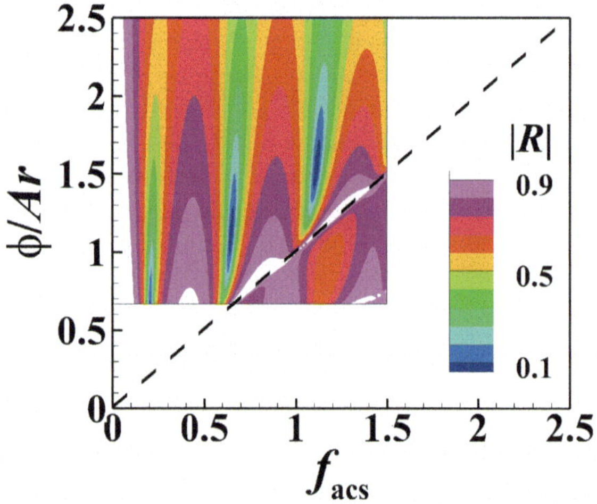

Fig. 3.3 Contours of reflection coefficient amplitude obtained from proposed theoretical model

The acoustic field of an arbitrary incident plane wave penetrating the coating surface can be expressed as (the time dependence $e^{-j\omega t}$ is omitted for simplicity)

$$\begin{aligned} p_i &= e^{jk_x x} e^{jk_z z} e^{-jk_y y}, \\ v_{y,i} &= \frac{1}{j\rho\omega}\frac{\partial p_i}{\partial y} = -\frac{k_y}{\rho\omega} e^{jk_x x} e^{jk_z z} e^{-jk_y y}, \end{aligned} \tag{3.17}$$

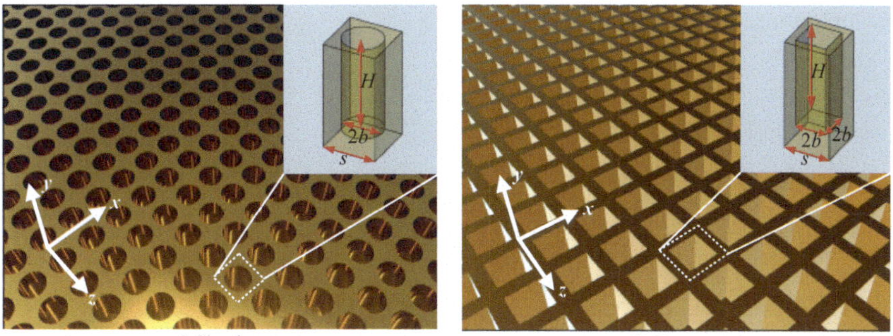

a) Circular pores b) Square pores

Fig. 3.4 Schematic illustrations of acoustic metasurfaces with pores of different cross-sections

where p_i is the incident pressure, $v_{y,i}$ is the y component of the particle velocity, and $j = \sqrt{-1}$. k_x and k_z are the parallel x- and z-momenta, respectively. $k_y = (k_0^2 - k_x^2 - k_z^2)^{1/2}$ is the perpendicular momentum, in which $k_0 = \omega/c$ is the wavenumber with ω being the angular frequency. The reflected pressure field $p_r^{(m,n)}$ and y-component particle velocity $v_{y,r}^{(m,n)}$ of the (m, n)-th order diffracted wave are expressed as

$$\begin{aligned} p_r^{(m,n)} &= R_{mn} e^{jk_x^{(m)}x} e^{jk_z^{(n)}z} e^{jk_y^{(m,n)}y}, \\ v_{y,r}^{(m,n)} &= \frac{k_y^{(m,n)}}{\rho\omega} R_{mn} e^{jk_x^{(m)}x} e^{jk_z^{(n)}z} e^{jk_y^{(m,n)}y}. \end{aligned} \tag{3.18}$$

Here, $k_x^{(m)} = k_x + \frac{2\pi m}{s}$, $k_z^{(n)} = k_z + \frac{2\pi n}{s}$, and $k_y^{(m,n)} = \sqrt{k_0^2 - (k_x^{(m)})^2 - (k_z^{(n)})^2}$, $m, n \in \mathbf{Z}$. R_{mn} is the reflection coefficient of the (m,n)-th order diffraction. Inside the pore, the fundamental wave mode dominates in the long- wavelength limit ($2b \ll \lambda_{\text{acs}}$, where λ_{acs} is the wavelength of the incident acoustic wave), and the sound pressure and particle velocity within the pore are given by

$$\begin{aligned} p_h &= C_1 e^{jk_h y} + C_2 e^{-jk_h y}, \\ v_{y,h} &= \frac{k_h}{\tilde{\rho}\omega}\left(C_1 e^{jk_h y} - C_2 e^{-jk_h y}\right). \end{aligned} \tag{3.19}$$

Here, the dynamic density $\tilde{\rho}$, compressibility $\tilde{C}$, and wavenumber k_h are complex and frequency-dependent quantities owing to the existence of thermal and viscous boundary layers inside the narrow pore:

$$\begin{aligned} \tilde{\rho} &= \rho / \Psi_v, \ \tilde{C} = \frac{\gamma - (\gamma - 1)\Psi_t}{\rho c^2}, \\ k_h^2 &= \omega^2 \tilde{\rho}\tilde{C} = k_0^2 \frac{\gamma - (\gamma - 1)\Psi_t}{\Psi_v}, \end{aligned} \tag{3.20}$$

in which, for a circular pore

$$\Psi_i = -\frac{\mathrm{J}_2(bk_i)}{\mathrm{J}_0(bk_i)}, \tag{3.21}$$

with J_0 and J_2 being Bessel functions of zeroth and second order, respectively, while, for a square pore,

$$\begin{aligned} \Psi_i &= 2k_i^2 \sum_{M=0}^{\infty} \left(\frac{1}{\alpha_M M'}\right)^2 \left(1 - \frac{\tan(\alpha_M b)}{\alpha_M b}\right), \\ \alpha_M &= \sqrt{k_i^2 - (M'/b)^2}, \quad M' = (M + \frac{1}{2})\pi, \quad (M = 0, 1, 2, \ldots). \end{aligned} \tag{3.22}$$

In both Eqs. (3.21) and (3.22), k_i has the same form

$$k_i^2 = \begin{cases} k_v^2 = j\omega \dfrac{\rho}{\mu}, & \text{viscous wave number} \\ k_t^2 = j\omega \dfrac{\rho C_p}{\kappa}, & \text{thermal wave number} \end{cases}. \tag{3.23}$$

The subscript i is either v or t, denoting the effect of the viscous or thermal boundary layer, respectively. In the above equations, κ is the thermal conductivity, μ is the viscosity, and $\gamma = C_p/C_v$ is the ratio of the specific heat at constant pressure C_p to the specific heat at constant volume C_v.

The bottom of the cavity is rigid $\left(v_{y,h}\big|_{y=-H} = 0\right)$ and thus we have $C_1 = C_2 e^{2jk_h H} \equiv C e^{2jk_h H}$. At the interface, the continuity of the sound pressure requires that the mean pressure over the opening area at $y = 0^+$ equals the pressure inside the pore at $y = 0^-$. Application of this condition to the case of a circular pore gives

$$\frac{1}{\pi b^2} \iint\limits_{x^2+z^2 \le b^2} \left(e^{jk_x x} e^{jk_z z} + \sum_{m,n=-\infty}^{+\infty} R_{mn} e^{jk_x^{(m)} x} e^{jk_z^{(n)} z} \right) \mathrm{d}x\mathrm{d}z = C(1 + e^{2jk_h H}). \tag{3.24}$$

After a simple process of deduction, we have

$$\sum_{m,n=-\infty}^{+\infty} (\delta_{mn,00} + R_{mn}) S_{mn} = C(1 + e^{2jk_h H}), \tag{3.25}$$

where $S_{mn} = \frac{1}{\pi b^2} \iint\limits_{x^2+z^2 \le b^2} e^{jk_x^{(m)} x} e^{jk_z^{(n)} z} \mathrm{d}x\mathrm{d}z = \frac{2\mathrm{J}_1\left(b\sqrt{\left(k_x^{(m)}\right)^2 + \left(k_z^{(n)}\right)^2}\right)}{b\sqrt{\left(k_x^{(m)}\right)^2 + \left(k_z^{(n)}\right)^2}}$ is the overlap integral between the (m,n)-th order diffracted mode and the fundamental mode inside the circular pore; $\delta_{mn,00}$ is the Kronecker delta function defined as $\delta_{mn,00} = 1$ for $(m, n) = (0, 0)$ and $\delta_{mn,00} = 0$ otherwise; J_1 is the Bessel function of first order.

The continuity requirement on the particle velocity v_y is that it should be continuous at the opening area and equal to zero elsewhere:

$$-\frac{k_y}{\rho\omega}e^{jk_x x}e^{jk_z z} + \sum_{m,n=-\infty}^{+\infty} \frac{k_y^{(m,n)}}{\rho\omega} R_{mn} e^{jk_x^{(m)}x} e^{jk_z^{(n)}z} = \begin{cases} \frac{k_h}{\tilde{\rho}\omega} C(e^{2jk_hH} - 1) & x, z \in \{x, z | x^2 + z^2 \leq b^2\} \\ 0 & x, z \notin \{x, z | x^2 + z^2 \leq b^2\} \end{cases}. \tag{3.26}$$

We multiply the above equation by $e^{-jk_x^{(r)}x}e^{-jk_z^{(q)}z}$ (where r and q are integers) and average over the unit cell area

$$\frac{1}{s^2} \sum_{m,n=-\infty}^{+\infty} \int_{x,z=-s/2}^{x,z=s/2} \frac{k_y^{(m,n)}}{\rho\omega} \left(\delta_{mn,00} - R_{mn}\right) e^{j(k_x^{(m)} - k_x^{(r)})x} e^{j(k_z^{(n)} - k_z^{(q)})z} \mathrm{d}x\mathrm{d}z = \frac{1}{s^2} \iint_{x^2+z^2 \leq b^2} \frac{k_h}{\tilde{\rho}\omega} C\left(1 - e^{2jk_hH}\right) e^{-jk_x^{(r)}x} e^{-jk_y^{(q)}z} \mathrm{d}x\mathrm{d}z. \tag{3.27}$$

Using the orthogonality of the exponential function, we can solve Eq. (3.27) to give

$$R_{rq} = \delta_{rq,00} - C(1 - e^{2jk_hH})\phi \frac{\rho k_h}{\tilde{\rho} k_y^{(r,q)}} S_{rq}. \tag{3.28}$$

Substituting Eq. (3.28) into the pressure continuity condition (3.25) yields

$$2S_{00} - C\left(1 - e^{2jk_hH}\right)\phi \frac{\rho}{\tilde{\rho}} \sum_{r,q=-\infty}^{+\infty} \frac{k_h}{k_y^{(r,q)}} S_{rq}^2 = C\left(1 + e^{2jk_hH}\right). \tag{3.29}$$

The coefficient C is then determined as

$$C = \frac{2S_{00}}{(1 + e^{2jk_hH}) + (1 - e^{2jk_hH})\phi \frac{\rho}{\tilde{\rho}} \sum_{r,q=-\infty}^{+\infty} \frac{k_h}{k_y^{(r,q)}} S_{rq}^2}, \tag{3.30}$$

and is substituted into Eq. (3.28). With the relation $-j\tan(k_hH) = \frac{1-e^{2jk_hH}}{1+e^{2jk_hH}}$, the reflection coefficients of Eq. (3.28) can then be expressed as

$$R_{mn} = \delta_{mn,00} + \frac{2j\tan(k_h H)\frac{\rho}{\bar{\rho}}\phi\frac{k_h}{k_y^{(m,n)}}S_{mn}}{1 - j\tan(k_h H)\frac{\rho}{\bar{\rho}}\phi\sum\limits_{r,q=-\infty}^{+\infty}\frac{k_h}{k_y^{(r,q)}}S_{rq}^2}. \quad (3.31)$$

Assuming normal incidence of the Mack second mode (i.e. $k_x = 0$, $k_z = 0$ and $k_y = k_0$) [14] and that the periodic spacing $s \ll \lambda_{acs}$, the effective impedance Z of the acoustic metasurface with circular pores can be derived as

$$Z = \frac{p}{v}\Big|_{y=0} = \frac{\frac{1}{s^2}\int_{x,z=-s/2}^{x,z=s/2}(p_i + \sum\limits_{m,n=-\infty}^{+\infty} p_r^{(m,n)})\mathrm{d}x\mathrm{d}z}{\frac{1}{s^2}\int_{x,z=-s/2}^{x,z=s/2}(v_{y,i} + \sum\limits_{m,n=-\infty}^{+\infty} v_{y,r}^{(m,n)})\mathrm{d}x\mathrm{d}z} = \rho c\frac{R_{00}+1}{R_{00}-1}, \quad (3.32)$$

where the reflection coefficient of zeroth-order diffraction (specular refraction) is

$$R_{00} = 1 + \frac{2j\tan(k_h H)\frac{\rho}{\bar{\rho}}\phi\frac{k_h}{k_0}}{1 - j\tan(k_h H)\frac{\rho}{\bar{\rho}}\phi\sum_{r,q=-\infty}^{+\infty}\frac{k_h S_{rq}^2}{\sqrt{k_0^2 - (\frac{2r\pi}{s})^2 - (\frac{2q\pi}{s})^2}}}. \quad (3.33)$$

For the case of a square pore, by applying the continuity requirements of pressure and velocity at the unit surface, a derivation similar to that from Eqs. (3.24)–(3.30) leads to expressions for the impedance and for the reflection coefficients that are similar to those in Eq. (3.32) and in Eqs. (3.31) and (3.33), respectively, except that $S_{mn} = \frac{1}{4b^2}\int_{x,z=-b}^{x,z=b} e^{jk_x^{(m)}x}e^{jk_z^{(n)}z}dxdz = \mathrm{sinc}(k_x^{(m)}b)\mathrm{sinc}(k_z^{(n)}b)$. Since the derivation includes the higher-order diffracted modes, the mutual wave coupling among neighboring pores is taken into account. In particular, Eq. (3.33) reduces to Fedorov's model if all higher-order modes are neglected and the porosity ϕ is allowed to tend to zero (so that the local oscillations inside each pore become independent), and reads

$$R_{00} = 1 + \frac{2j\tan(k_h H)\frac{\rho}{\bar{\rho}}\phi\frac{k_h}{k_0}}{1 - j\tan(k_h H)\frac{\rho}{\bar{\rho}}\phi\frac{k_h}{k_0}}. \quad (3.34)$$

In order to mimic the operational environment as much as possible, a steady flow calculation of a Mach 6 hypersonic flat-plate flow was first conducted [18]. The freestream flow conditions are referred to the wind tunnel experiment of Bountin et al. [19]: Mach number $Ma_\infty = 6.0$, unit Reynolds number $Re_\infty = 10.5 \times 10^6\ \mathrm{m}^{-1}$, and temperature $T_\infty = 43.18$ K. The wall is isothermal with a temperature of $T_w = 293$ K. The flow parameters at the wall ($\rho_w = 5.59 \times 10^{-3}$ kg/m^3 and $p_w = 470.1$ Pa) are obtained at 75% of the plate length where the Mack second mode dominates [18], and are provided to test the acoustic characteristics of the acoustic metasurfaces. The following assumptions are also hypothesized (1) a continuum without rarefaction effect, (2) a perfect gas with Prandtl number $Pr = 0.72$ and $\gamma = 1.4$, and (3) Sutherland-law temperature-dependent viscosity/conductivity. Two corresponding acoustic Reynolds numbers $Re_{acs} = \rho_w c_w b/\mu_w$ of 10 and 100 are chosen. The aspect ratio Ar, porosity ϕ, and the normalized incident-wave frequency ($f_{acs} = fH/c = H/\lambda_{acs}$) are varied in the range of 0.06–0.5, 0.2–0.8, and 0–1.5, respectively. These porous parameters are also relevant for practical applications in high altitude hypersonic flight [10]. Figure 3.5 compares the reflection coefficient contours obtained from the proposed model and Fedorov's model. It is observed that the results of two models are consistent at lower f_{acs}, and diverge when f_{acs} increases for both kinds of coatings (circular and square pores). As addressed in Ref. [15], when f_{acs} increases, the interactions among adjacent pores become strong and the higher-order terms in the proposed model take effect, yielding the differences from the Fedorov's model. By including the data from a finite element solver (COMSOL Multiphysics), the reflection coefficient distributions at different combinations of Ar, ϕ and Re_{acs} are compared in Fig. 3.10. In the previous investigations of the acoustic metasurface with 2D slits [10, 15], a resonant mode of $f_{res} = \phi/Ar$ with $|R|$ approaching unity was found. For coatings with 3D pores, when λ_{acs} approaches s, resonant modes can still be found at $f_{res} = \frac{2}{\sqrt{\pi}}\frac{\sqrt{\phi}}{Ar}$ for circular pores (in particular, $f_{res} = 1.26$ in Fig. 3.10c)), and at $f_{res} = \sqrt{\phi}/Ar$ for square pores (in particular, $f_{res} = 1.12$ in Fig. 3.10d)). The predicted reflection coefficients by the proposed model are in good agreement with the numerical results up to the limit $\lambda_{acs} = s$, whereas that by the Fedorov's models tend to be shifted to the higher f_{acs}. Notably, when the interaction of the scattered waves at the porous surface become strong, i.e., when f_{acs} approaches f_{res}, the predictions by Fedorov's model deviate from the numerical results. In addition, a possible coupling between diagonal pores ($\lambda_{acs} = \sqrt{2}s$) is not found. The acoustic Reynolds number, Re_{acs}, mainly affects the oscillatory behavior of the reflection coefficient. At lower Re_{acs} (e.g., $Re_{acs} = 10$ in Fig. 3.2a)), the viscosity dissipates the acoustic energy in deep pores (e.g. $Ar = 0.2$) without additional reflection from the pore bottoms, and the reflection coefficient monotonically decreases with frequency in Fig. 3.10a. At higher Re_{acs} (e.g., $Re_{acs} = 100$ in Fig. 3.2b, c)), reflections from the pore bottoms lead to destructive/constructive reinforcement at some specific frequencies [14], causing the resonant-like curves in Fig. 3.10b–d.

Notably, Sousa et al. [20] have made comparisons of measured and predicted absorption coefficients, using the inverse Helmholtz solver, the homogeneous absorber theory and the model for acoustic impedance of a porous surface with random porosity used by Fedorov et al. [1] for the classical C/C versus base pressure. The discrepancies between the measured and predicted absorption coefficients may be partially attributed to the lack of considering of the scattering effect in the models. More in-depth investigations of the scattering effects on the C/C composite with an irregular porous surface will be conducted in the future (Figs. 3.6, 3.7, 3.8, 3.9 and 3.10).

According to previous research [21], the stabilization effects of coatings with different slit or pore shapes could be collapsed in good agreement for the same hydraulic diameter $d_h = 4A_p/C$, where A_p is the slit or pore area and C is the circumference. In line with this, and for the same values of ϕ and H, the predicted reflection

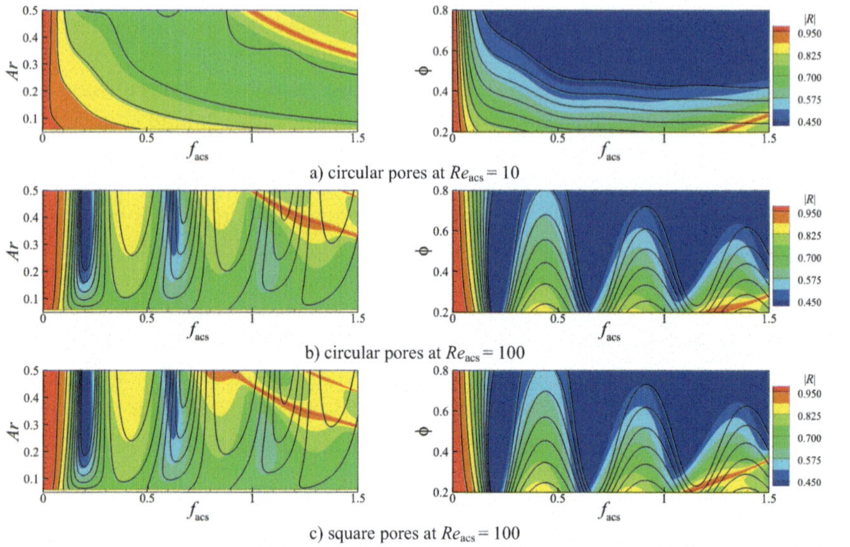

Fig. 3.5 Reflection coefficient contours from the proposed model (color contours) and Fedorov's model (black solid lines). Left column: ϕ is kept constant and $\phi = 0.2$, Right column: Ar is kept constant and Ar $= 0.4$

coefficients are generally consistent with each other up to the resonant frequency of the slits (Fig. 3.11).

3.2.3 Metasurface Corrugated with Random Microstructures

Since the majority of TPS materials have random microstructures, many studies were performed to model the acoustic characteristics of this type of metasurface [22–27]. The formations resemble those in Eq. (3.2). However, there is no rigorous theory to obtain the exact complex dynamic density $\tilde{\rho}$ and the complex dynamic compressibility $\tilde{C}$. Delany and Bazley [28] provided empirical relationships for $\tilde{\rho}$ and $\tilde{C}$ for fibrous absorbent materials that are widely used in various applications such as sound attenuation in dust, room acoustics, and transmission loss through walls. Fedorov et al. [22, 23] derived semi-empirical relations for a felt-metal material (Fig. 3.12). Their model requires additional parameters because of their random nature, including tortuosity a_∞, flow resistivity σ, and characteristic pore size r_p. a_∞ is equivalent to the structure-form factor [29] and it is coupled with the dynamic density $\tilde{\rho}\ (\omega)$ as $a_\infty = \lim_{\omega\to 0} \tilde{\rho}\ (\omega)$; it adopts a value of unity. Further, σ should be measured and fitted in the laboratory, for example, in an impedance tube experiment; σ has a large measurement uncertainty of up to 12.6%, although this uncertainty weakly affects the second-mode amplification [22]. For isotropic porous materials,

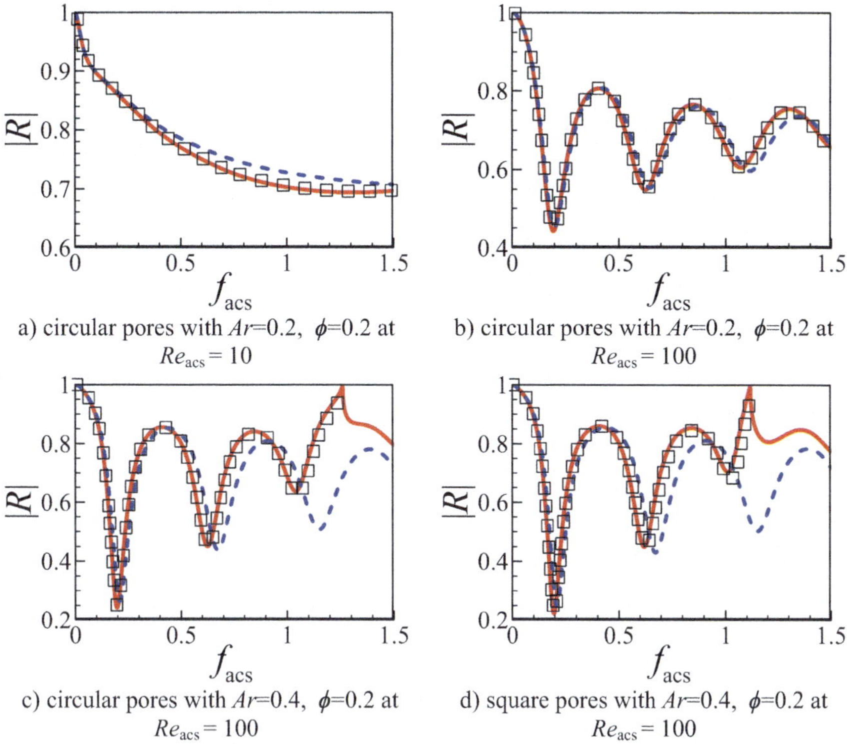

Fig. 3.6 Comparison of reflection coefficient amplitude from COMSOL (black square), the proposed model (red solid curve) and Fedorov's model (blue dashed curve)

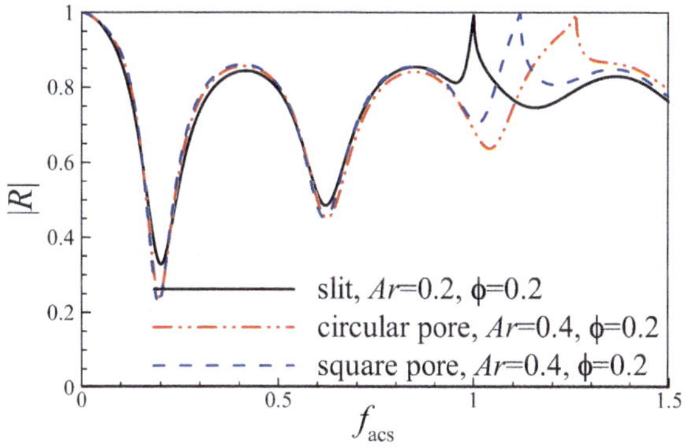

Fig. 3.7 Comparison of the reflection coefficient amplitude distributions for slits or pores with the same hydraulic diameter

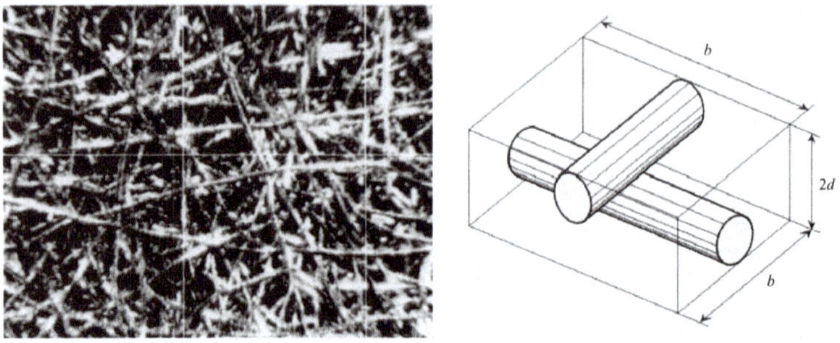

Fig. 3.8 Elementary cell of the felt-metal microstructure; b × b × 2d parallelepiped including two adjoining sections of fibers with diameter d [22]

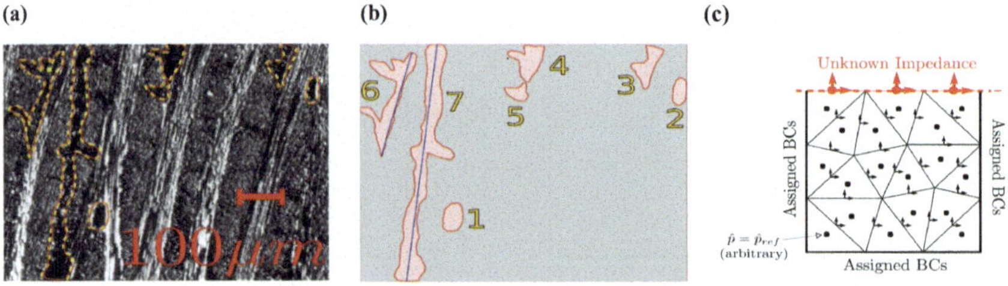

Fig. 3.9 Inverse Helmholtz solver procedure used to calculate the impedance of cavities typical of blind-hole porous surfaces, **a** high-definition image of the C/C block, **b** detected large cavity edges, and **c** simplified two dimensional computational setup [26, 27]

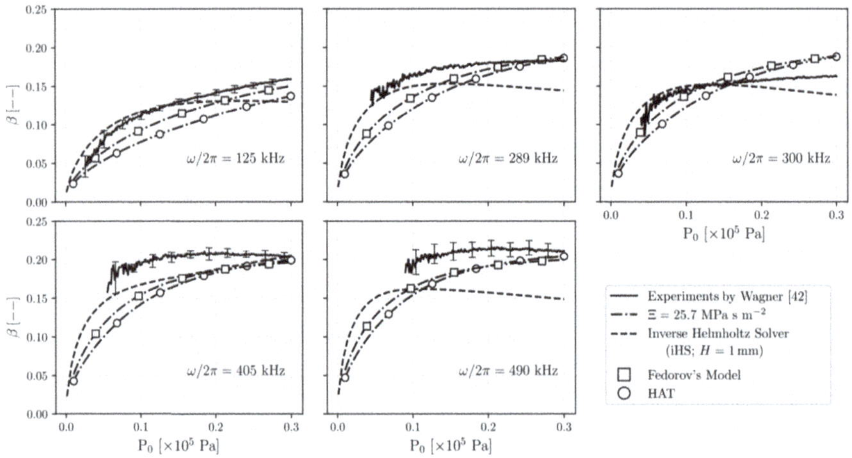

Fig. 3.10 Comparison of measured and predicted absorption coefficient β for the classical C/C versus base pressure [20]

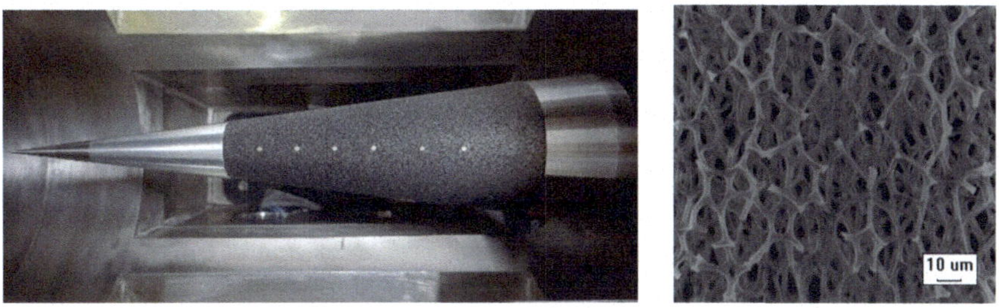

Fig. 3.11 Left: sharp cone model with porous carbon surface; right: microstructures in X-ray microscopy

r_p can be expressed as $r_p = s_h\sqrt{8\mu a_\infty/\sigma\phi}$, where s_h denotes the dimensionless shape factor [30]. The felt-metal is highly anisotropic, and Fedorov et al. [22, 23] treated r_p as a hydraulic radius. They considered an average statistical elementary cell simulating the topology of the felt metal microstructure (Fig. 3.12), and they represented r_p as a function of the porosity ϕ and fiber diameter d, which can be measured experimentally.

$$r_p = \frac{\pi d}{2(1-\phi)(2-\phi)}. \tag{3.35}$$

In contrast to the above analytical impedance models for metasurfaces with regular and felt-metal microstructures, the German DLR and Purdue University adopted the inverse Helmholtz solver (IHS) to estimate the C/C impedance. The IHS is a computational methodology that allows the evaluation of the spatial distribution of acoustic impedance at the open surface of an arbitrarily shaped cavity for a given

Fig. 3.12 Statistical porosity

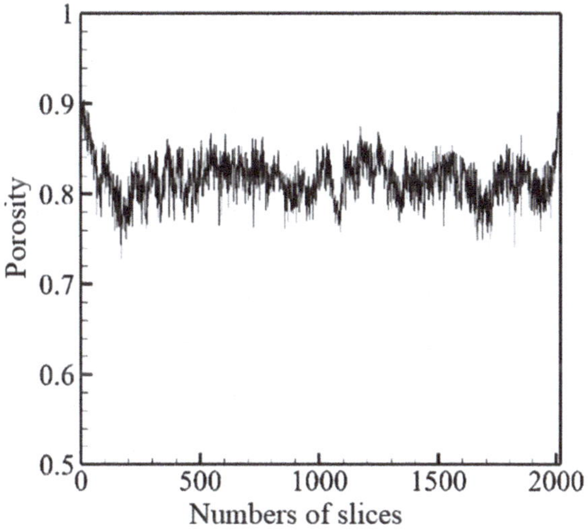

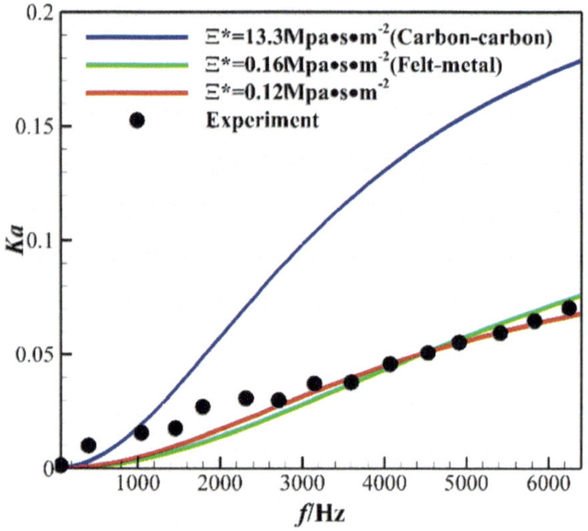

Fig. 3.13 Comparison of absorption coefficient of foam carbon materials

frequency [25–27]. In the application, a high-resolution image of the C/C block (Fig. 3.13) auto-detects the edges of large surface cavities depending on the grayscale value (Fig. 3.13b), makes an automatic approximation to a rectangular slot or a cylindrical pore based on the extent of the pores in each direction, and meshes their volumes to serve as an input geometry to the IHS (Fig. 3.13c). Finally, the broadband impedance of each individual surface cavity is evaluated. These impedances are combined to yield the effective surface-averaged impedance for the C/C block.

Sousa et al. [31] compared analytical impedance models and the IHS method in the prediction of the acoustic characteristics of C/C materials at various base pressures. The analytical impedance models used the homogeneous absorber theory (HAT) [32] and the felt-metal model proposed by Fedorov et al. [22, 23]. Figure 3.14 shows that the IHS can predict a consistent absorptive coefficient with the experiment by Wagner et al. [27], especially in an environment of lower base pressure. When the test base pressure increases or the incident wave frequency increases, the prediction accuracy of the IHS degrades significantly. The discrepancy in the IHS results was attributed to the underestimated volume porosity, given that only a limited selection of the largest visible pores in Fig. 3.13a was used for the direct impedance estimate [31]. Compared to the HAT model, the felt-metal model had a higher accuracy and showed a better tendency with the experiment.

Our team has theoretically and experimentally analyzed the stabilization effect of foam carbon surface in a Mach 6 sharp-cone BL flow. This kind of metasurface is made of foam carbon material by powder sintering, and there are lots of irregular pore structures inside, which has the characteristics of high porosity, low density and high temperature resistance.

In the theoretical analysis, the felt-metal model was applied. In Eq. (3.2)

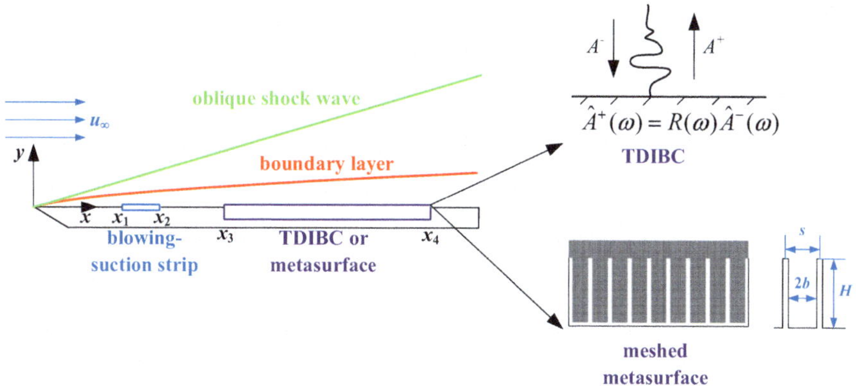

Fig. 3.14 Schematic drawing of the physical problem

$$
\begin{aligned}
&\Lambda = -\frac{\mathrm{i}\omega M_e}{\sqrt{T_w}}\sqrt{\tilde{\rho}\tilde{C}}, \quad Z_0 = \frac{\sqrt{\tilde{\rho}/\tilde{C}}}{M_e\sqrt{T_w}} \\
&\tilde{\rho} = a_\infty\left[1 + \frac{g(\lambda_1)}{\lambda_1}\right], \quad \tilde{C} = \gamma - \frac{\gamma - 1}{1 + g(\lambda_2)/\lambda_2} \\
&g(\lambda) = \sqrt{1 + \frac{4a_\infty\mu_w^*\lambda}{\sigma\phi r_p^{*2}}} \\
&\lambda_1 = -\mathrm{i}a_\infty\rho_w^*\omega^* / (\phi\sigma), \lambda_2 = 4\,\mathrm{Pr}\,\lambda_1
\end{aligned}
\tag{3.36}
$$

where ϕ is statistical porosity, a_∞ is equivalent to the structure-form factor and has a value of unity, and σ is the flow resistivity, usually measured experimentally. The characteristic size $r_p{}^*$ can be calculated from porosity and flow resistivity:

$$
r_p^* = s_h\sqrt{\frac{8\mu a_\infty}{\sigma\phi}}
\tag{3.37}
$$

where r_p^* is a shape parameter. Since the pore structure of foam carbon material is approximately a sphere, s_h adopts unity. Because of the natural irregularity of the pore structures of foam carbon, it is necessary to test ϕ and σ through experiments to provide input for its acoustic impedance model. As shown in Fig. 3.15 (right map), the 3D reconstruction of foam carbon materials was carried out with Zeiss Xradia520 high-resolution 3D X-ray microscope. We cut the foam carbon material into 2016 slices, and divide the pore area on each slice by the slice area, and get the porosity of each slice. As shown in Fig. 3.12, the porosity of each slice is mainly between 76 and 86%. The average porosity of the foam carbon material is 83.3% according to the statistics of each slice (Fig. 3.16).

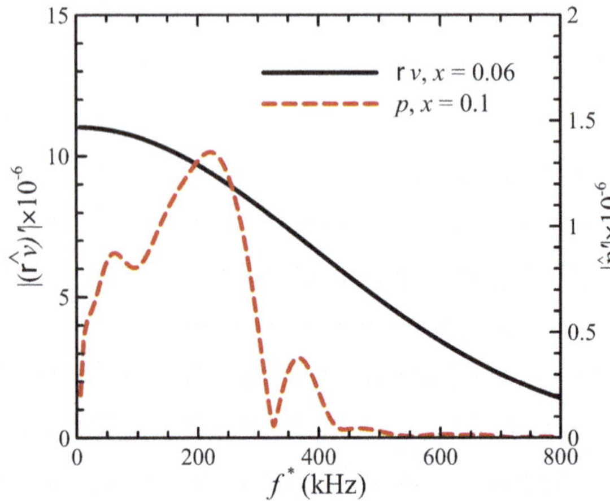

Fig. 3.15 Spectra of the wall-normal mass flow rate at x = 0.06 in the blowing-suction strip and of the pressure at x = 0.1 downstream

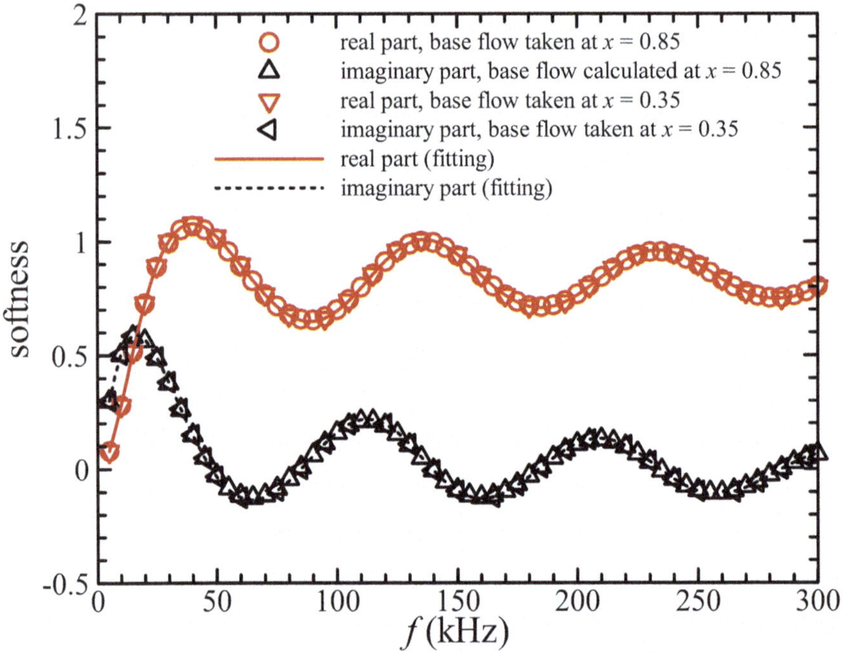

Fig. 3.16 Comparison of wall softness W(ω) between the impedance model (symbols) and the multi-pole fitting results (lines)

The flow resistivity σ is obtained by fitting the absorption characteristics with the data measured in s impedance tube. As shown in Fig. 3.13, the calculated absorption characteristics could predict a more consistent results with experimental data when σ is chosen 0.12 Mpa·s·m^{-2}.

3.3 Implementation in DNS

3.3.1 Implementation Method

The impedance $Z(\omega)$ is originally defined in the frequency domain and should be solved in the time domain when coupling with the (N–S) equations. For a single disturbance frequency ω, this relationship is written

$$v_{\mathrm{w}}=p'_{\mathrm{w}}\mathrm{Real}(1/Z) - \frac{1}{\omega}\frac{\partial p_{\mathrm{w}}{}'}{\partial t}\mathrm{Imag}(1/Z) \tag{3.38}$$

It can deal with the metasurface stabilization effect on the most unstable mode, but fails to simulate the response of the broadband disturbances in the real flight or most wind tunnel conditions. Mathematically, the time-domain equivalent of $Z(\omega)$ can be derived by taking its inverse Fourier transform. However, the inverse Fourier transform of the frequency-domain impedance boundary condition leads to a convolution integral. The computation of the convolution integral is time-consuming, especially when coupling with DNS studies. Fund and Ju [33] established an elegant formulation of the time-domain impedance boundary condition (TDIBC) to avoid repeated Fourier transforms and save computational cost.

Given the renormalized pressure fluctuation $p'_a = p' \big/ (\overline{\rho}_w \overline{a}_w)$, the amplitude of the up-traveling reflection and down-traveling incident characteristic waves are given by:

$$A^+ = v' + p'_a,\ A^- = v' - p'_a, \tag{3.39}$$

respectively. The two waves are causally correlated by the reflection coefficient in the frequency domain as $\hat{A}^+(\omega) = R(\omega)\hat{A}^-(\omega)$. Based on the specific impedance $Z_a = -\hat{p}_a(\omega)/\hat{v}(\omega)$, the reflection coefficient is computed by $R(\omega) = (1 - Z_a)\big/(1 + Z_a)$. The wall softness W is then obtained as $W(\omega) = 1 + R(\omega) = 2/(1 + Z_a)$. Accordingly, it yields:

$$\hat{A}^+(\omega) = -\hat{A}^-(\omega) + W(\omega)\hat{A}^-(\omega). \tag{3.40}$$

Fung and Ju [33] proposed to approximate the softness by a multi-oscillator model:

$$W(\omega) \approx \sum_{k=1}^{n_k} \left(\frac{\beta_k}{i\omega - p_k} + \frac{\beta_k^\dagger}{i\omega - p_k^\dagger} \right), \tag{3.41}$$

where (p_k, β_k) represent the pole and residue, the superscript † denotes complex conjugate, and n_k is the total number of conjugate pairs. By applying the Laplace transform and assuming Re(p_k) < 0, Eq. (3.40) is transformed into the time domain at the instant $t + \Delta t$:

$$A^+(t + \Delta t) = -A^-(t + \Delta t) + \sum_{k=1}^{n_k} G_k^-(t + \Delta t) + \text{c.c.}, \tag{3.42}$$

where.

$$G_k^-(t + \Delta t) = \int_0^\infty \beta_k e^{p_k \tau} A^-(t + \Delta t - \tau) d\tau, \tag{3.43}$$

and c.c. denotes complex conjugate. The above equation naturally constitutes a recursive scheme due to

$$G_k^-(t + \Delta t) = \theta_k G_k^-(t) + \int_0^{\Delta t} \beta_k e^{p_k \tau} A^-(t + \Delta t - \tau) d\tau; \, \theta_k = e^{p_k \Delta t}. \tag{3.44}$$

Numerically, a second-order trapezoidal quadrature rule is applied to evaluate Eq. (3.44). The incident wave $A^-(t + \Delta t)$ is given by the first-order extrapolation at $y = 0$ suggested by Scalo et al. [34]. Given the incident wave amplitude $A^-(t + \Delta t)$, the reflection wave $A^+(t + \Delta t)$ is obtained by Eq. (3.42). Subsequently, the wall-normal velocity at the next time step is updated by

$$v'(t + \Delta t) = \left(A^+ + A^-\right)/2 = \text{Re}\left(\sum_{k=1}^{n_k} G_k^-(t + \Delta t) \right), \tag{3.45}$$

while p' is calculated via Eq. (3.39). Finally, the instantaneous wall-normal velocity and pressure on the wall boundary are obtained by superimposition on the corresponding base-flow variables.

Our team has carried a comprehensive comparative study between the TDIBC and an acoustic metasurface case with meshed micro-cavities. As shown by Fig. 3.18, a Mach 6 two-dimensional (2-D) flat-plate boundary layer with a sharp leading edge is directly simulated. The free-stream conditions, identical to the wind-tunnel ones by Bountin et al. [19] are given as follows: Mach number $Ma_\infty = 6.0$, static temperature $T_\infty^* = 43.18$ K and unit Reynolds number $\text{Re}_\infty^* = 1.05 \times 10^7$ m^{-1}, where the asterisk denotes the dimensional quantity. The wall is assumed to be isothermal with a temperature of $T_w^* = 293$ K. Besides, a no-slip velocity boundary condition is given on the wall boundary, and a no-penetration condition is imposed expect for the blowing-suction strip and impedance boundaries. A perfect gas assumption is made with the specific heat ratio being $\gamma = 1.4$. The Prandtl number is assumed to be $Pr = 0.72$, and the dynamic viscosity is computed by the Sutherland's law. The primitive variables are nondimensionalized by the corresponding freestream base-flow quantities, except that the pressure p is by $\rho_\infty u_\infty^2$, where ρ and u are the density and streamwise velocity, respectively. The reference length scale L^* ref = 0.2 m is used for nondimensionalization. A single-frequency or Gaussian broadband disturbance is introduced through the wall blowing-suction technique between $x_1 = 0.0523$ and $x_2 = 0.0773$ in Fig. 3.18. The surface wall-normal momentum density is disturbed, whose dimensionless form is given by

$$(\rho v)'_w(x,t) = \begin{cases} \epsilon \sin\left(2\pi \dfrac{x - x_1}{x_2 - x_1}\right) \sin(\omega t), & \text{for single - frequency disturbance,} \\ \epsilon \sin\left(2\pi \dfrac{x - x_h}{x_2 - x_1}\right) \exp\left\{-\dfrac{(t - t_a - [t/T]T)^2}{2\sigma_b^2}\right\}, & \text{for broadband disturbance.} \end{cases} \quad (3.46)$$

Here, the amplitude $\epsilon = 0.001$ is given, and t denotes time. For the single-frequency case, the angular frequency is $\omega = 220.6$ and the dimensional frequency is $f^* = 138.74$ kHz, which finally evolve into a second mode in around $x > 0.5$. For the broadband disturbance case in Eq. (3.46) the parameters are given by: $x_h = (x_1 + x_2)/2$, $t_a = 0.0119$, $T = 1.6$ and $\sigma_b = 0.0016$. Furthermore, the operator "[·]" in Eq. (3.46) represents a floor function, i.e., the greatest integer function. Thus, the broadband disturbance is generated in the form of a pulse wavepacket with a period of $T = 1.6$. The spectra of the broadband mass-flow pulse in the blowing-suction strip and the response of the pressure right downstream are shown in Fig. 3.15. Throughout this paper, fast Fourier transform (FFT) is performed to obtain spectra of signals after taking sampling every 100 time steps during a period of the broadband pulse.

Downstream of the forcing region in Fig. 3.18, the TDIBC or the acoustic metasurface with subwavelength grooves (2-D cavities) is imposed between x_3 and x_4, where $x_3 = 0.2$ and $x_4 = 1$. Sponge zones are placed in the range $x > 1$ to minimize the reflection of disturbances near the outflow boundary. In the schematic drawing of the TDIBC, A^+ and A^- represent the up-traveling reflection and down-traveling incident characteristic waves, respectively, and R is the reflection coefficient to be modeled. With regard to the metasurface case, the dimensionless half width $b = 9.8 \times 10^{-4}$, the unit-cell period $s = 2.6 \times 10^{-3}$ and depth $H = 8.21 \times 10^{-3}$ are identical to those in Ref. [35], as shown in Fig. 3.14. The geometric setting corresponds to a porosity of $\phi = 2b / s \approx 0.75$.

The determination of the conjugate pairs of poles and residues in Eq. (3.41) is not unique. Algorithms, such as the iterative least square fitting in Ref. [36], enable a satisfactory fitting to the discrete dataset provided by either experimental results or reliable impedance models. When the frequency exceeds 300 kHz in the

Table 3.1 Conjugate pairs of dimensionless poles and residues

Pair number k	Poles p_k	Residues β_k
1	(− 111.980, 595.313)	(109.678, 20.631)
2	(− 203.277, 3871.378)	(269.167, 14.133)
3	(− 48.977, − 209.187)	(32.931, − 7.710)
4	(− 10.673, − 0.033)	(0.798, − 259.963)
5	(− 17.116, − 997.046)	(146.984, − 2.523)
6	(− 1145.867, 297.759)	(58.582, 225.439)
7	(− 325.950, − 2670.483)	(204.960, − 25.017)
8	(− 329.014, 2729.641)	(248.601, 29.965)
9	(− 88.475, − 360.592)	(55.609, − 13.644)
10	(− 52.805, 28.276)	(55.756, 104.123)

considered broadband disturbance, these spectral components decay rapidly for both metasurface and TDIBC cases (shown in Fig. 3.22 later). Thus, we only concentrate on the fitting of the acoustic wall softness below 300 kHz. Figure 3.16 shows an excellent fitting agreement within the frequency range of interest. Meanwhile, in the considered flat-plate boundary layer, Fig. 3.16 illustrates that the wall softness $W(\omega)$ calculated based on the impedance model is entirely insensitive to the selection of the streamwise location for base-flow quantities. Ten conjugate pairs of poles and residues are applied in the final fitting function, where the information is detailed in Table 3.1.

3.3.2 *Single-Frequency-Disturbance Case*

Starting from a steady solution at $t = 0$, the propagating single-frequency instability wave finally penetrates the metasurface region $0.2 < x < 1$. Figure 3.17 shows that the TDIBC model is entirely consistent with the direct coupling method for a constant wall impedance. Compared to the real metasurface case, these models slightly underestimate the pressure fluctuation. It is also observed that the maximum pressure fluctuation of the impedance models approaches the lower bound of the streamwise oscillating fluctuation of the real metasurface case. This is consistent with the observation in Ref. [35] under the same condition (see Fig. 3.7 in Ref. [35]). The streamwise oscillation of p' in the real metasurface case has also been reported in Ref. [35].

It is preliminarily deduced from Fig. 3.17 that the minor deviation between the impedance models and the metasurface does not result from the numerical error by the TDIBC procedures, such as the extrapolation or convolution manipulation, since the direct coupling method does not involve these effects. Besides, as shown in Fig. 3.18, the micro-cavities of the metasurface can distort the near-wall base flow

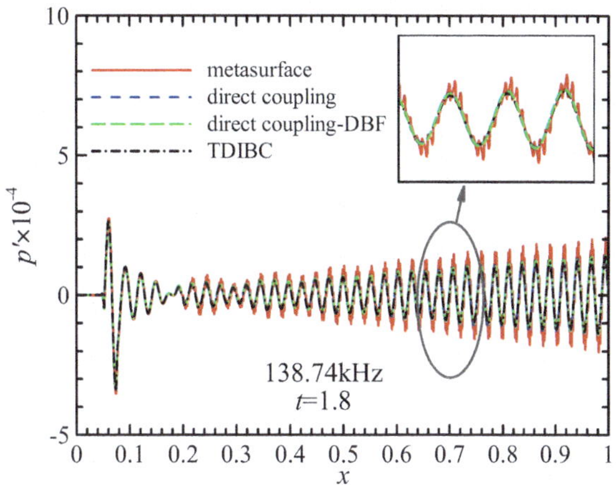

Fig. 3.17 Comparison of the pressure fluctuation along y = 0 for the single-frequency case at t = 1.8

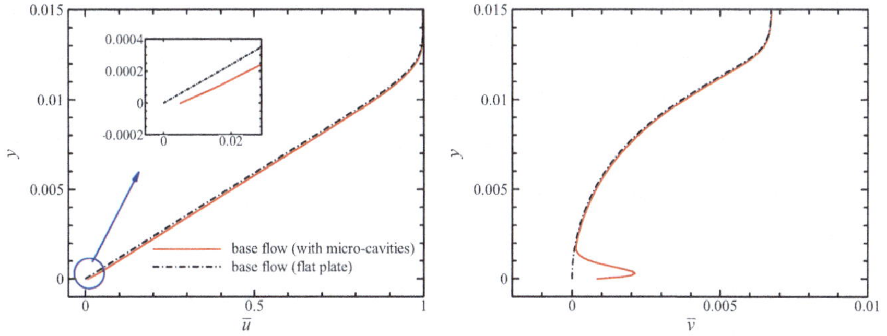

Fig. 3.18 The cavity-induced distortion in the base flow streamwise velocity (left) and wall-normal velocity (right) at x = 0.8, which is at the three-quarter length of the cavity

profiles via mass transfer across the shear layer. To consider this effect, the 2-D base flow of the direct-coupling-method case, i.e., the flat-plate boundary layer, is replaced with that of the real metasurface corresponding to the solid lines in Fig. 3.18. The wall boundary condition is adjusted to allow the velocity slip and penetration as is in the real metasurface case. The corresponding result is given in Fig. 3.17 labeled as "direct coupling-DBF" (i.e., distorted base flow), which nearly overlaps with the counterpart of the original direct coupling method. Therefore, the deviation between the TDIBC and the real metasurface is not attributed to the base-flow-distortion effect induced by micro-cavities. The responsible effect is thus inferred to be the minor impedance modeling error.

3.3.3 Broadband-Disturbance Case

Figure 3.19 compares the maximum pressure fluctuation in history between the TDIBC and the real metasurface cases for broadband disturbances. Result of a smooth-wall boundary layer case without any metasurface or impedance boundary is also shown. The metasurface starting from $x = 0.2$ effectively suppresses the peak pressure fluctuation. Although streamwise oscillation occurs in the real metasurface case (also reported in Ref. [35]), favorable agreement is reached between TDIBC and metasurface cases. Figure 3.20 further displays the 2-D pressure and wall-normal velocity fluctuations between the metasurface (contour) and TDIBC (line) cases at two instants. The near-field wave packet as well as the far-field propagating wave interacting with the leading-edge shock wave is well captured. The peak and valley characteristics are accurately given by TDIBC, particularly for the downstream wave front region. Quantitatively, Fig. 3.21 compares the instantaneous pressure fluctuation along the boundary $y = 0$ at $t = 0.4$ and $t = 0.8$. The amplitude and phase information is satisfactorily predicted by TDIBC. The main discrepancy is observed in the tail region of the wave packet.

For clarification purposes, Fig. 3.22 further shows wall pressure spectra in the metasurface or TDIBC region. We aim to identify what frequency bands are stabilized or destabilized, and what bands belong to the first-mode or second-mode instabilities. The two selected probes at $x = 0.35$ and at $x = 0.65$ are located in the cavity region of the metasurface, while that at $x = 0.45$ belongs to the solid-wall region of the metasurface. For the smooth-wall case, the components concentrated mostly between 100 and 200 kHz are amplified, while the high-frequency band is generally damped. With the installation of the metasurface, the stabilization band compared to the smooth-wall case is gradually enlarged as the increase of the streamwise coordinate. Take the location $x = 0.45$ for example, a wide range of [93 kHz, 300 kHz] covering the main energetic band is effectively stabilized by the metasurface compared to the smooth-wall case. The low-frequency range below 93 kHz, subsequently proved to belong

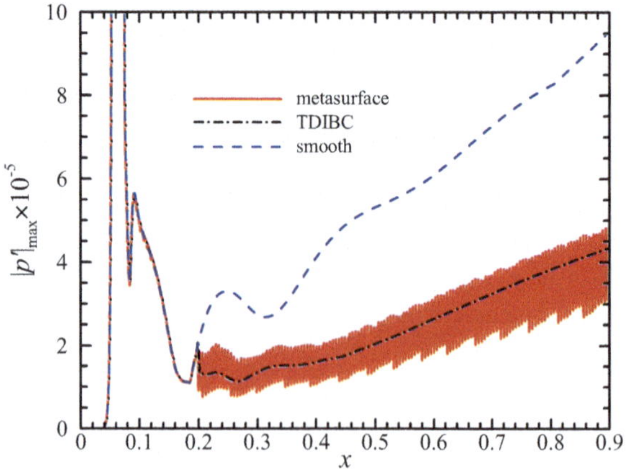

Fig. 3.19 Comparison of the maximum pressure fluctuation along y = 0 for the broadband disturbance case

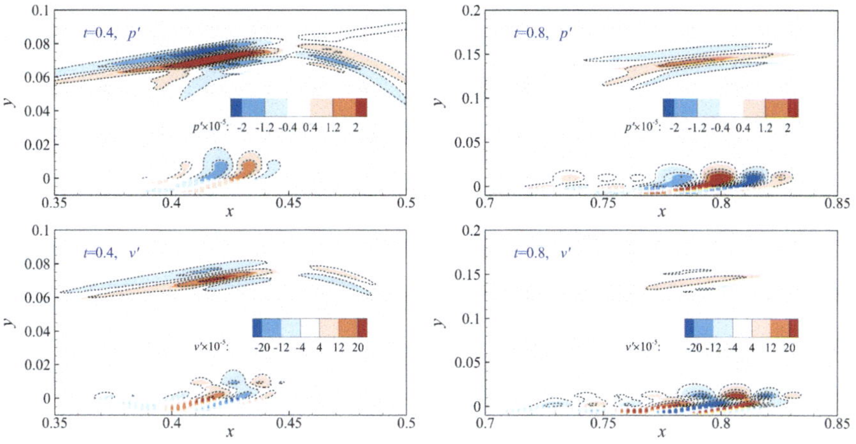

Fig. 3.20 Contours of the pressure and wall-normal velocity fluctuations at t = 0.4 (left) and t = 0.8 (right) for the metasurface (contour) and TDIBC (dashed line) broadband-disturbance cases

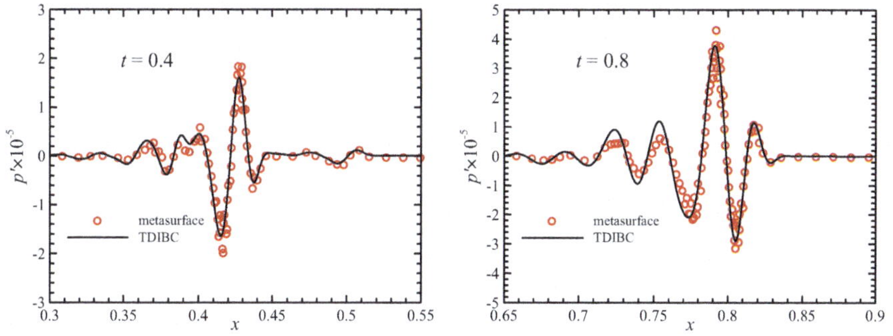

Fig. 3.21 Comparison of the pressure fluctuation for the broadband-disturbance case at t = 0.4 (left) and t = 0.8 (right) along y = 0 between the metasurface and different TDIBC models

to the first mode instability, is destabilized at $x = 0.45$. Besides, the range between 100 and 150 kHz reports the majority of the discrepancy between the TDIBC and metasurface cases, which will be discussed later.

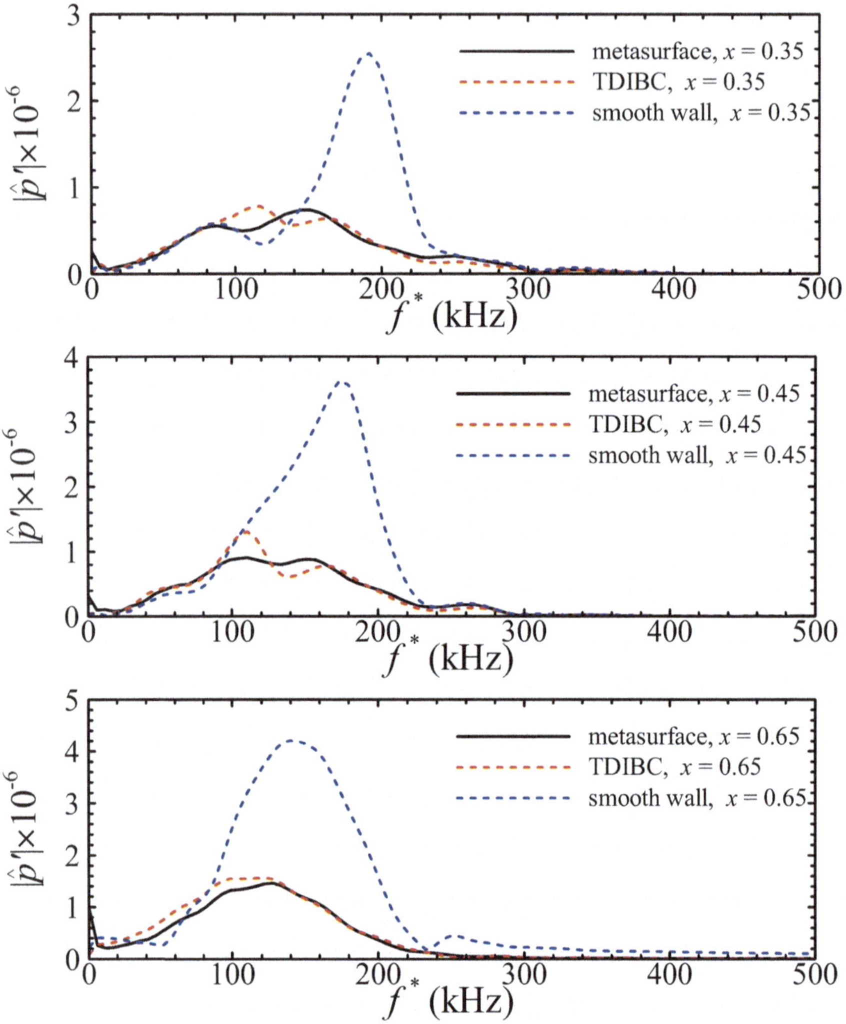

Fig. 3.22 Comparison of wall pressure spectra for the broadband-disturbance case at x = 0.35 (top), x = 0.45 (middle) and x = 0.65 (bottom)

3.4 Conclusion

The proposed impedance models consider high-order diffracted modes and therefore incorporate scattered-wave interactions among adjacent slits/pores. In contrast with the results of Fedorov's models, the predicted reflection coefficients have been shown to be consistent with numerical results from a finite element solver (COMSOL Multiphysics), and the coupling modes induced by neighboring slits/pores are reproduced well. The distributions of the reflection coefficient show the oscillatory behavior at higher acoustic Reynolds numbers, otherwise monotonically decrease especially for deep pores. Therefore, a straightforward optimization strategy involving program loops is introduced to achieve a maximum absorptive performance. The Mack second mode is strongly suppressed by the optimized porous coating, with the maximum fluctuating pressure decreasing by about 88%. Additionally, the acoustic characteristics are in general agreement for coatings with different slit or pore shapes, provided these have the same hydraulic diameter d_h, porosity ϕ and depth H.

Through a careful comparative study with an acoustic metasurface with micro-cavities, the conventional TDIBC model is shown to predict the propagation of both single-frequency and broadband disturbances accurately. Based on the case studies, the minor discrepancy in the wall fluctuation is found to arise from the impedance modeling error rather than the effects of numerical aspects or cavity-induced base flow distortion. For a broadband pulse, the discrepancy between TDIBC and the real metasurface is mainly observed in the tail of the wave packet, which corresponds to the spectral deviation. The spectra indicate that the impedance model becomes less applicable either in the vicinity of the switching band from the first to the second mode or in the pure first-mode band.

References

1. Federov AV, Malmuth ND, Rasheed A, Hornung HG (2001) Stabilization of hypersonic boundary layers by porous coating. AIAA J 39(4):605–610. https://doi.org/10.2514/3.14776
2. Gaponov SA (1971) Influence of porous layer on boundary-layer stability, Izevestia SO AN SSSR. Seria Technicheskich Nauk, Vyp 1:21–23 (in Russian)
3. Gaponov SA (1975) Influence of gas compressibility on stability of boundary layer on porous surface at subsonic speeds. Zhurnal Prikladnoi Mechaniki i Technicheskoi Fiziki 1:121–125 (in Russian)
4. Gaponov SA (1977) Stability of supersonic boundary layer on porous wall with heat conductivity. Izvestia AN SSSR, Mechanika Zhidkosti I Gaza 1:41–46 (in Russian)
5. Daniels FB (1950) On the propagation of sound waves in a cylindrical conduit. J Acoustic Soc Am 22:563–564. https://doi.org/10.1121/1.1906650
6. Benade AH (1968) On the propagation of sound waves in a cylindrical conduit. J Acoustic Soc Am 44:616–623. https://doi.org/10.1121/1.1906650
7. Johnson DL, Koplik J, Dashen R (1987) Theory of dynamic permeability and tortuosity in fluid saturated porous media. J Fluid Mech 176:379–402. https://doi.org/10.1017/s0022112087000727
8. Maslov AA (2003) Experimental and theoretical studies of hypersonic laminar flow control using ultrasonically absorptive coatings (UAC). Int Sci Technol Cent, Rept 2172–2001, Moscow, May 2003
9. Kozlov VF, Fedorov AV, Malmuth ND (2005) Acoustic properties of rarefied gases inside pores of simple geometries. J Acoust Soc Am 117(6):3402–3412
10. Brès GA, Inkman M, Colonius T, Fedorov AV (2013) Second-mode attenuation and cancellation by porous coatings in a high-speed boundary layer. J Fluid Mech 726:312–337. https://doi.org/10.1017/jfm.2013.206
11. Wu T, Cox TJ, Lam YW (2000) From a profiled diffuser to an optimized absorber. J Acoust Soc Am 108(2):643–650. https://doi.org/10.1121/1.429596

12. Schwan L, Geslain A, Romero-García V, Groby JP (2017) Complex dispersion relation of surface acoustic waves at a lossy metasurface. Appl Phys Lett 110(5):051902. https://doi.org/10.1063/1.4975120
13. Liu T, Liang S, Chen F, Zhu J (2018) Inherent losses induced absorptive acoustic rainbow trapping with a gradient metasurface. J Appl Phys 123(9):091702. https://doi.org/10.1063/1.4997631
14. Brès GA, Colonius T, Fedorov AV (2010) Acoustic properties of porous coatings for hypersonic boundary-layer control. AIAA J 48(2):267–274. https://doi.org/10.2514/1.40811
15. Zhao R, Liu T, Wen CY, Zhu J, Cheng L (2018) Theoretical modeling and optimization of porous coating for hypersonic laminar flow control. AIAA J 56(8):2942–2946. https://doi.org/10.2514/1.J057272
16. Zhao R, Zhang XX, Wen CY (2020) Theoretical modeling of porous coatings with simple microstructures for hypersonic boundary-layer stabilization. AIAA J 58(2):981–986. https://doi.org/10.2514/1.J058403
17. Brès GA, Colonius T, Fedorov AV (2008) Interaction of acoustic disturbances with micro-cavities for ultrasonic absorptive coatings. AIAA Paper 2008–3903
18. Zhao R, Wen CY, Tian XD, Long TH, Yuan W (2018) Numerical simulation of local wall heating and cooling effect on the stability of a hypersonic boundary layer. Int J Heat Mass Transf 121:986–998. https://doi.org/10.1016/j.ijheatmasstransfer.2018.01.054
19. Bountin D, Chimitov T, Maslov A, Novikov A, Egorov I, Fedorov A, Utyuzhnikov S (2013) Stabilization of a hypersonic boundary layer using a wavy surface. AIAA J 51(5):1203–1210. https://doi.org/10.2514/1.J052044
20. Sousa VCB, Patel D, Chapelier J-B, Wartemann V, Wagner A, Scalo C (2019) Numerical investigation of second-mode attenuation over carbon/carbon porous surfaces. J Spacecr Rocket 56(2):319–332. https://doi.org/10.2514/1.A34294
21. Sandham ND, Lüdeke H (2009) A numerical study of mach 6 boundary-layer stabilization by means of a porous surface. AIAA J 47(9):2243–2252. https://doi.org/10.2514/1.43388
22. Fedorov AV, Shiplyuk AN, Maslov AA, Burov E, Malmuth ND (2003) Stabilization of a hypersonic boundary layer using an ultrasonically absorptive coating. J Fluid Mech 479:99–124. https://doi.org/10.1017/S0022112002003440
23. Fedorov A, Shiplyuk A, Maslov A, Burov E, Malmuth N (2015) Stabilization of a hypersonic boundary layer using an ultrasonically absorptive coating - corrigendum. J Fluid Mech 769:725–728. https://doi.org/10.1017/jfm.2015.155
24. Tritarelli RC, Lele SK, Fedorov AV (2015) Stabilization of a hypersonic boundary layer using a felt-metal porous coating. J Fluid Mech 769:729–739. https://doi.org/10.1017/jfm.2015.156
25. Patel D, Gupta P, Scalo C, Rothermel T, Kuhn M (2017) Towards impedance characterization of carbon-carbon ultrasonically absorptive cavities via the inverse Helmholtz problem. AIAA Paper 2019–2151. https://doi.org/10.2514/6.2017-0460
26. Patel D, Gupta P, Scalo C (2017) Surface impedance determination via numerical resolution of the inverse Helmholtz problem. AIAA Paper 2017-3695. https://doi.org/10.2514/6.2017-3695
27. Wagner A, Martinez Schramm J, Dittert C, Sousa V, Patel DI, Scalo C (2018) Experimental and numerical acoustic characterization of ultrasonically absorptive porous materials. AIAA Paper 2018-2948. https://doi.org/10.2514/6.2018-2948
28. Delany MA, Bazley EN (1970) Acoustic properties of fibrous absorbent materials. Appl Acoust 3:105–116. https://doi.org/10.1016/0003-682X(70)90031-9
29. Zwikker, Kosten CW (1949) Sound absorbing materials. Elsevier, New York
30. Allard J-F, Champoux Y (1992) Empirical equations for sound propagation in rigid frame porous materials. J Acoust Soc Am 91:3346–3353. https://doi.org/10.1121/1.402824
31. Sousa VCB, Patel DI, Chapelier JB, Wagner A, Scalo C (2018) Numerical investigation of second mode attenuation over carbon/carbon surfaces on a sharp slender cone. AIAA Paper 2018-0350
32. Möser M (2009) Engineering acoustics: an introduction to noise control, 2nd edn. Springer
33. Fung KY, Ju H (2004) Time-domain impedance boundary conditions for computational acoustics and aeroacoustics. Int J Comput Fluid Dyn 18(6):503–511. https://doi.org/10.1080/10618560410001673515

34. Scalo C, Bodart J, Lele SK (2015) Compressible turbulent channel flow with impedance boundary conditions. Phys Fluids 27(3):035107. https://doi.org/10.1063/1.4914099
35. Zhao R, Wen CY, Long TH, Tian XD, Zhou L, Wu Y (2019) Spatial direct numerical simulation of the hypersonic boundary-layer stabilization using porous coatings. AIAA J 57(11):5061–5065. https://doi.org/10.2514/1.J058467
36. Douasbin Q, Scalo C, Selle L, Poinsot T (2018) Delayed-time domain impedance boundary conditions (D-TDIBC). J Comput Phys 371(15):50–66. https://doi.org/10.1016/j.jcp.2018.05.003

Chapter 4
Acoustic Metasurface Design

4.1 Introduction

As mentioned above, acoustic metasurfaces are planar metamaterial structures constructed with monolayer or multilayer stacks of subwavelength building blocks, which have significantly broadened the horizon of acoustic wave manipulation from wave-front modulation, to sound insulation, and absorption. This chapter describes the design process of optimum absorptive metasurface, impedance-near-zero metasurface, and reflection-controlled metasurface. The acoustic characteristics can be obtained by searching the geometrical parameter space of the microstructures using the impedance model proposed in Chapter 3. Also, the dominant mode of transition may change as the flight Mach number and attitude change. The suppression of transition over a wide frequency range by acoustic metasurfaces is an important research direction. This chapter also respectively introduces the design methods in the transitions where the first mode dominates, and where the second mode dominates. Furthermore, it should be noted that the joint active control technologies would be required when applied to the real complex flight vehicles, and the micro-blowing/suction/metasurface joint control technology is presented. Additionally, the possible machining flaw effect is discussed. In the final, a realistic metasurface caused by the thermal ablation of C/C composites is introduced. The represent volume cells of C/C composites with different mesostructures are designed and established, and the surface morphology evolutions are numerically solved. It hopes the desired acoustic metasurface can be "naturally" formed to inhabit transition in the cruise stage after the controlled thermal ablation of the protection material during the reentry stage in the future.

R. Zhao et al., *Stabilization of Hypersonic Boundary-Layer: Acoustic Metasurfaces*,
https://doi.org/10.1007/978-981-97-8623-7_4

4.2 Design Particular Acoustic Characteristics

4.2.1 Design of Optimum Absorptive Metasurface

Brès et al. [1] deduced an expression of optimum cavity depth to ensure the phase opposition between the reflection from the solid wall and from the porous surface. However, R is a nonlinear function of the porosity parameters and flow quantities (Eq. (4.1)). A numerical solution seems to be an effective way to obtain the minimum $|R|$ by applying triple program loops in the following order: loop 1: $0.2 \le \phi \le 0.8$, loop 2: $0.06 \le Ar \le 0.3$, loop 3: $0 < f_{\text{acs}} < \min(\phi/Ar, 2.0)$. These parameters span the range of interest for practical applications [2] and also ensure the accuracy of the proposed model.

$$R = 1 + \frac{2j\tan(k_c H)\phi\frac{\rho_w}{\bar{\rho}}\frac{k_c}{k_0}}{1 - j\tan(k_c H)\phi\frac{\rho_w}{\bar{\rho}}\sum\limits_{r=-\infty}^{+\infty}\frac{k_c}{\sqrt{k_0^2-\left(\frac{2\pi r}{s}\right)^2}}S_r^2}. \tag{4.1}$$

To confirm the proposed optimization strategy, we numerically study the stabilization problem of a hypersonic flow along a two-dimensional flat plate at zero angle of attack (Fig. 4.1). The freestream flow conditions are the same as in the experiment of Bountin et al. [3]: Mach number $Ma_\infty = 6.0$, unit Reynolds number $\text{Re}_\infty = 10.5 \times 10^6\ \text{m}^{-1}$, and temperature $T_\infty = 43.18$ K. The wall is isothermal with temperature $T_w = 293$ K. An unsteady disturbance is introduced at the beginning of the plate by a slot of periodic suction-blowing at the fixed frequency of 138.74 kHz. The metasurface covers the second half of the plate, and its effect is modeled by the boundary condition $v_w' = \text{A}p_w'$ at the wall. Because the sound speed c_w is determined by T_w, the remaining input flow parameter for Eq. (4.13) is ρ_w. In this case, we use the value of ρ_w at $x/L_{\text{ref}} = 0.75$, and the optimized shape parameters are $\phi = 0.66$, $Ar = 0.28$, and $f_{\text{acs}} = 0.67$. With these shape parameters, the streamwise minimum $|R|$ is 3.4×10^{-5} at $x/L_{\text{ref}} = 0.75$, and the maximum $|R|$ is 2.9×10^{-3} at $x/L_{\text{ref}} = 0.5$. For comparison, the performance of a conventional acoustic metasurface [2] with relatively deep cavities ($\phi = 0.66$, $Ar = 0.1$, and $f_{\text{acs}} = 1.87$) is also presented.

Figure 4.2 shows the instantaneous contours of fluctuating pressure, and Fig. 4.3 compares the amplitude distributions of fluctuating pressure for the three cases. Taking the baseline case as an example (Fig. 4.2a), two-cell structures form downstream ($x/L_{\text{ref}} > 0.5$) with longitudinal wavelength approximately equal to twice the boundary-layer thickness, which corresponds to the typical Mack second-mode structure. Upon installing the acoustic metasurface, the Mack second mode is strongly suppressed, especially for the optimized case (Fig. 4.2b). Notably, upstream of the acoustic metasurface, the strong oscillations in Fig. 4.3 result from the coexistence of multiple waves (including mode F and mode S of acoustic waves, and entropy/vorticity waves) in the boundary layer. Compared with the baseline case and the conventional acoustic metasurface, the maximum fluctuating pressure for

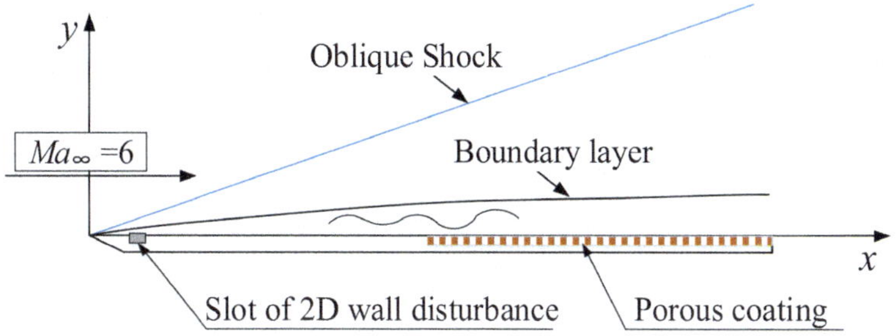

Fig. 4.1 Schematic drawing of problem formulation. Acoustic metasurface is at $0.5 \le x/L_{\text{ref}} \le 1.0$, $L_{\text{ref}} = 0.2$ m is the reference length

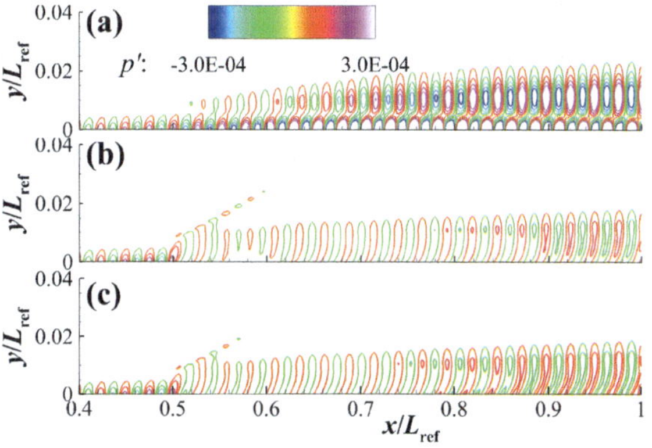

Fig. 4.2 Instantaneous fluctuating pressure fields (p' is normalized by $\rho_\infty u_\infty{}^2$, and ρ_∞ and u_∞ are the freestream density and velocity, respectively): **a** baseline case without acoustic metasurface; **b** optimized acoustic metasurface; **c** conventional porous coating with relatively deep cavities

the optimized case decreases by 88% and 30%, respectively. Also, the optimized shallow cavity is easier to manufacture. For the second-mode wave packets having a frequency band between 88.28 and 176.58 kHz, the calculated $|R|$ of the optimized acoustic metasurface varies more obviously than those of the conventional one, and its advantage can only be found in a narrow frequency band between 124.5 and 153.8 kHz (Fig. 4.4). It suggests a acoustic metasurface consisting of varied-depth cavities, such as a gradient coating, will be a potential candidate to suppress the disturbance with broadband frequencies. Once again, the Fedorov's model shifted the predicted frequency at the optimized shape parameters since it neglects the higher-order diffracted modes (Fig. 4.4).

4.2.2 Design of Impedance-Near-Zero Metasurface

As the Mack second mode exhibits acoustic-wave-like behavior, one straightforward way to suppress its development while minimally disturbing the mean flow is to dissipate the acoustic wave energy by implementing an absorptive boundary. Contrary

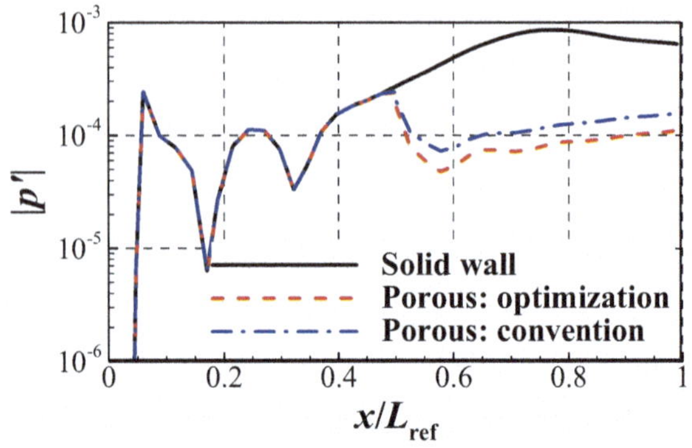

Fig. 4.3 Amplitude distributions of pressure perturbation along the wall (p' is normalized as the way in Fig. 4.5)

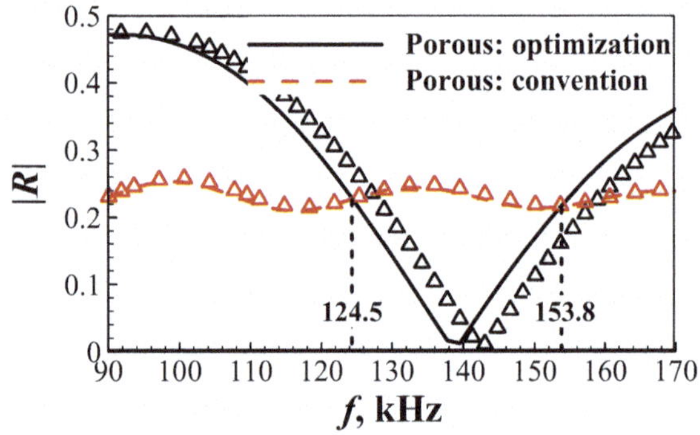

Fig. 4.4 Calculated reflection coefficients for different disturbance frequencies. Line: calculated by the proposed model, Symbol: calculated by Fedorov's model

to this intuitive thinking, here we show that an effective boundary with hardly any absorptive capability can still suppress the Mack second mode provided that its surface acoustic impedance approaches zero. The performance is even improved over that of the absorptive boundary.

For hypersonic boundary-layer flow problems, the boundary condition at the wall surface is generally modeled as $v_w' = p_w'/Z$, where v_w' and p_w' are the vertical flow velocity and acoustic pressure at the wall, respectively. Z is the surface acoustic impedance, which depends on the properties of the wall material, mean flow characteristics at the wall surface, and flow-perturbation parameters such as wave frequency. Z becomes infinite for a rigid wall, and is often designed to match the specific acoustic impedance of the background medium to maximize the absorption effect. Note that, as a trapped mode within a waveguide, the sound ray of the Mack second mode is almost normal to the wall surface [1], which means the absorption reaches a maximum for the matched impedance. The stability characteristics of hypersonic boundary layers can be theoretically evaluated using linear stability analysis theory (LST). In our theoretical modeling, as depicted in Fig. 4.5a, a slot of periodic suction-blowing at a fixed frequency of $f = 138.74$ kHz is introduced at the beginning of the surface. At this frequency, the typical Mack second mode will be excited along the second half of the surface [4]. The surface is 0.2 m long with an isothermal wall temperature of T_w

= 293 K. Its first half is rigid, and an impedance boundary condition is implemented along the second half (highlighted area in Fig. 4.5a), in which we define a normalized surface acoustic impedance $Z^* = Z/Z_0$, where Z_0 is the background flow impedance. The freestream conditions are as follows: Mach number $Ma_\infty = 6.0$, unit Reynolds number $Re_\infty = 1.05 \times 10^7$ m^{-1}, and temperature $T_\infty = 43.18$ K. As mentioned above, the maximum sound absorption corresponds to the matched surface acoustic impedance, namely, $Z^* = -1$, suggesting the most prominent dissipation of the Mack second mode. As shown in Fig. 4.5b, by varying the surface impedance from $Z^* \rightarrow -\infty$ (rigid) to $Z^* = -2$ and $Z^* = -1$, the growth rate of the Mack second mode is indeed largely reduced. Surprisingly, the growth rate continues to decrease as Z^* approaches zero. In other words, the boundary tends to become non-dissipative, yet the Mack second mode can still be effectively suppressed in an even better manner. This is not in line with our conventional physics understanding of hypersonic boundary-layer stabilization.

Actually, our findings can be evidently explained by examining the pressure distributions of the Mack second mode using a high-order accurate DNS. As shown in the upper subplot of Fig. 4.6a, a typical Mack second mode (two-cell structure) is formed when the perturbation propagates downstream along a rigid surface. This is similar to a surface acoustic mode confined to the interface and evanescent along the vertical direction. The amplitude of its pressure perturbation reaches a maximum at the surface, corresponding to an antinode. An absorptive boundary efficiently dissipates the acoustic wave energy because of the small angle of incidence, but cannot fully eliminate the Mack second mode as the flow field keeps offering energy to the boundary layer (middle subplot of Fig. 4.6a). In contrast, the impedance-near-zero

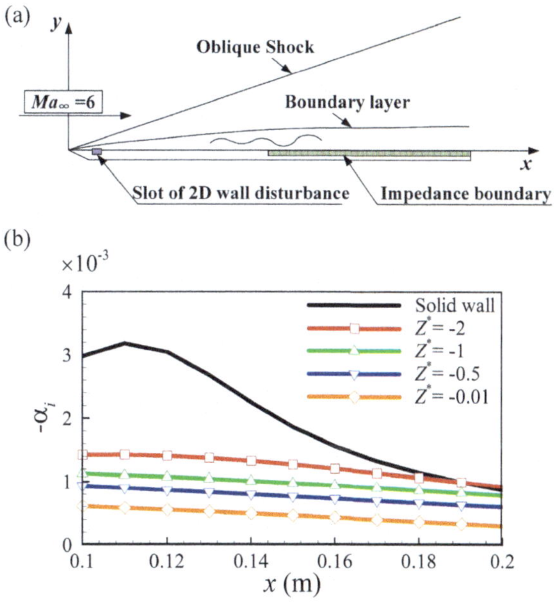

Fig. 4.5 Linear stability analysis of hypersonic boundary-layer flow along an impedance boundary. **a** Schematic drawing of problem formulation. **b** Growth rate distributions of the Mack second mode for different surface impedance values

boundary forces the acoustic pressure fluctuation at the antinode to approach zero (lower subplot in Fig. 4.6a). According to the relation $v'_w = p'_w/Z$, the excitation of the Mack second mode is fundamentally suppressed. In this case, the pressure perturbation in the hypersonic flow field can no longer be effectively coupled into the Mack second mode. As can be seen in Fig. 4.6b, the pressure fluctuation above the impedance-near-zero boundary (red line) is much weaker than that of the absorptive boundary (blue line).

Despite the potential of impedance-near-zero boundary to stabilize the hypersonic boundary layer flow, realizing it with traditional materials and structures is extremely challenging. The recent emergence of acoustic metasurface provides a practical way to bring such unusual artificial boundary condition to life. Among the wide variety of unconventional functionalities, effective impedance engineering enabled by acoustic metasurfaces [5–9] makes it possible to tune the boundary impedance to near-zero for the above-mentioned hypersonic boundary layer problem. In order to validate our idea unambiguously, here we consider the simplest structural design: a rigid

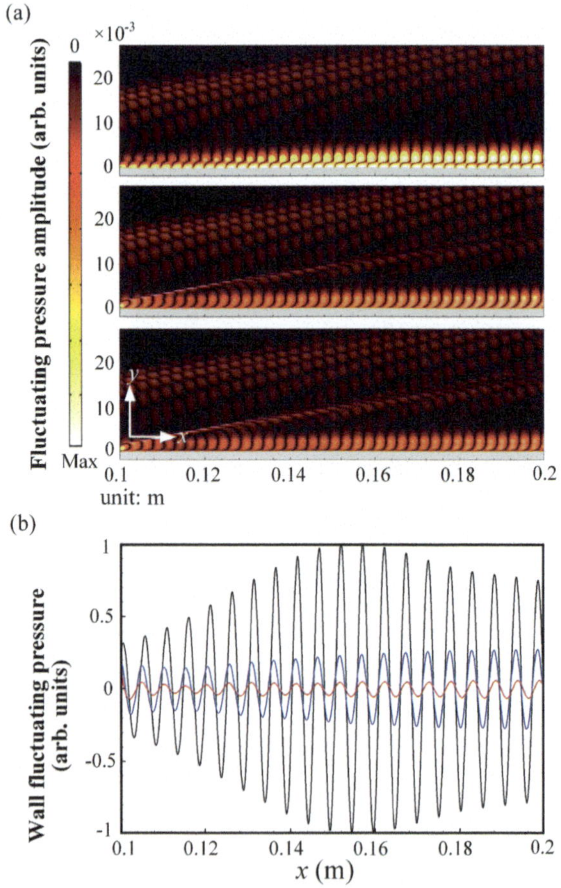

Fig. 4.6 Direct numerical simulation of the fluctuating pressure field in the hypersonic flow. **a** Fluctuating pressure amplitude contours for the rigid boundary $Z^* \to -\infty$ (upper), the absorptive boundary $Z^* = -1$ (middle), and the impedance-near-zero boundary $Z^* = -0.01$ (lower). **b** Comparison of fluctuating pressure distributions for different boundary conditions: black, rigid; blue, $Z^* = -1$; red, $Z^* = -0.01$. All curves are normalized to the maximum fluctuating pressure over the rigid boundary

surface periodically corrugated with subwavelength grooves. As Fig. 4.7a shows, the acoustic metasurface extends to infinity in the x and z directions. The half-width and depth of the cavities are b and H, respectively, with the unit-cell period being s. The porosity and aspect ratio are $\phi = 2b/s$ and $Ar = 2b/H$, respectively. Assuming that the periodicity s is much smaller than the acoustic wavelength λ_{acs}, the plane wave expansion method [8, 9] can be used to derive the normalized effective surface acoustic impedance of the proposed metasurface as

$$Z^* = \frac{1}{\rho_w c_w} \frac{p}{v}\Big|_{y=0} = 1 + \frac{1}{j\tan(k_c H)\phi \frac{\rho_w}{\tilde{\rho}} \frac{k_c}{k_0}} - \sum_{n=-\infty}^{+\infty} \frac{k_0}{\sqrt{k_0^2 - \left(\frac{2\pi n}{s}\right)^2}} S_n^2, \tag{4.2}$$

where ρ_w, c_w and $k_0 = \omega/c_w$ are the local density, speed of sound, and wavenumber at the wall, respecitively; $\tilde{\rho}$ and k_c are the complex dynamic density and wavenumber inside the cavity. $S_n = \text{sinc}(k_x b + \phi\pi n)$ is the overlap integral between the nth-order diffracted mode and the fundamental mode inside the cavity, where k_x is the parallel momentum. At the resonance frequency, Z^* is a purely real number governed by the geometrical parameters of the unit cell as well as the viscosity and thermal conductivity inside the cavities. Note that, despite the non-zero nature of Z^* in a passive system (owing to the inherent visco-thermal losses), we may still obtain a relatively low value of Z^* through a careful selection of the geometrical parameters. To this end, a numerical optimization process was applied to determine the geometrical parameters required by a near-zero surface acoustic impedance, through solving Eq. (4.2) with program loops in the following order: loop 1: $0.2 \leq \phi \leq 0.8$, loop 2: $0.06 \leq Ar \leq 1.5$, loop 3: $0 < f_{acs} < \min(\phi/Ar, 2.0)$, where $f_{acs} = fH/c_w$ is the normalized frequency.

Before taking into account the hypersonic fluid field, we perform full-wave finite element simulations to examine the acoustic characteristics of our design in the first place. As shown in the lower subplot of Fig. 4.7b, the absolute acoustic pressure field generated by a near-surface point source is extremely weak at the interface $y = 0$, as a result of the out-of-phase behavior between the incident and reflected waves. Sound waves are localized within the central cavity and hardly transmit to neighboring ones, corresponding to a bandgap of the surface acoustic wave. This is in stark contrast to the sound field of an impedance-matched metasurface (middle subplot of Fig. 4.7b), namely, an optimized UAC with $|Z^*| = 1$, in which most of the incident waves are absorbed compared with the case of rigid surface (upper subplot of Fig. 4.7b) but the acoustic pressure oscillation at the interface remains strong.

With this carefully designed impedance-near-zero metasurface, the aforementioned stabilization problem of the two-dimensional hypersonic boundary-layer flow (Fig. 4.5a) is reconsidered by employing DNS to simulate the spatial development of perturbations that interact with the metasurface. The corresponding cavity shape parameters are as listed in the caption of Fig. 4.7, and the excitation source located in the left-hand side is a slot of periodic suction-blowing at a fixed frequency of $f = 138.74$ kHz. Unlike the typical two-cell Mack second mode structures along the rigid surface (Fig. 4.8a, upper subplot), the upper cells join the lower part when

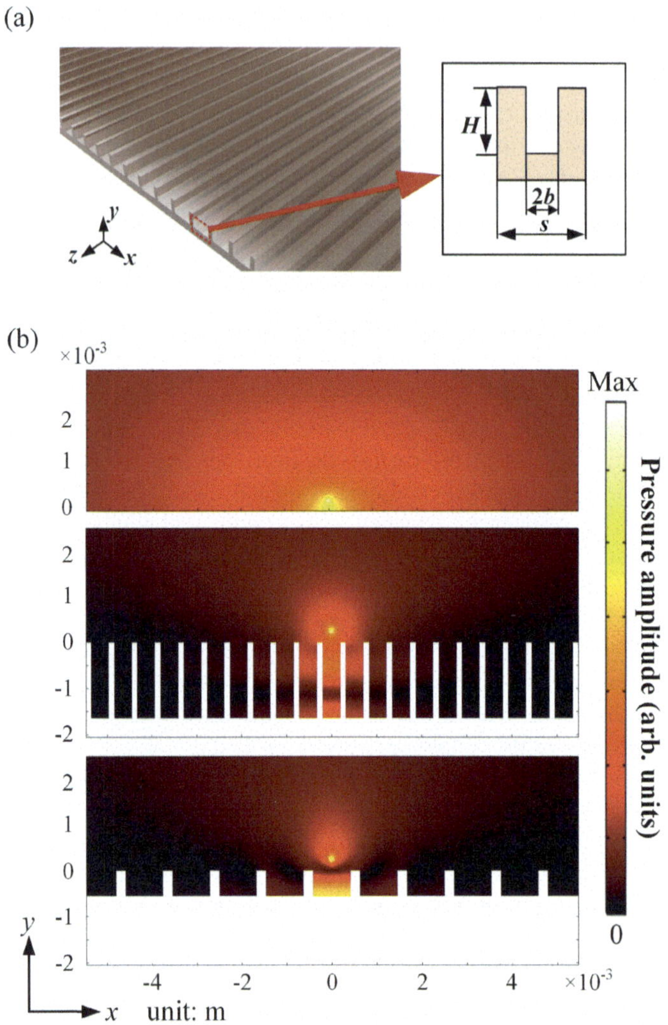

Fig. 4.7 Acoustic metasurface with subwavelength grooves. **a** Schematic illustration of the metasurface. **b** Simulated absolute acoustic pressure fields for different types of surfaces in the absence of hypersonic fluid flow. Upper: rigid surface; middle: absorptive metasurface with matched impedance $|Z^*| = 1$ and the corresponding shape parameters $\phi = 0.76$, $Ar = 0.24$ and $b = 1.96 \times 10^{-4}$ m, lower: impedance-near-zero metasurface with $|Z^*| = 0.15$ and the corresponding shape parameters $\phi = 0.8$, $Ar = 1.5$ and $b = 4.16 \times 10^{-4}$ m. During all simulations, a point sound source was located at $y = 2.48 \times 10^{-4}$ m (one-tenth of the incident wavelength) and immersed in the quiescent air of density $\rho_w = 5.59 \times 10^{-3}$ kg/m^3 and speed of sound $c_w = 343.11$ m/s, with the operating frequency being 138.74 kHz. The above ρ_w and c_w are equal to the flow parameters at wall location $x = 0.15$ m from steady flow calculation

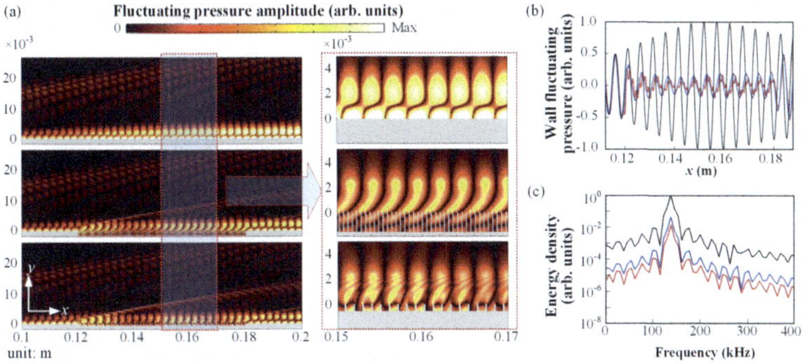

Fig. 4.8 Direct numerical simulation of fluctuating pressure fields when hypersonic flow passes the metasurface. Problem formulation is as for Fig. 4.2a except that the metasurface is only located at $x = 0.12$–0.18 m to save computational resource. **a** Fluctuating pressure amplitude contours for rigid boundary (upper), impedance-matched metasurface (middle), and impedance-near-zero metasurface (lower), with the corresponding zoom-in views shown in the right column. **b** Comparisons of fluctuating pressure along the surface ($y = 0$) for different boundaries. **c** Comparisons of energy density at $(x, y) = (0.16$ m, 0 m$)$ for different boundaries. In **b** and **c**, the black line represents data for the rigid boundary, the blue line for the impedance-matched metasurface, and the red line for the impedance-near-zero metasurface

moving through the metasurface region ($x = 0.12 - 0.18$ m), and then separate to form the typical two-cell structures downstream beyond $x > 0.18$ m (Fig. 4.8a, middle and lower subplots). For the impedance-matched metasurface (Fig. 4.8a, middle subplot), the trailing leg moves within the cavity. The intensity is greatly weakened due to the internal visco-thermal dissipation. Whereas for the impedance-near-zero metasurface (lower), a minimum value appears as a result of the phase opposition of the instability waves at the surface, and the mode structure is seen divided in the cavity and above the surface. The change in mode structures inhibits the amplification process of the Mack second mode. Figure 4.8b and c compare the fluctuating pressure distributions along the wall and the energy density spectrum at $x = 0.16$ m, respectively. Together, they confirm that the proposed impedance-near-zero acoustic metasurface can effectively suppress the Mack second mode instability, as predicted by our theoretical analysis. Compared with the optimized absorptive metasurface (viz. the impedance-matched metasurface), the overall performance is slightly improved, and the shallow-cavity-design would also be an advantage for manufacture and maintenance.

4.2.3 *Design of Reflection-Controlled Metasurface*

In this subsection, we further propose a reflection-controlled acoustic metasurface that can suppress the Mack second mode by controlling the reflection directions of disturbance waves and consequently stabilize hypersonic BL flow. We first discuss

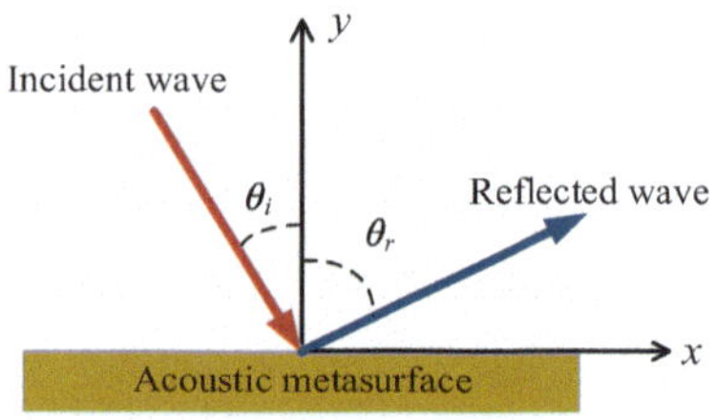

Fig. 4.9 Schematic diagram of reflection angle control

the theory of reflection direction control based on the generalized Snell's law and then demonstrate a practical implementation of such a reflection-controlled boundary based on the concept of acoustic metasurfaces. Finally, we perform DNS to validate our predictions.

According to the generalized Snell's law [10, 11], the relationship between the reflection and incident angles in terms of the gradient of the reflection phase along the wall is

$$\sin\theta_r - \sin\theta_i = \frac{\lambda}{2\pi}\frac{\mathrm{d}\Phi(x)}{\mathrm{d}x} \tag{4.3}$$

where θ_r is the reflection angle, θ_i is the incident angle, λ is the wavelength of the incident wave, and $\Phi(x)$ is the reflection phase, which is equal to the reflection coefficient phase when the phase of the incident wave is zero (Fig. 4.9). In principle, the general Snell's law holds for all monochromatic waves, which is first applied to optical waterfront engineering [10], and further used to manipulate acoustic waves [11]. For second-mode instability waves, a normal incidence can be hypothesized, i.e., $\theta_i = 0$. As a result, Eq. (4.3) can be simplified as

$$\sin\theta_\mathrm{r} = \frac{\lambda}{2\pi}\frac{\mathrm{d}\Phi(x)}{\mathrm{d}x}. \tag{4.4}$$

The next step is to carefully design the microstructures of the acoustic metasurface to produce a constant gradient of $\Phi(x)$ along the x-direction in order to meet the required θ_r. Here, we consider the simplest structural design: a rigid surface corrugated with subwavelength grooves (Fig. 4.10). Unlike traditional absorptive or impedance-near-zero acoustic metasurfaces with uniform grooves, reflection-controlled surfaces consist of periodic groove groups. As Fig. 4.10 shows, each group has N grooves with the same half-width b and unit-cell period s, but different depths H_i ($i = 1 \sim \mathrm{N}$). Because of the varied H_i values, the reflection coefficient phase $\Phi(x)$ can be designed to change linearly from 0 to 2π in each group based on a discrete approximation:

$$\frac{\mathrm{d}\Phi_\mathrm{r}(x)}{\mathrm{d}x} \approx \frac{2\pi}{Ns} \tag{4.5}$$

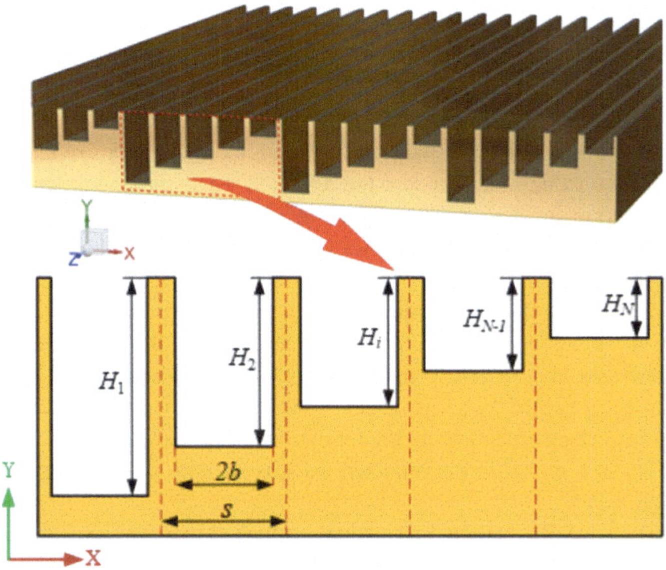

Fig. 4.10 Schematic illustration of reflection-controlled metasurfaces with subwavelength grooves

Substituting Eq. (4.5) into Eq. (4.4) allows the relation between the unit-cell period s and the wavelength λ to be unveiled

$$s = \frac{\lambda}{N \sin \theta_\mathrm{r}} \tag{4.6}$$

Using the designed θ_r, a given disturbance frequency f, and an appointed N, Eq. (4.6) thus provides the period s for Eq. (4.3). Then, upon choosing a particular b, the porosity ϕ is determined and the relationship between the phase Φ and depth H can be plotted using Eq. (4.3). The H_i value ($i = 1 \sim$ N) can consequently be chosen to construct the groove groups.

In Sect. 2.1, we has proposed an optimized absorptive acoustic metasurface through a careful selection of the geometrical parameters for Eq. (4.1) to achieve $|R|$=0, suggesting the most prominent dissipation of the Mack second mode. To enable direct comparisons with the optimized absorptive acoustic metasurface, the disturbance frequency and the half-width are kept the same values of $f = 138.74$ kHz and $b = 1.96 \times 10^{-4}$ m. N is chosen to make the resulting porosity ϕ approach 0.76. Taking the designed angle $\theta_r = 90°$ as an example, N is set to 5 to achieve the porosity $\phi = 0.79$ (Eq. (4.6)). The distribution of the reflection phase Φ is plotted for various depths H by calculating Eq. (4.3) using the sound speed $c_\mathrm{w} = 343.11$ m/s and density $\rho_\mathrm{w} = 5.59 \times 10^{-3}$ kg/m^3 as flow parameters. As documented in Part III, these are taken from DNS results at 75% of the plate surface to mimic the operational environment as much as possible. As shown in Fig. 4.11a, Φ increases from 0° to 360° in an approximately linear manner at the beginning of the curve, and the H_i (i

= 1 ~ 5) values can be chosen every $\Delta\Phi = 72°$. Also, all five grooves have much larger $|R|$ values than that of the optimized absorptive acoustic metasurface, which has a depth of $H = 1.63 \times 10^{-3}$ m and $|R| \approx 0$, as shown in Fig. 4.11b. The same procedure is also performed to design another two θ_r cases: $\theta_r = -90°$ and $\theta_r = -45°$, with the goal of comparing the stabilization effects of various reflection angles, where a negative value indicates that the reflected waves propagate upstream. The corresponding porosity and groove depths in one group are listed in Table 4.1. It should be noted that the groove depths for $\theta_r = \pm 90°$ are the identical but arranged in opposite manner.

Before considering the hypersonic flowfield, we perform full-wave finite element simulations to examine the acoustic characteristics of the designed reflection-controlled metasurfaces in quiescent air using COMSOL Multiphysics®. Compared to Fig. 4.12a, the reflected waves propagate downstream and upstream in the direction parallel to the surface, thus turning the reflection angles to $\theta_r = \pm 90°$ in Fig. 4.4b and Fig. 4.4c. In Fig. 4.12d, the reflected wave is seen propagating upstream in an oblique angle of 45°.

With these carefully designed reflection-controlled metasurfaces, the aforementioned stabilization problem of the 2D hypersonic BL flow (Fig. 4.13) is reconsidered by employing DNS to simulate spatial development of perturbations that interact with

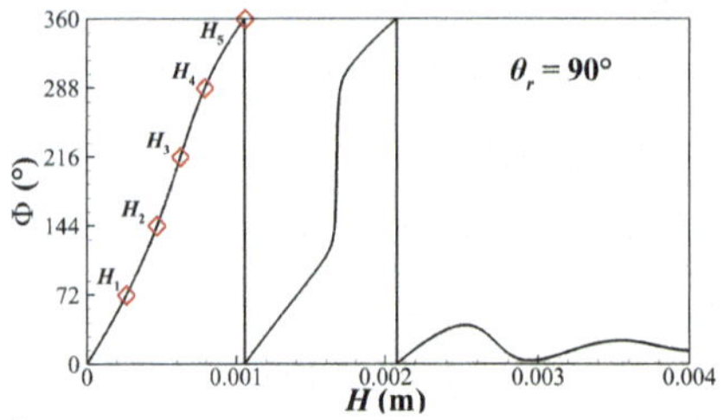

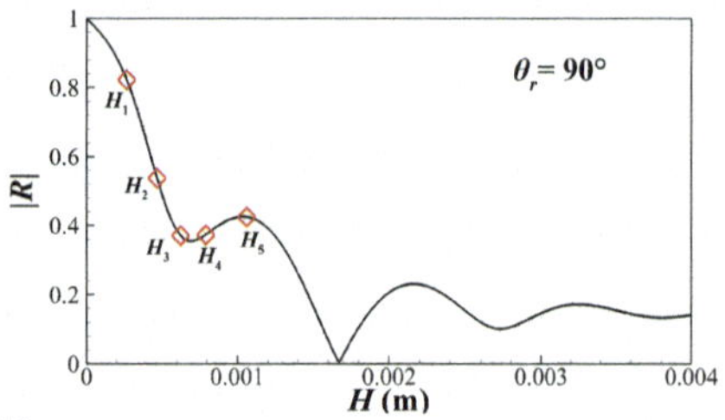

Fig. 4.11 The distribution of the reflection coefficient R along the groove depth H: **a** the phase and **b** amplitude distributions

Table 4.1 Geometric parameters for the three types of reflection-controlled metasurfaces

Cases	ϕ	$H_1 \times 10^{-3}$ m	$H_2 \times 10^{-3}$ m	$H_3 \times 10^{-3}$ m	$H_4 \times 10^{-3}$ m	$H_5 \times 10^{-3}$ m	$H_6 \times 10^{-3}$ m	$H_7 \times 10^{-3}$ m
Case 1: $\theta_r = 90°$	079	0.26	0.47	0.63	0.79	1.06	-	-
Case 2: $\theta_r = -90°$	0.79	1.06	0.79	0.63	0.47	0.26	-	-
Case 3: $\theta_r = -45°$	0.78	1.06	0.85	0.71	0.60	0.49	0.36	0.20

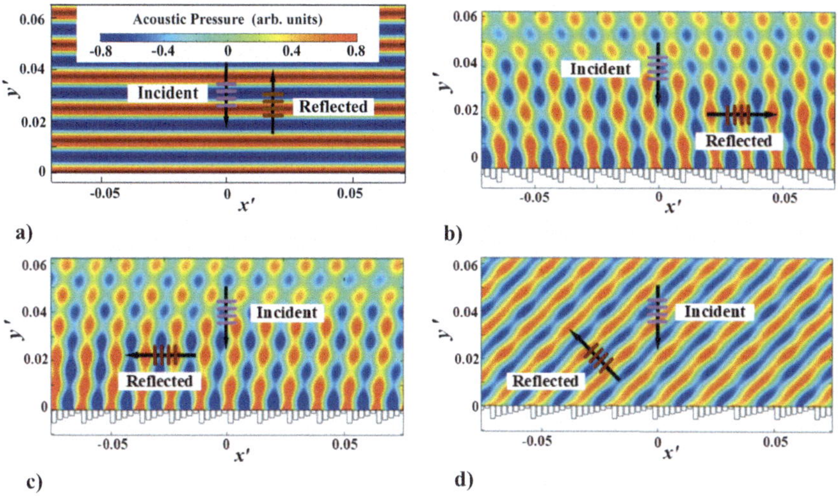

Fig. 4.12 Simulated scattered pressure fields for various metasurface reflection angles in the absence of hypersonic fluid flow: **a** rigid surface, **b** Case 1: $\theta_r = 90°$, **c** Case 2: $\theta_r = -90°$, and **d** Case 3: $\theta_r = -45°$. The x' coordinate in the streamwise direction and the y' coordinate normal to the wall are normalized by the reference flat-plate length of 0.2 m

the designed metasurfaces. The freestream flow conditions reference the wind tunnel experiment of Bountin et al. [3]: a Mach number $Ma = 6.0$, unit Reynolds number $= 10.5\ \mathrm{m}^{-1}$, and freestream temperature $T_\infty = 43.18$ K. The wall is isothermal with $T_w = 293$ K and the flat-plate length is $L = 0.2$ m. A periodic suction-blowing disturbance of frequency $f = 138.74$ kHz is introduced at the leading edge of the plate. At this frequency, the excited Mack second mode has a wavelength $\lambda = 5.7 \times 10^{-3}$ m, which indicates there are about 10 grooves in one wavelength. The metasurface region is located at $x' = 0.6$–0.9, where development of the Mack second mode is greatly excited. A total of 24 groove groups are meshed for Cases 1 and 2, and 17 groove groups are meshed for Case 3. According to the grid independence study in Ref. [12], the shallowest groove has a grid dimension of 20 × 15 (width × depth) in Case 3 ($H_7 = 0.2 \times 10^{-3}$ m in Table 4.1), while the deepest groove ($H_1 = 1.06 \times 10^{-3}$ m in Table 4.1) has a grid dimension of 20 × 55 in Cases 1 –3.

As shown in Fig. 4.14, the profiles of velocity and temperature above the metasurfaces are consistent with those above the rigid surface, which indicate the designed metasurfaces have little effect on the mean flowfield. Figure 4.15 compares instantaneous pressure fluctuation contours for all cases, including the fields of rigid surfaces and optimized absorptive metasurfaces. Here the fluctuating pressure p' is normalized by $\rho_\infty u_\infty{}^2$, and ρ_∞ and u_∞ are the freestream density and velocity, respectively. Unlike typical "two-cell" structures associated with the Mack second mode along the rigid surface in Fig. 4.15a, the upper mode cells tend to fuse with the lower parts and the trailing legs sink into the microgrooves when passing along the metasurfaces (Fig. 4.6e and 15b). Case 1 ($\theta_r = 90^{\circ}$, Fig. 4.15c) has the most similar structures

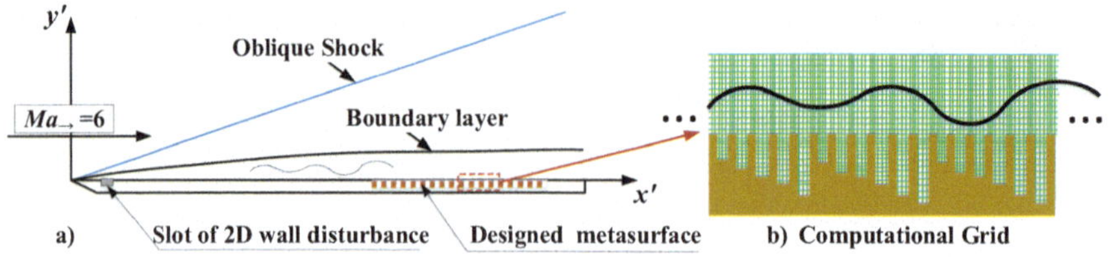

Fig. 4.13 Schematic drawing of the problem formulation: **a** hypersonic BL flow and **b** computational grids among the groove groups

as those along the optimized absorptive metasurface (Fig. 4.15b), except that some oscillation can be observed. For Cases 2 and 3, the mode structures are divided into several segments in the normal direction. The change in mode structures might inhibit the Mack second mode amplification process, leading to an intensity that is greatly weakened compared to that on the rigid surface. As shown in Fig. 4.16, the amplitudes of the fluctuating wall pressure on the three types of reflection-controlled metasurfaces all decrease to the level of the optimized absorptive pressure. The averaged magnitudes between $x' = 0.6$–0.9 for Cases 1–3 are respectively 1.59×10^{-4}, 1.63×10^{-4} and 1.67×10^{-4}, compared to 1.63×10^{-4} of the optimized absorptive one. This might indicate that the better inhibition strategy is to control the reflection wave as it propagates downstream ($\theta_r = 90^\circ$). However, it needs more designs and calculations to determine the optimum reflection angle. Behind the metasurface region ($x' > 0.9$), the amplitude rises quickly. Nevertheless, it remains smaller than that of the rigid surface. In particular, the recovered fluctuating pressure profiles of Cases 1–3 are all slightly below that of the optimized absorptive metasurface, which shows the potential priority over the conventional absorptive design.

In conclusion, reflection-controlled acoustic metasurfaces were designed to restrain the Mack second mode based on the generalized Snell's law. The surfaces were periodically corrugated with subwavelength groove groups. The DNS results reveal that the typical "two-cell" structure is divided into several segments when

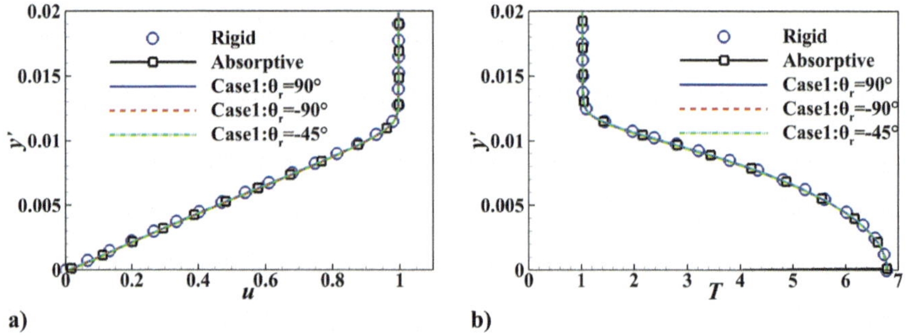

Fig. 4.14 Comparisons of streamwise velocity and temperature profiles at $x'= 0.7$. **a** Streamwise velocity profile, and **b** Temperature profiles

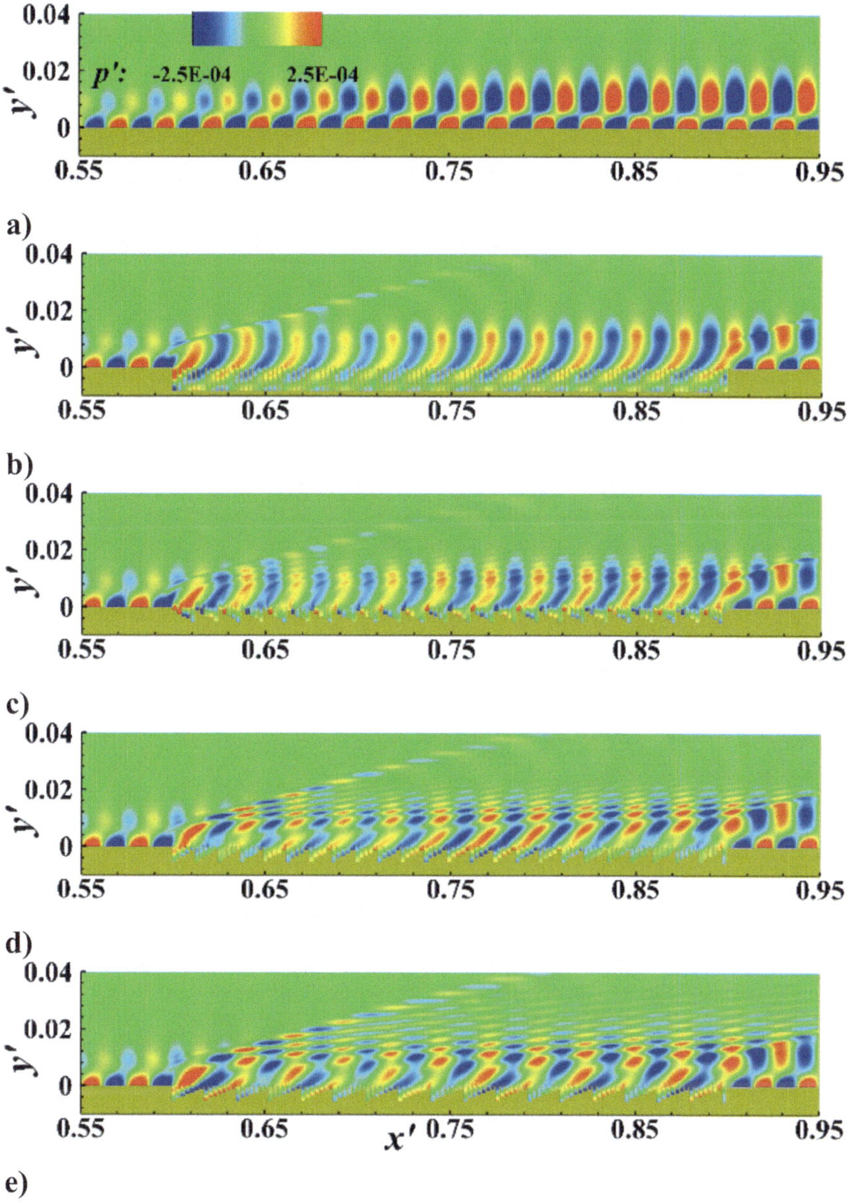

Fig. 4.15 Fluctuating pressure contours for: **a** a rigid surface and **b** the optimized absorptive metasurface, as well as three reflection-controlled surfaces: **c** Case 1: $\theta_r = 90^\circ$, **d** Case 2: $\theta_r = -90^\circ$, and **e** Case 3: $\theta_r = -45^\circ$

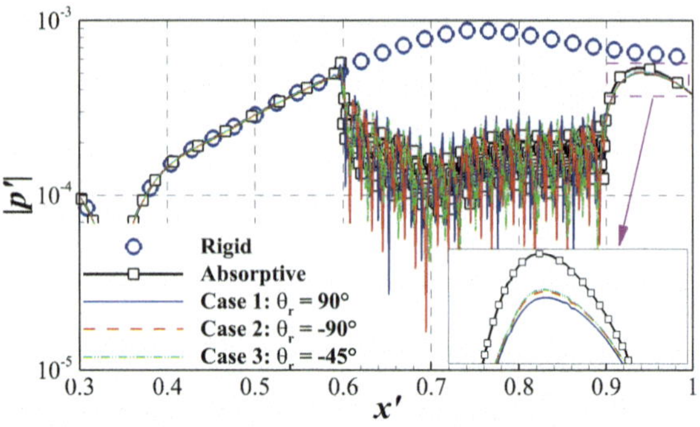

Fig. 4.16 Comparison of fluctuating pressure amplitudes along the wall

moving through the designed metasurface, and this change is thought to break down the Mack second mode propagation pattern in the BL. Among the three metasurfaces with different reflection angles, namely $\theta_r = \pm 90°$ and $\theta_r = -45°$, the most effective way to suppress disturbance amplification is to tune reflection wave propagation downstream in the direction parallel to the surface ($\theta_r = 90°$). Careful comparison of fluctuating pressure distributions indicates that its performance is slightly better than that of the optimized absorptive metasurface. In addition, further study will be carried to determine the optimum reflection angle with more DNS calculations. This finding reveals an alternative paradigm that can stabilize the Mack second mode and delay hypersonic transitions using porous metamaterials.

4.3 Optimization Methodology of Broadband Frequencies

It has been widely acknowledged that the first mode, which always coexists with the second mode, has been observed to be slightly destabilized by the presence of acoustic metasurfaces.[13, 14] A few recent studies have attempted to optimize metasurfaces to improve the stabilization efficiency on broadband disturbances. Tian et al. [15] discovered that the first mode is amplified when the admittance phase θ (admittance $A = |A|e^{i\theta}$) varies from 0.5 to π, while the second mode is destabilized when θ tends to 0.5π and suppressed as $\theta \geq 0.75\pi$. Furthermore, they designed a metasurface corrugated by gradient micro-slits to stabilize the second mode in a wide frequency range with the first mode slightly amplified in a Mach 4 BL flow via linear stability theory (LST) analysis. However, the first mode should be the primary concern in Mach 4 flows, and their LST analysis neglected the metasurface roughness effect caused by the recirculation zones inside the micro-slits as well as the alternating expansion and compression waves induced at the slit edges. Zhao et al. [16] focused on metasurface optimization to suppress the first mode while marginally amplifying the second mode in a Mach 4 BL flow. Tian et al. [17] designed a metasurface comprising two layers of perforated plates to suppress the second mode over a relatively wide bandwidth in a Mach 6 BL flow, but they did not consider its stabilization effect on the first mode of low frequencies. Our team designed acoustic metasurfaces with comprehensive admittance response to the local frequencies of the first and second modes in supersonic and hypersonic BL flows, respectively.

4.3.1 Broadband Stabilization in Supersonic BL Flows

In supersonic BL flows, the first mode is generally considered as the dominant instability to promote transition. Compared to the abundance of studies for stabilizing the Mack second mode, there seem to be limited options for controlling the first mode. Kimmel [18] noted that this mode could be stabilized by wall cooling, suction and a favorable pressure gradient. Another way to damp the first mode is to use a very thin perforated sheet stretched over a plenum chamber. When the TS waves propagate along the BL, the fluctuating pressure forces air in and out of the plenum chamber that produces a near-wall region of negative Reynolds stress. Carpenter and Porter [19] conducted LST analyses for the incompressible BL on a flat plate and showed that the TS waves are stabilized when the wall admittance phase is close to $\pi/2$. Whether the experiments will confirm the theoretical predictions remains to be seen. They indicated that this theory leads to severe constraints on the perforated sheet thickness, and the pressure variations may lead to areas of quasi-steady inflow and outflow, thereby modifying the mean flow and its stability characteristics. Wang and Zhong [20] analyzed the porous coating admittance and noticed that its phase angle was closely related to the destabilization of the first mode, i.e., the destabilization effect decreased as the phase angle of admittance decreased. However, they used an artificial porous coating boundary for numerical demonstration instead of realistic coating structures. Tian et al. [15] found that the first mode is amplified as the admittance phase θ varies from 0.5π to π, whereas the second mode is destabilized when θ tends to 0.5π and suppressed as $\theta \geq 0.75\pi$; furthermore, the stabilization or destabilization effects on Mack modes are intensified by the increase in the amplitude of admittance. Following this principle, they theoretically designed an acoustic metasurface that could damp the second mode in a wide frequency range with the first mode slightly destabilized. Zhao et al. [21] designed a metasurface that suppressed the first mode in a Mach 4 BL flow with the Mack second mode not considered. Furthermore, we proposed a broadband design strategy that damps the first mode and Mack second mode in the high-frequency region through LST analyses. In the present section, a piecewise acoustic metasurface aimed at suppressing the first mode while marginally exciting the Mack second mode is designed by LST and subsequently validated by DNS.

The stabilization problem is simplified as a high-speed flat-plate BL flow at zero angle of attack, as shown in Fig. 4.17a. The freestream conditions are as follows: Mach number of 4, temperature of 70 K and unit Reynolds number of $1.39 \times 10^7\ \mathrm{m}^{-1}$. The wall is isothermal with temperature of 295 K. The acoustic metasurface is placed at $0.05\ \mathrm{m} < x^* < 0.4\ \mathrm{m}$ in both LST and DNS calculations. The computation domain above the surface is $0.48\ \mathrm{m} \times 0.1\ \mathrm{m}$, which has a grid dimension of 8618×200 in the

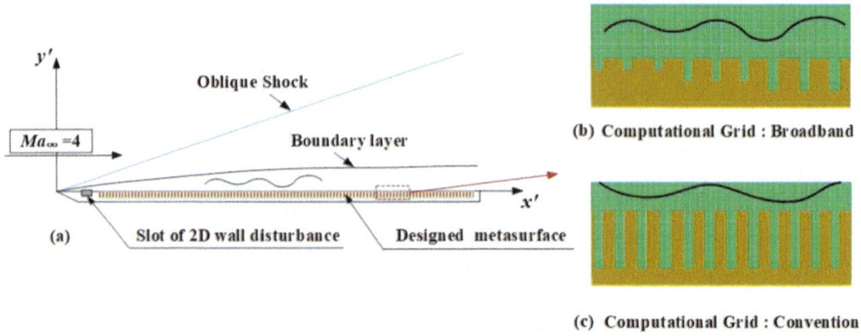

Fig. 4.17 Schematic drawing of the problem formulation: **a** high-speed boundary-layer flow, **b** computational grids among the cavities of broadband design, **c** computational grids among the cavities of convention design

streamwise and vertical directions. We elongated the domain to $x^* = -0.04$ m with symmetry conditions at $y^* = 0$ m to avoid numerical instability caused by the singular point at the tip of the flat plate. The computational grid clusters at the leading edge and plate surface with approximately 50 grids within the BL to resolve the disturbance accurately. To achieve global optimum stabilization, the broadband acoustic metasurface is cut into 35 pieces with different sets of geometrical dimensions of microslits (Fig. 4.17b). The shallowest slit has a grid dimension of 10 × 15 (width × depth), while the deepest slit has a grid dimension of 15 × 20. The detailed dimensions of broadband acoustic metasurface will be introduced in the following part. For comparison, a conventional metasurface with microslit depth $H^* = 1.0 \times 10^{-3}$ m, width $2b^* = 1.0 \times 10^{-4}$ m (aspect ratio, $Ar \equiv 2b^*/H^* = 0.1$), and porosity n ($\equiv 2b^*/s^*$) $= 0.5$ is also meshed with a uniform grid dimension of 10 × 30 (Fig. 4.17c).

An unsteady disturbance source is introduced in the calculation of DNS by a slot of periodic blowing and suction between $x_1{}^* = 0.02$ m and $x_2{}^* = 0.03$ m with the expression

$$q_w^*(x^*, t^*) = \varepsilon \rho_\infty^* u_\infty^* \sin\left(2\pi \frac{x^* - x_1^*}{x_2^* - x_1^*}\right) \sin(2\pi f^* t^*), \quad x_1^* \le x^* \le x_2^*, \tag{4.7}$$

where $q_w{}^*$ is the mass flow flux in the normal wall direction. A small forcing amplitude $\varepsilon = 1.0 \times 10^{-3}$ is adopted, and the forcing frequency f^* is fixed at 60 kHz. At this frequency, the first mode is gradually amplified along the surface.

Figure 4.18 shows the growth rates of unstable modes in the BL over the smooth wall analyzed by LST. Both the first mode ($\omega < 0.12$) and second mode ($\omega > 0.23$) coexist in the BL. Since the dimensionless angular frequencies ω of their maximum growth rates basically remain constant in the streamwise direction, ω can be chosen to characterize the BL instabilities. As shown in Fig. 4.19, the stabilization or destabilization effect on the unstable modes was closely dependent on the metasurface admittance phase θ, and the effect was intensified by an increase in the admittance

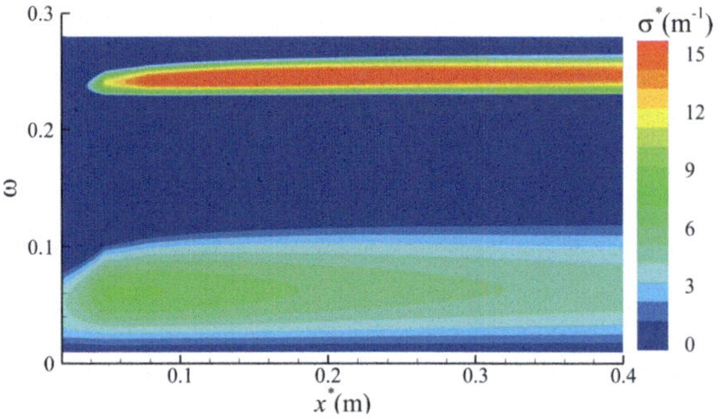

Fig. 4.18 Contour of the growth rates of the unstable modes as a function of the dimensionless angular frequencies ω and the coordinates along the streamwise x^* for a smooth solid wall

magnitude $|A|$. The first mode is marginally stabilized if θ approaches 0.5π and is greatly destabilized near $\theta = \pi$ as $|A|$ increases. In contrast, the second mode is restrained when $\theta \geq 0.75\pi$ and amplified as θ approaches 0.5π. For the present study, the metasurface was cut into 35 pieces in the streamwise direction to enable engineering realization. The admittance A of each piece should simultaneously meet the following three principles through a careful selection of the geometrical parameters in Eq. (4.8); that is: (1). As the dimensionless angular frequency $\omega < 0.12$, the admittance phase θ should be designed to approach 0.5π with a small $|A|$ to ensure the stabilization effect on the first mode (Fig. 4.19a); (2). in the frequency region where the first and second modes overlap ($0.12 < \omega < 0.23$), for all phases, amplitude $|A|$ should be designed as small as possible to minimize the destabilization effect on the second mode; (3). When the frequency $\omega > 0.23$, the admittance phase should satisfy $\theta > 0.75\pi$, and the amplitude effect on the growth rate is negligible (Fig. 4.19b and c).

As shown in Fig. 4.20, the required geometry parameters are achieved by solving Eq. (4.8) with program loops in the following order: loop 1 is 0.1×10^{-3} m $\leq H^* \leq 1 \times 10^{-3}$ m, loop 2 is $0.1 \leq Ar \leq 1.5$, and loop 3 is $0.1 \leq n \leq 0.8$. These parameters span the range of interest for practical applications. The selected parameters should first satisfy the admittance phase θ requirement (Step I in Fig. 4.20), and then the amplitude $|A|$ should be small enough to marginally excite the second mode (Step II in Fig. 4.20). After optimization, the final microstructure parameters at different streamwise directions x^* are shown in Table 4.1. The increase in H^* along the streamwise ensures the stabilization effect on the first and the high-frequency second modes. Ar increases to 0.7 to detune the admittance phase θ to approach 0.5π for first mode, whereas a further increasement will lead to the increase in the amplitude $|A|$. The porosity n increases gradually along the streamwise direction, but restrains small values to obtain a small $|A|$. For the sake of simplification, the optimization processes are only performed at 7 locations in the streamwise direction, and the optimal parameters are then interpolated into the midpoints of 35 pieces of the designed metasurface (Table 4.2).

The accuracy of DNS and the self-similarity compressible Blasius solution used in LST are first verified in the BL flow. As shown in Fig. 4.21, the profiles of velocity and temperature of DNS are consistent with the self-similar solutions at $x^* = 0.24$ m.

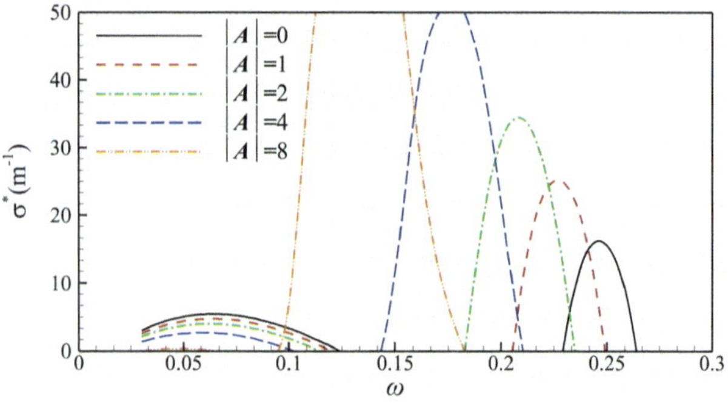

(a) phase $\theta = 0.5\pi$

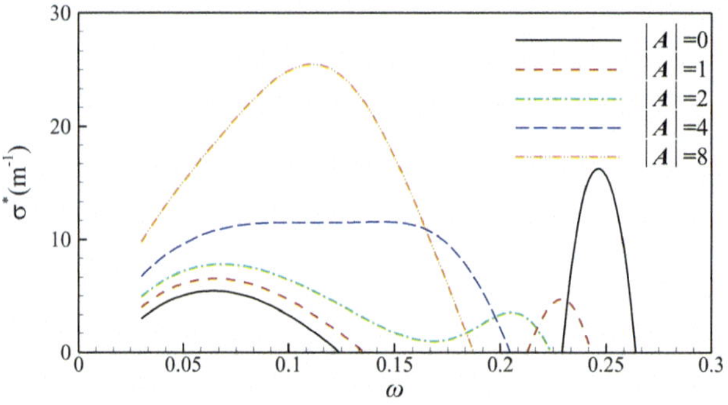

(b) phase $\theta = 0.75\pi$

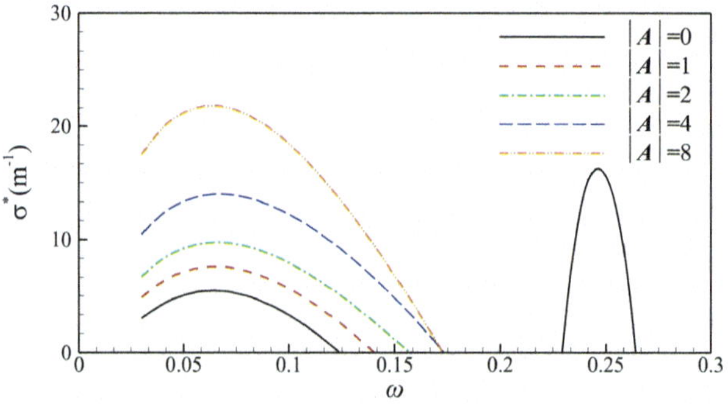

(c) phase $\theta = \pi$

Fig. 4.19 Effects of the admittance phase and amplitude of the acoustic metasurface on the growth rates of unstable modes (|*A*|=0 corresponds to the smooth wall)

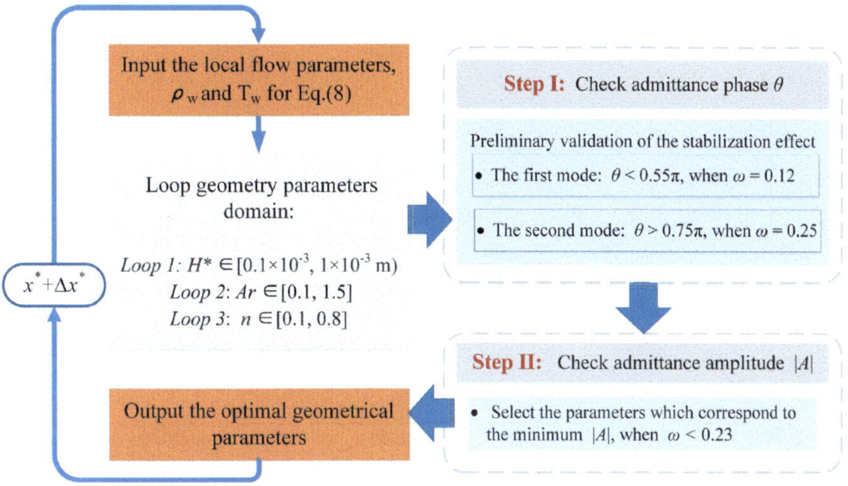

Fig. 4.20 Optimal geometric parameters selection method for the broadband metasurface design

Table 4.2 Optimal geometric parameters along the streamwise direction

x^*(m)	$H^*(10^{-3}$ m)	n	Ar
0.10	0.28	0.12	0.7
0.15	0.37	0.12	0.7
0.20	0.44	0.13	0.7
0.25	0.51	0.13	0.7
0.30	0.57	0.14	0.7
0.35	0.63	0.14	0.7
0.40	0.68	0.14	0.7

There are no visible discrepancies between the velocity and temperature profiles of smooth walls and the two above-mentioned acoustic metasurfaces, which indicates that acoustic metasurfaces have minimal effects on the mean flow.

As shown in Fig. 4.22a, the broadband acoustic metasurface successfully damps wide frequencies of first mode disturbances compared with the smooth wall. For Mack second mode disturbances with low frequencies, as depicted in Fig. 4.22b, the designed metasurface weakly enhances the Mack second mode and promotes the position of the maximum growth rate upstream. The growth rates of the high-frequency second mode disturbances become less than zero when two metasurfaces are used, which indicates that the high-frequency disturbance is completely suppressed (Fig. 4.22c). For comparison, the conventional acoustic metasurface damps all the second mode perturbations but strongly excites the first mode.

As shown in Fig. 4.23, the e^N method is employed to verify the transition delay performance of the acoustic metasurface. In the frequency range of 40 ~ 80 kHz, the broadband acoustic metasurface damps the first mode disturbances, while the

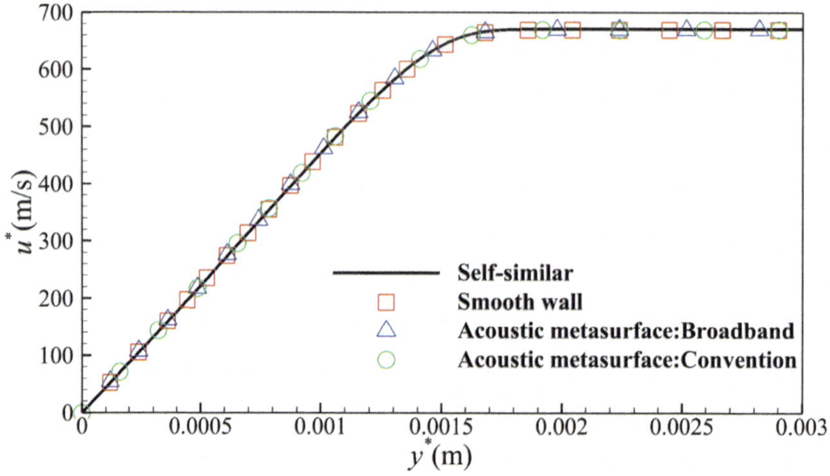

(a) Streamwise velocity profile

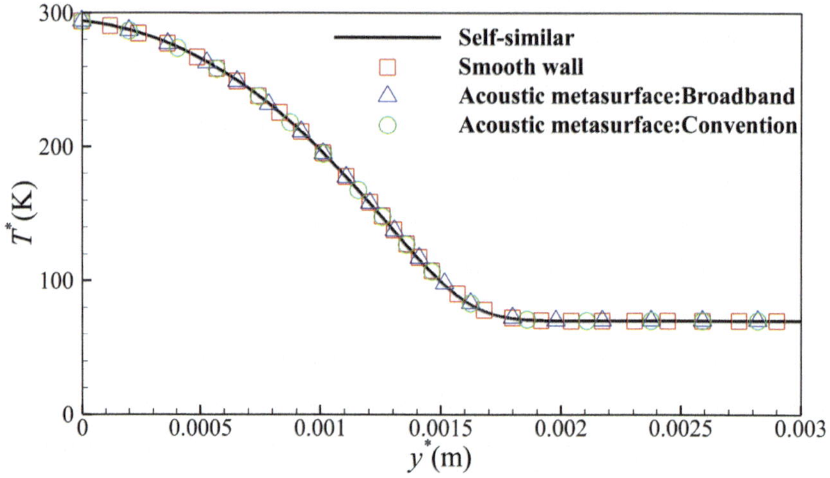

(b) Temperature profiles

Fig. 4.21 Comparisons of streamwise velocity and temperature profiles calculated from DNS and compressible Blasius solution at $x^* = 0.24$ m above different surfaces

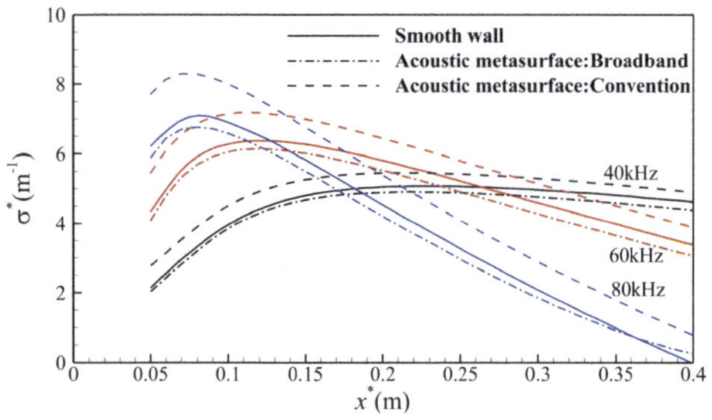

(a) the first mode

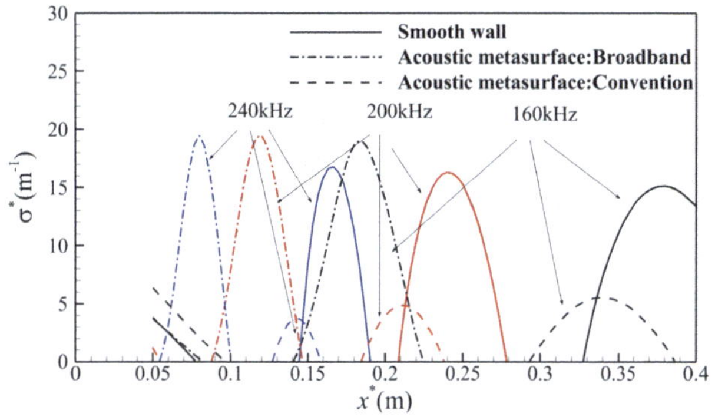

(b) the second mode of low frequencies

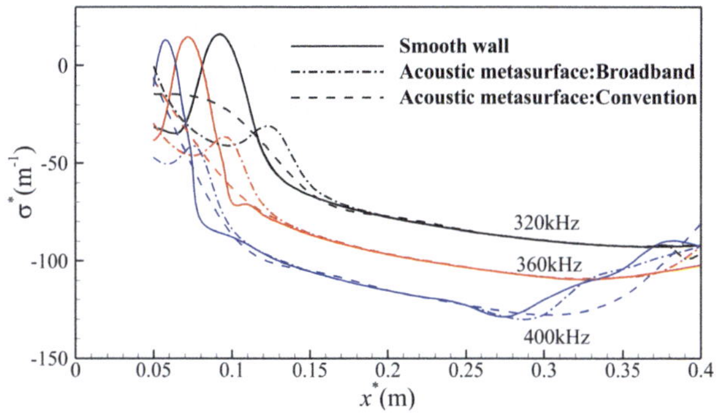

(c) the second mode of high frequencies

Fig. 4.22 Dimensional growth rates of unstable modes in the BL along different surfaces conventional metasurface amplifies them. It can also be observed that 60 kHz is the dominant frequency, of which the disturbance reaches the largest N at the end of the flat plate. With respect to the Mack second mode of low frequencies (160 ~ 240 kHz), the N factor curves of both acoustic metasurfaces move upstream, while the maximum N factor of the broadband metasurface is only slightly reduced compared to that of the smooth wall. The N factor in the high-frequency region (320 ~ 400 kHz) of both metasurfaces is less than zero, which indicates that both metasurfaces could completely suppress the high-frequency second mode. Figure 4.24 depicts the N-factor envelopments of different surfaces. Although the N factor of the broadband acoustic metasurface in the upstream region ($x^* = 0.05 \sim 0.1$ m) is higher than that of the smooth wall, the transition would not be triggered because of the small N factor. Moving downstream, the N factor envelopment of the broadband acoustic metasurface is lower than that of the smooth wall, which indicates its potential to delay the transition. In contrast, the conventional one tends to prompt transition with a higher N factor envelopment under the Mach 4 inflow condition

In the above LST analyses, the impedance boundary condition was applied to represent the acoustic effect of the real microstructures on the BL but ignored the recirculation flows inside the slits and the alternating expansion and compression waves induced at the slit edges (Fig. 4.25). To take into account this simplification, we perform DNS of the broadband and conventional acoustic metasurfaces (see Fig. 4.17) to verify the stabilization effect at the dominant frequency of 60 kHz. As shown in Fig. 4.26, compared with the smooth wall, the designed broadband acoustic metasurface actually destabilizes the first modes with larger fluctuating pressure amplitudes at the surface, which is in contrast to the analysis of LST (Fig. 4.22a). The conventional metasurface largely excites the first mode, as expected. For clarification, the impedance boundary condition for the conventional one is also implemented in DNS for comparison. The impedance model could generally represent the effect

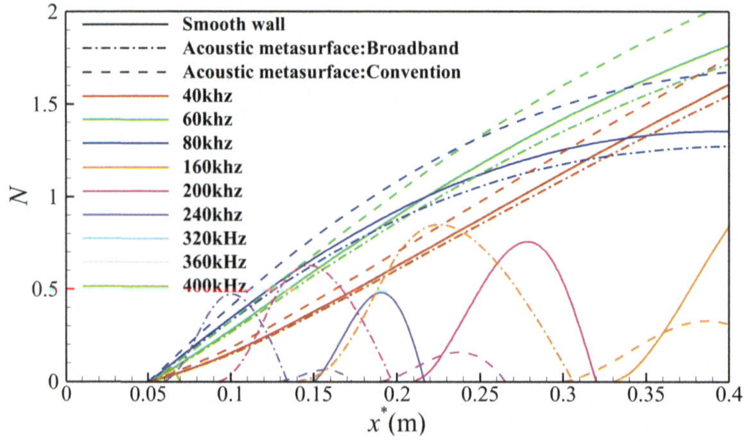

Fig. 4.23 Comparison of N-factor curves of different surfaces

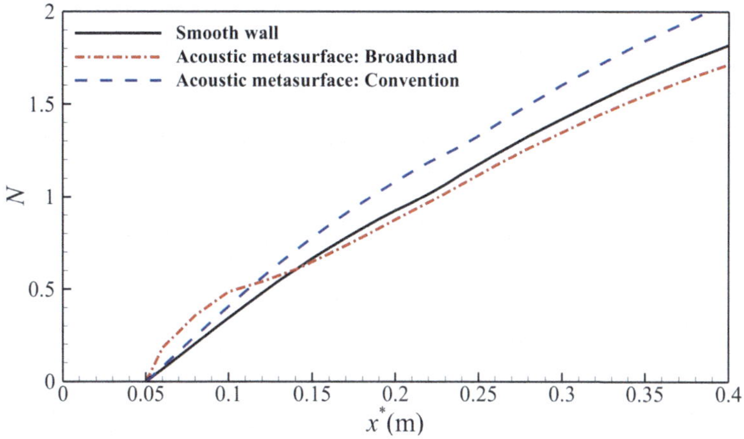

Fig. 4.24 Comparison of N-factor envelopments of different surfaces

of realistic metasurface microstructures for the first mode, but the results are not consistent with the comparisons for the Mack second mode. The fluctuating pressure amplitude is obviously larger in the CFD simulation on the realistic metasurfaces with micro-slits, which indicates that the local flow perturbations would represent a 'roughness' effect and excite the first mode. Therefore, it can be concluded that the acoustic metasurface has two opposite influences on the first mode: the acoustic characteristics could be designed to stabilize the first mode wave, but the very minute local flow perturbance induced by the microstructures will amplify it. This is different from the design of the metasurface for the stabilization of the Mach second mode at higher Mach numbers, wherein the flow perturbances are generally not influential.

To prove the stabilization effect of the acoustic metasurface on the first mode by DNS, the wall temperature increases to $T_w = 400$ K to further amplify the first mode

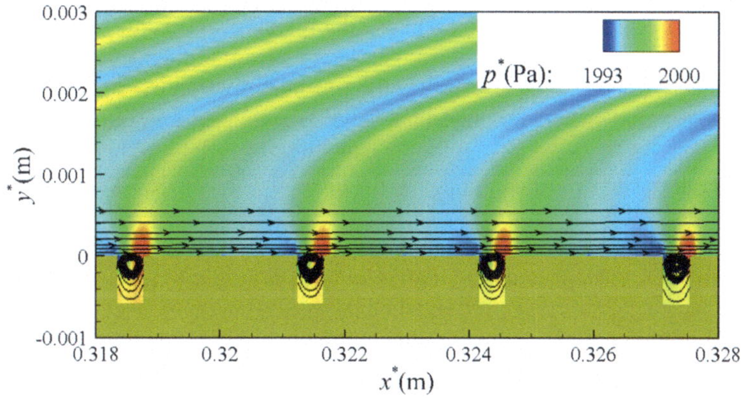

Fig. 4.25 Local pressure contour around the cavities of the broadband acoustic metasurface

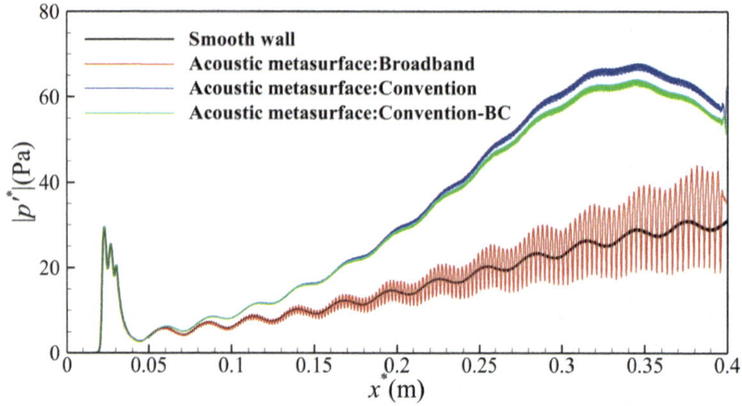

Fig. 4.26 Comparisons of wall fluctuating pressure amplitude distributions. (BC: calculated with the impendence boundary condition)

under the same freestream flow conditions as above. The periodic suction-blowing disturbance is chosen to be 80 kHz in the unsteady calculation, which is the dominant frequency of the first mode for this wall temperature condition after LST analysis. The microslits are designed with uniform parameters: depth $H^* = 4.0 \times 10^{-4}$ m, width $2b^* = 6.0 \times 10^{-4}$ m, and porosity $n = 0.536$. These parameters ensure that the metasurface admittance phase is equal to 0.52π to stabilize the first mode at 80 kHz. Figure 4.27 shows the disturbance growth rates predicted by LST. The hot wall condition approximately doubles the growth rate of the first mode compared to that in Fig. 4.22a. With the designed metasurface installed along the plate surface (0.05 m $< x^* <$ 0.4 m), the first mode is seen to be significantly suppressed. In the DNS study, the metasurface is only installed at the maximum growth rate region, i.e., 0.08 m $< x^* <$ 0.2 m to avoid the 'roughness' effect on the downstream flow. As shown in Fig. 4.28, the wall fluctuating pressure amplitudes at the metasurface region are larger than those on the smooth wall, which is contrary to the LST prediction again. The 'roughness' effect of the metasurface plays a dominant role in exciting the first mode. However, the amplitudes gradually decrease downstream of the metasurface, which indicates that the acoustic metasurface could substantially stabilize the amplification process of the first mode. On the other hand, 'roughness' is limited to the local influence. The results indicate that the acoustic metasurface should be installed in a confined region where the maximum growth ratio of the first mode occurs to minimize the 'roughness' effect.

In conclusion, a piecewise acoustic metasurface is designed to suppress the first mode across a wide frequency band without amplifying the second mode distinctly in a Mach 4 flat-plate flow. The LST proves its stabilization effect, and the e^N method demonstrates the trend of delay transition. However, the DNS simulations show contrary results. The discrepancies lie in the neglect of the local flow perturbation caused by the recirculation flows inside the microslits and the alternating expansion and compression waves induced at the slit edges, which behave as 'roughness' to

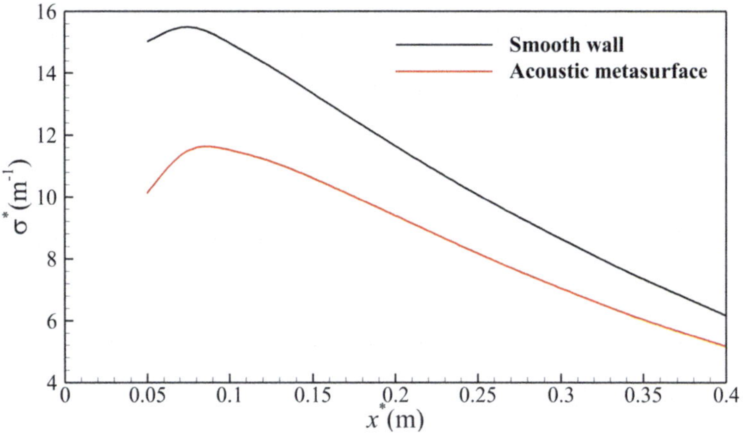

Fig. 4.27 Dimensional growth rates along the hot wall with a temperature of 400 K and disturbance frequency of 80 kHz

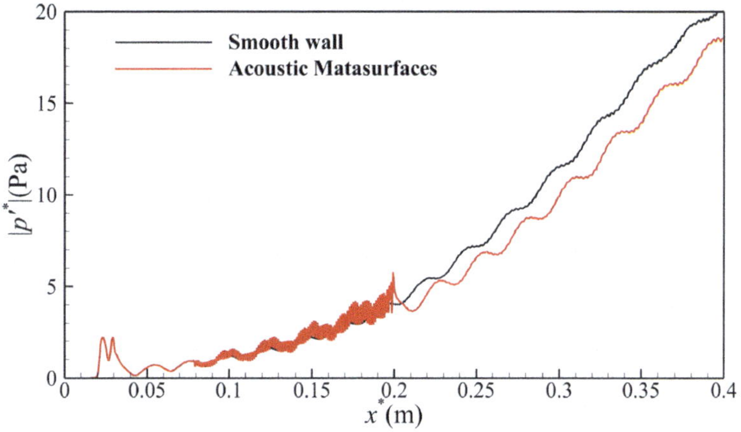

Fig. 4.28 Comparisons of fluctuating pressure amplitude distributions along the hot wall with a temperature of 400 K and disturbance frequency of 80 kHz

excite the first mode. The competitive mechanism between the acoustic stabilization and 'roughness' destabilization effect of the metasurface on the first mode was also found, which was different from that for stabilizing Mack second mode. In the metasurface region, 'roughness' plays a leading role in exciting the first mode, while downstream, the disturbance amplification process is inhibited because of acoustic modulation by the metasurface. This not only means that the metasurface microstructures should be carefully designed but also that the installment location should be considered when applying the metasurface to delay the BL transition triggered by the first mode. However, the 'roughness' mechanism on the first mode should be further clarified. Additionally, the impedance model should be improved by including the 'roughness' effect to reach a consistent result between LST and DNS.

4.3.2 Broadband Stabilization in Hypersonic BL Flows

This study serves as a supplement to absorptive acoustic metasurfaces in Sect. 2.1; it proposes a metasurface optimization strategy to stabilize the broadband disturbances in a Mach 6 BL flow. Without loss of generality, the *Ma* 6.0 hypersonic flat-plate flow is analyzed. Figure 4.29 displays the growth rates and N-factors of the unstable modes of different frequencies on the smooth plate calculated via the LST and e^N methods. The figure shows that the second mode of high frequencies, especially 130 kHz, dominates the maximum N value at the end of the plate. If one only considers the stabilization of acoustic second mode to delay transition, the intuitive way is to design the metasurface to absorb disturbances of all frequencies, or achieve the optimum absorption performance for the most unstable modes of 130 – 150 kHz. As shown in Fig. 4.30, the conventional metasurface with relatively deep micro-slits (M1: deep-slits design in Fig. 4.29b, with width b = 0.39 mm, depth H = 3.9 mm, aspect ratio $Ar = b/H = 0.1$, porosity $n = 0.8$) can monotonously dissipate acoustic waves, and the optimized absorptive metasurface (M2: $|R|_{\min}$ design in Fig. 4.29b, with b = 0.39 mm, H = 1.64 mm, $Ar = 0.24$, porosity $n = 0.8$; the details of the optimization process are present in Ref. [9] exhibits absorptive advantage in the frequency band between 130 and 150 kHz).

When considering the first mode of relative low frequencies, the metasurface admittance phase θ and amplitude $|A|$ should be carefully designed. As shown in Fig. 4.31, the metasurface generally decreases the growth rate of the second mode but slightly excites the first mode when its acoustic admittance phase θ varies from 0.75π to 1.25π. Furthermore, the metasurface partially destabilizes the second mode of high frequencies (>150 kHz) when $\theta > 0.75\pi$, and the destabilization effect on

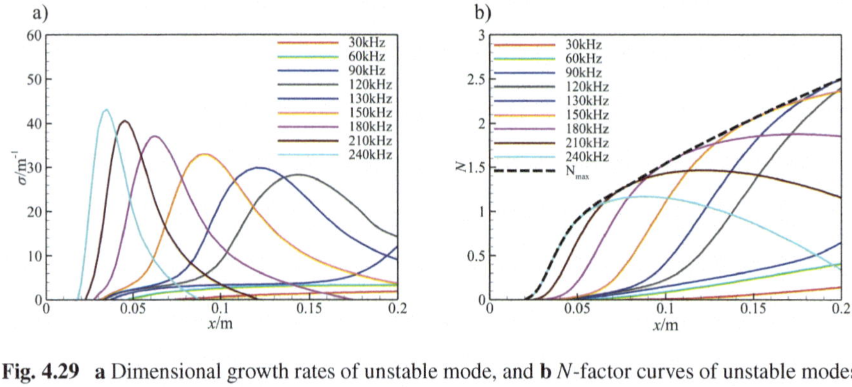

Fig. 4.29 **a** Dimensional growth rates of unstable mode, and **b** N-factor curves of unstable modes along a smooth wall of 0.2 m long

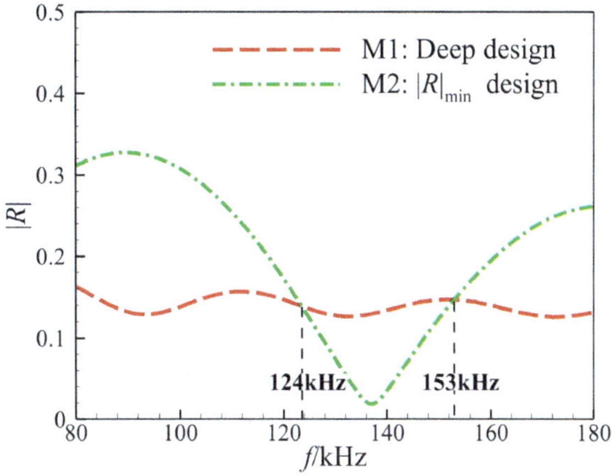

Fig. 4.30 Reflection coefficients for different disturbance frequencies

the first mode alleviates when θ approaching 1.25π. The stabilization or destabilization effect is promoted by a large admittance amplitude |A|. Therefore, to design a broadband metasurface to stabilize the Mach 6 BL flow, θ should approach 0.75π and the amplitude |A| should be large to stabilize the dominant second mode in the high frequency band. Moreover, θ should approach 1.25π and |A| should be small to alleviate the destabilization effect of the first mode of relative low frequencies.

Since the frequency band of the first or second mode varies along the streamwise direction, the metasurface is divided into several equal pieces in the streamwise direction, and the micro-slits of each piece are carefully designed to meet the local required admittance [16]. It can be predicated that one designed with more pieces could achieve better stabilization effects but increase the manufacture difficulties. In the considered Mach 6 BL flow over a flat plate, the metasurface region is divided into five pieces as an example. In the first piece ($x = 0.04 - 0.072$ m), the phase speeds are first predicated by LST at the midpoint $x = 0.056$ m (Fig. 4.5a). The three horizontal dashed lines represent dimensionless phase velocities of fast acoustic wave, entropy and vorticity waves, and slow acoustic wave, respectively. Modal F originates from fast acoustic spectrum, while Modal S originates from slow acoustic spectrum. The two modes synchronizes with each other at $f = 187$ kHz, and Modal

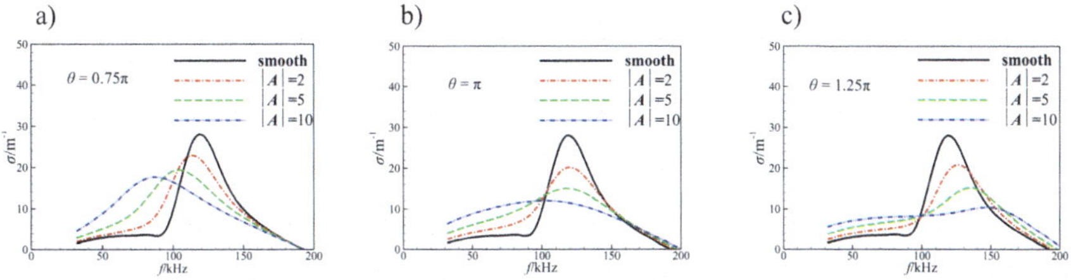

Fig. 4.31 Effects of the admittance phase and amplitude of the acoustic metasurface on the growth rates of unstable modes: **a** $\theta = 0.75\pi$, **b** $\theta = \pi$, **c** $\theta = 1.25\pi$

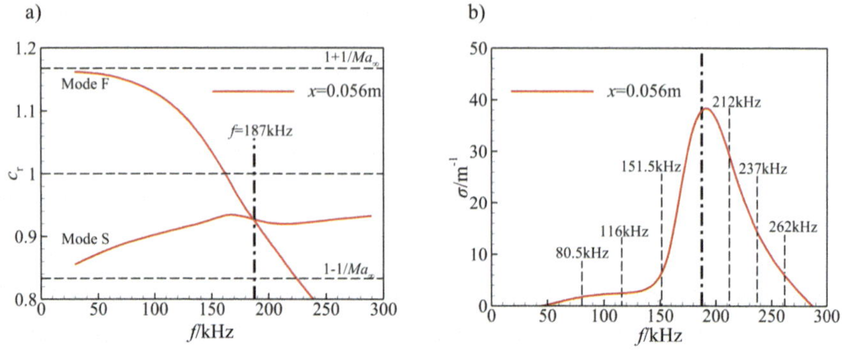

Fig. 4.32 **a** S/F modal phase velocity curves and **b** growth rate curve of S mode at $x = 0.056$ m. c_r is normalized by u_∞

S is either termed as the first or second mode depending on whether the frequency is smaller or greater than the synchronization frequency, respectively. As shown in Fig. 4.32b, three frequencies with an equally-spaced Δf from each modal band should be first evaluated to afford an optimum comprehensive phase response (Eq. (4.8)), and subsequently, the optimum amplitude $|A|_{opt}$ should be simply chosen as the minimum or maximum value depending on whether the most amplified frequency, 130 kHz, belongs to first or second mode at this piece, respectively.

$$g(b, H, n, f_i, f_j) = \min\left(\underbrace{\sum_{i=1}^{3}(1.25\pi - \theta_i)}_{\text{First mode band}} + \underbrace{\sum_{j=1}^{3}(\theta_j - 0.75\pi)}_{\text{Second mode band}}\right), \quad 0.75\pi \le \theta_i, \theta_j \le 1.25\pi \tag{4.8}$$

The optimization process of Eq. (4.8) is performed by solving Eq. (4.1) with program loops at every chosen frequency in the following order: loop 1 is 1×10^{-4} m $\le H \le 2 \times 10^{-3}$ m, loop 2 is 1×10^{-4} m $\le b \le 1 \times 10^{-3}$ m, and loop 3 is $0.2 \le n \le 0.8$.

The left four pieces of the metasurface are optimized following the above procedure, and the geometrical parameters of the designed broadband metasurface are listed in Table 4.3. Figure 4.33 compares the maximum N value envelopes of the three kinds of metasurfaces: M1—deep-slits design, M2—$|R|_{min}$ design, and M3—broadband design. M3 has largely decreased N_{max} at the end of the plate, signifying its superiority in the potential to delay the transition. It also could be observed that N_{max} on M3 is slightly larger than those on M1 and M2 at around $x = 0.1$ m, because M3 is more focused on weakening the excitation of the first mode at $x < 0.104$ m. The optimized absorption design M2 actually does not exhibit a better stabilization effect than the conventional M1, which consequently emphasizes the importance of considering the broadband disturbance.

As discussed in Ref. [16], the above LST analyses utilize the impedance boundary condition to represent the acoustic effect of real microstructures on the BL but ignore perturbed flows induced by the metasurface roughness. The ignored roughness effect actually affords results that are contrary to the results of DNS verification in the Mach 4 BL flow. In the present Mach 6 BL flow, a slot of Gaussian pulse is added at the leading edge to evaluate the broadband stabilization effect of metasurfaces M1–M3 via DNS. The equation for the Gaussian mass flux of the slot is

$$\rho v(x,t)'_w = \varepsilon_b(\rho u)_\infty \exp\left\{-\frac{(t-\mu_b)^2}{2\sigma_b^2}\right\} \sin\left\{\frac{2\pi(x-x_b)}{l_b}\right\} \tag{4.9}$$

where $l_b = 0.002$ m is the length of the slot, $x_b = 0.01$ m is the center of the slot measured from the leading edge of the flat plate, $\epsilon_\mathrm{b} = 1.0 \times 10^{-4}$ scales the function, $\mu_\mathrm{b} = 3.0 \times 10^{-6}$ shifts the Gaussian component to avoid negative times, and $\sigma_\mathrm{b} = 3.0 \times 10^{-7}$ adjusts the spectral content of the function. The Fourier transform of the Gaussian pulse includes most of the frequency content less than 500 kHz (Fig. 4.34).

The computational mesh resolutions are referred to Ref. [16]. The plate is extended to $x = 0.24$ m to study the modal development after the metasurface region. The wall fluctuating pressure amplitudes are compared in Fig. 4.35a. In contrast to the results of the Mach 4 BL flow [16], the DNS results are consistent with the e^N

Table 4.3 Broadband geometric parameters along the streamwise direction

Locations /m	0.04–0.072	0.072–0.104	0.104–0.136	0.136–0.168	0.168–0.2
$H \times 10^{-3}$ m	0.86	1.07	1.25	1.39	1.52
$b \times 10^{-3}$ m	0.48	0.48	0.48	0.44	0.44
n	0.2	0.2	0.8	0.8	0.8

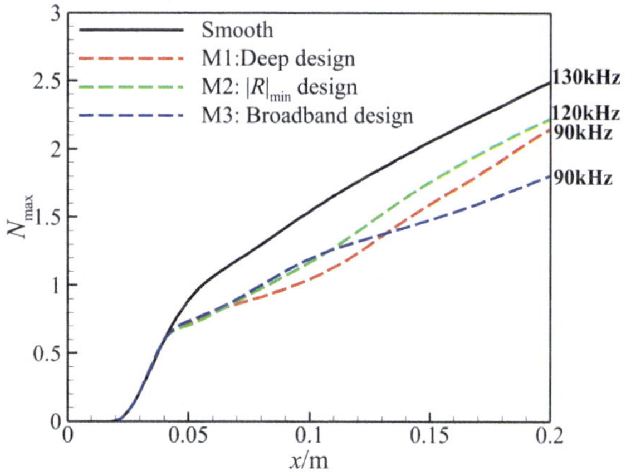

Fig. 4.33 Comparison of N-factor curves among three different acoustic metasurfaces and the smooth wall

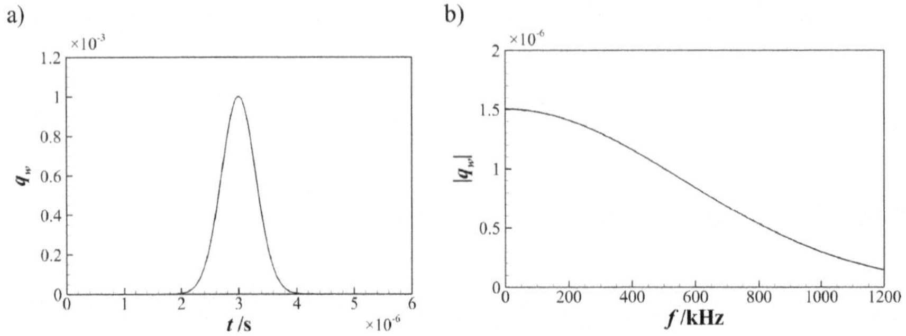

Fig. 4.34 Disturbance distribution over **a** time and **b** frequency for the Gaussian pulse

predications (Fig. 4.33) in the metasurface region. The broadband design M3 attempts to inhibit the excitation of the first mode in the first two pieces, which helps achieve a comprehensive stabilization effect in the downstream three pieces. Furthermore, the roughness effect does not significantly promote the unstable modes at least in the metasurface region. This denotes that M1 and M2 without considering the first mode are also suitable for stabilizing the second-mode dominant BL flow. However, the fluctuating pressure amplitudes rapidly increase downstream of the metasurface region especially for the M1 and M2 designs. Figure 4.36 shows the amplitude distributions in the frequency domain calculated through Fast Fourier transform. Compared to the contour on the smooth plate, the unstable frequencies extend to lower bands on the metasurfaces, which is consistent with the results of the wind tunnel experiment [13, 22]. Lukashevich et al. [23] determined that the presence of a metasurface in the region of the first mode results in the enhancement of the measured pressure amplitude in the downstream smooth part of the cone model in a Mach 5.8 wind tunnel experiment. In this study, the disturbances of $f < 120$ kHz are largely promoted on the downstream smooth part for M1 and M2 (Fig. 4.36), which causes the wall fluctuating pressure amplitude to quickly exceed that on the smooth plate (Fig. 4.35a). Additionally, the broadband design of M3 helps inhibit the quick modal growth behind the metasurface. The wall fluctuating pressure amplitude of the most amplified frequency of 100 kHz is also extracted and compared in Fig. 4.35b. Notably, the amplitudes at the metasurface region are not obviously larger than those on the smooth plate, although the metasurface roughness should excite the first mode of 100 kHz. This result is different from the observation in the Mach 4 BL flow, wherein the wall fluctuating pressure is largely amplified on the metasurface of the deep-slits design. This may be due to the coexistence of mode *F*, mode *S*, acoustic waves, and entropy/vorticity waves in the BL. The metasurface stabilization effect on other modes/waves might compensate the excitation of the first mode (low frequency band of mode *S*), resulting in the overall stabilization performance.

In conclusion, this study continued the optimization of the metasurface to stabilize high-speed BL flows. A broadband design strategy was proposed to suppress the dominant second mode while marginally destabilizing the first mode in a Mach 6 BL

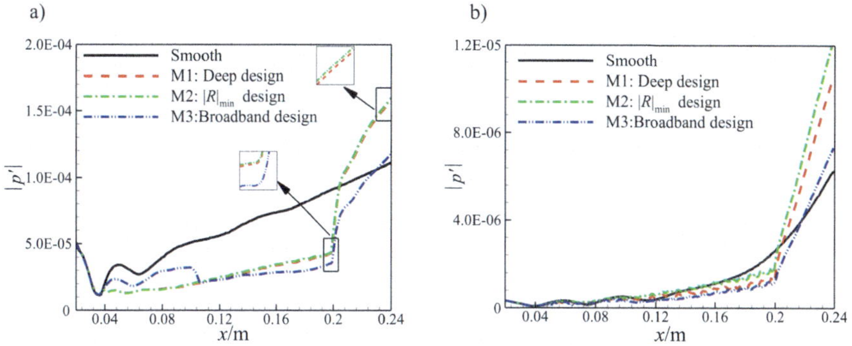

Fig. 4.35 Comparisons of **a** wall fluctuating pressure amplitude distributions on different surfaces ($|p'|$ is normalized by $\rho_\infty u_\infty{}^2$) perturbed by Gaussian pulse and **b** $|p'|$ of the most amplified frequency of 100 kHz

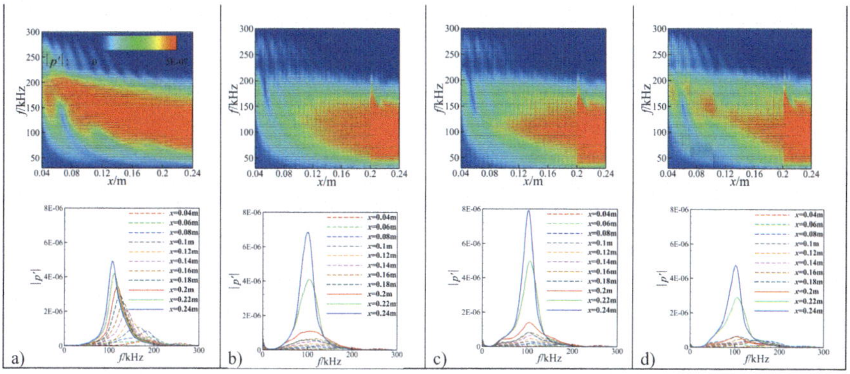

Fig. 4.36 Fourier analysis results of wall fluctuating pressure: **a** smooth wall, **b** M1: deep-slits design, **c** M2: $|R|_{\min}$ design, **d** M3: broadband design

flow. The results show that the disturbance of low frequencies, belonging to the first mode, are actually not obviously excited at the metasurface region but quickly grow in the downstream smooth part. The designed broadband metasurface can partially inhibit the downstream growth trend. The reasons should be clarified in future studies.

4.4 Combination with Micro-blowing

A number of studies have investigated the effects of blowing on this transition because of the extensive involvements of blowing in ablation [24–26] and transpiration cooling [27]. Schneider et al [28] reviewed experiments on the effects of blowing on the transition of the HBL and claimed that blowing promotes its transition upstream. They showed that a higher rate of mass flow in the presence of lighter gases results in an earlier transition. The effects of blowing on the stabilization of the HBL have also been investigated. Johnson et al. [29] used linear stability theory (LST) and parabolized stability equations (PSE) to study the effects of blowing and suction on a Mach 4.5 flat-plate boundary layer. Their results indicated that blowing destabilizes the boundary layer while suction stabilizes it. The effects of stabilization or destabilization are more prominent in the second mode than in the first mode. Ghaffari et al. [30] studied the effect of the stabilization of an HBL due to blowing by using the direct numerical simulation (DNS) and LST. The results of both showed that the growth rate of the second mode increases with the rate of blowing. According to the research of Pagella et al. [31]. they assumed that instability in the case of blowing results from an increase in the thickness of the boundary layer, which reduces the stabilizing influence of viscosity. In contrast to the aforementioned studies regarding the destabilization effect of blowing, Li et al. [24] reported that strong blowing could weakly stabilize the HBL. They studied steady-wall blowing on a Mach 7.4 sharp cone model by using PSE, and observed that the maximum N factors in the case of blowing was lower than that in the case of a smooth wall. However, Marvin and Akin [32] came to the opposite conclusion based on wind tunnel experiments on the same cone model and inflow condition. It has been claimed that this discrepancy was obtained owing to a potentially destabilizing factor that had been neglected in the stability analysis by Li et al. [24]

Most relevant numerical studies have simply used vertical velocities as the boundary condition at the wall and ignored the blowing medium that normally contains micropores to inject the gas. If the length scales of the pores are much smaller than the wavelength of the second mode, the blowing medium could server as metasurfaces (with infinitely deep micropores) that have been widely investigated for suppressing the second mode [33–39]. Fedorov et al. [40] applied LST to a Mach 5.3 sharp cone with porous coatings by injecting CO_2 into them. Both the vertical velocities and the acoustic admittance were combined as the boundary conditions of the wall. The results of a stability analysis indicated that CO_2 injection destabilized the acoustic modes, while the metausrface stabilized the HBL flow with CO_2 injection. Miró et al. [41] experimentally and theoretically investigated the effects of a combination of the blowing and metasurface on the stability of the HBL of a sharp cone at Mach 6. The results indicated that blowing, when used by itself, increased the maximum rate of amplification of disturbances while the pores helped stabilize them and delayed the transition used together with blowing. Highly accurate numerical simulations of blowing through real micropores remain elusive. LST neglects complex interactions of flow among the microstructures that might excite disturbances in the HBL [16]. Moreover, the locations of implementing steady blowing are critical to stabilization-related performance, and thus require further investigation.

This study uses both the DNS and LST to investigate the effects of combining blowing and a metasurface with respect to the stabilization of a Mach 6 flat-plate flow (Fig. 4.37). The stability-related performance of blowing is discussed with an emphasis on the effects of the locations of blowing. The additional stabilization afforded by the blowing medium, i.e., the acoustic metasurface, is also investigated.

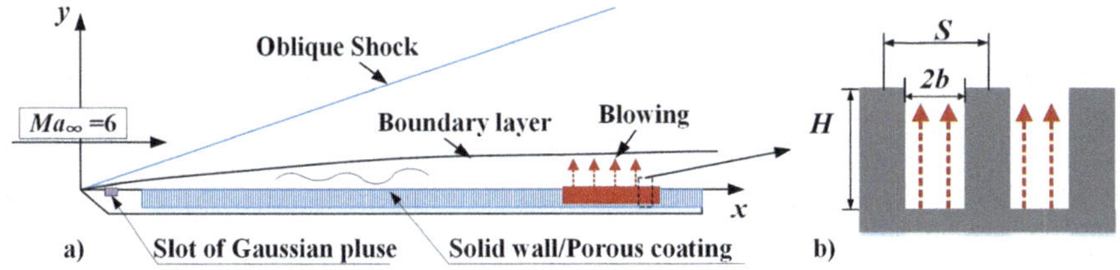

Fig. 4.37 Schematic drawing of the problem formulation: **a** hypersonic boundary-layer flow, **b** blowing medium

The inflow condition is the same as that in Sect. 3.2. We focused on the stabilization effects of different blowing positions. As shown in Table 4.4 Cases B1 ~ B3 featured locations of blowing upstream, middle, and downstream of the plate, with the same length of the blowing strip of 0.04 m, while the other part of the plate is smooth. Case B4 assumed blowing over the entire plate (0.02 ~ 0.2 m). All cases of blowing in LST had the same mass flux rate of blowing $\dot{m}=\rho_w v_w / \rho_\infty v_\infty = 0.0004$, which is consistent with that in the Ref. [32]. Cases B1-Porous ~ B4-Porous considered the metasurface as the medium of blowing, and the coating was installed along the entire plate (0.02 ~ 0.2 m). The metasurface was composed of deep microslits with a width $2b = 0.39$ mm, a depth $H = 3.9$ mm, and a porosity $\phi = 0.5$. The aspect ratio was relatively low ($2b/H = 0.1$), and the stabilization effect of the microslits resembled that of infinitely deep slits [33]. In DNS simulations, only Cases B1-Porous and B3-Porous are chosen to verify the corresponding LST studies. The microslits in the coating are meshed, and blow the gas into the HBL from the bottoms (Fig. 4.1b). The mass flow rate of blowing was $\dot{m}/\phi$, and caused the total mass flow entering the boundary layer to be equal to that in Cases B1 and B3.

The base flows are provided by DNS calculations by applying the vertical velocity as the wall boundary condition. When considering the acoustic metasurfaces (Cases B1-Porous ~ B4-Porous), the admittance wall boundary is included in LST calculations. In DNS verifications (Cases B1-Porous and B3-Porous), a converged steady flow with blowing at the microslit bottoms is first solved. Subsequently, unsteady calculations start up to simulate the evolutions of broadband disturbances excited by adding the Gaussian pulse (Eq. (4.9)) at the leading edge.

With the numerical base flows calculated by the DNS, the characteristics of stability are interpreted by using LST. Figure 4.38 shows the distributions of the thickness of the boundary layer as determined by $u^*(y)/u^*_\infty = 0.99$. Its thickness

Table 4.4 Setups of the locations for wall blowing in LST and DNS

Cases	B1/B1-Porous	B2/B2-Porous	B3/B3-Porous	B4/B4-Porous
Location of blowing (m)	0.02 ~ 0.06	0.08 ~ 0.12	0.14 ~ 0.18	0.02 ~ 0.2
LST analysis	✓	✓	✓	✓
DNS calculation	✓	–	✓	–

increased quickly from the beginning of the blowing strip. The extent of thickness increase became larger when the blowing strip was placed further downstream. Figure 4.39 shows the growth rates and N-factors of the disturbances at different frequencies on the smooth plate, calculated via LST and the e^N method. It shows that the second mode in the frequency region of 120 ~ 150 kHz dominated the maximum value of N at the end of the plate. Figure 4.40 shows the curves of phase velocity of the F/S mode and the curves of the growth rate of mode S for the baseline (smooth plate), Cases B1 and B3, calculated at the dominant frequency $f = 120$ kHz. In the leading-edge region of the plate, the wave number of mode S approaches that of a slow acoustic wave with the phase speed $c_r \equiv \omega/\alpha = 1 - 1/M$, whereas the wave number of mode F approaches that of a fast acoustic wave with $c_r \equiv \omega/\alpha = 1 + 1/M$. Further downstream, the modes F and S synchronized at $x = 0.142$ m for the baseline case, and mode S was considered to be either the first or the second mode depending on whether the location was upstream or downstream of the synchronization point, respectively. When the blowing strip was located further upstream (Case B1), the synchronization point moved forward to $x = 0.116$ m. The first mode ($x < 0.116$ m) was then significantly excited, and the maximum growth rate of the second mode ($x > 0.116$ m) increased weakly as it moved upstream, compared to the baseline. When the blowing strip was behind the synchronization point (Case B3), the phase velocity of mode F decreased rapidly right upstream of the blowing strip, thereby leading to the forward movement of the synchronization point. Moreover, downstream of the synchronization point, the growth rate of the second mode first increased quickly to a maximum and then deceased significantly.

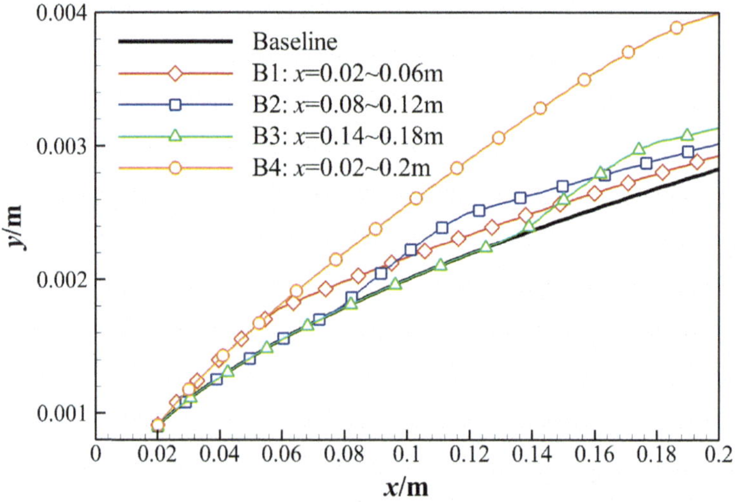

Fig. 4.38 Comparisons of boundary-layer thickness distributions in cases of blowing at different locations

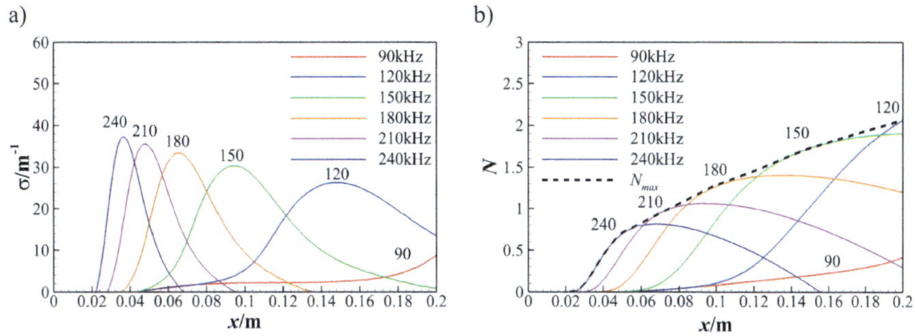

Fig. 4.39 Baseline (Smooth plate case): **a** dimensional growth rates of the unstable mode. **b** *N*-factor curves of the unstable modes

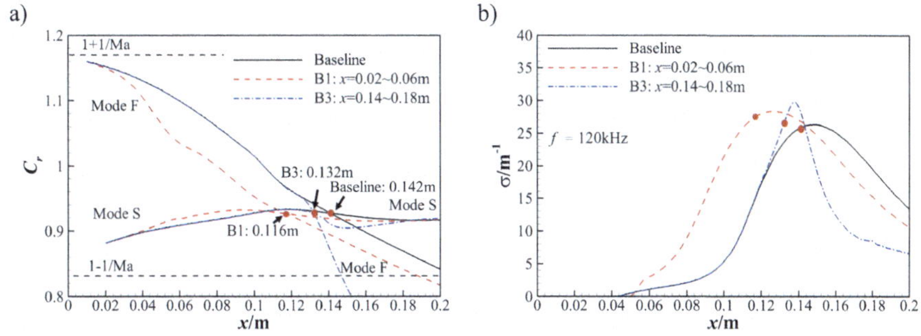

Fig. 4.40 **a** Curves of the *S/F* modal phase velocity and **b** the growth rates of mode *S* at f = 120 kHz. c_r is normalized by u_∞

As shown in Fig. 4.41, the maximum growth rates of the disturbances increased significantly compared with the baseline (Fig. 4.39a). The region of unstable disturbances shrank for all locations of the blowing strips. Take the 240 kHz of Case B1 for example, the unstable region is only 83% of that in the baseline case. This can be attributed to two reasons: The forward movement of the synchronization point reduced the region of the first mode, and the growth rate of the second mode decreased quickly at the strips. As shown in Fig. 4.41a, disturbances at frequencies lower than ~ 180 kHz were all excited because the strip was located upstream of their synchronization points. When the blowing strip was installed in the middle of the plate (Fig. 4.41b), the disturbance at the dominant frequency of 120 kHz also increased because the blowing excited its first modal part. By moving the blowing strip further downstream in Case B3 (Fig. 4.41c), the dominant disturbances at frequencies of 120 ~ 150 kHz were suppressed but lower-frequency disturbances, such as those at 90 kHz, were excited. If we consider blowing over the entire plate (Case B4 in Fig. 4.41d), the maximum growth rates of the disturbances at nearly all frequencies increased and those at lower frequencies were excited. Moreover, the second peaks

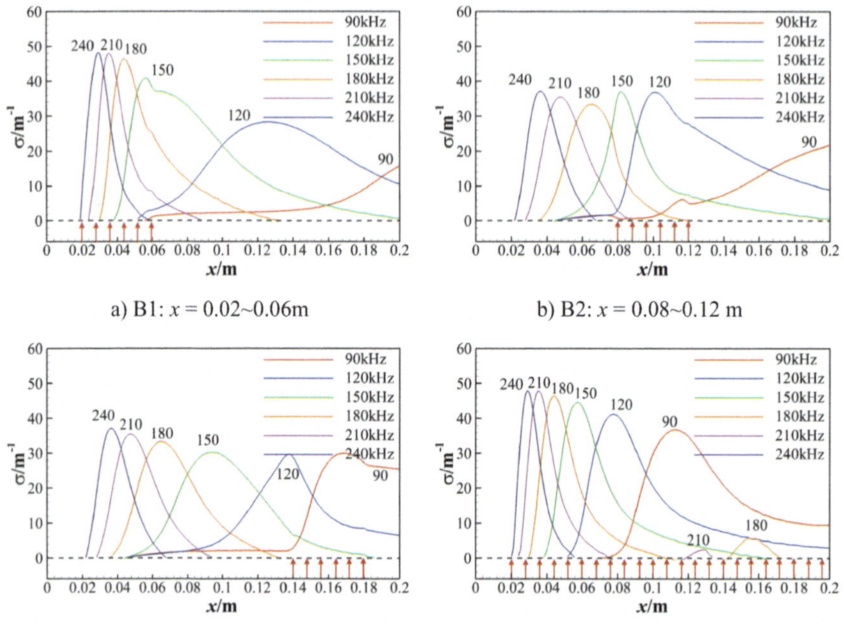

a) B1: $x = 0.02\sim0.06$m

b) B2: $x = 0.08\sim0.12$ m

c) B3: $x = 0.14\sim0.18$ m

d) B4: $x = 0.02\sim0.2$ m

Fig. 4.41 Dimensional growth rates of unstable modes in cases of blowing at different locations

of the high-frequency (180, 210 kHz) disturbances appeared downstream, which correspond to the Mack third mode reported by Li et al. [24]

As shown in Fig. 4.42, the e^N method was used to check the overall stabilization effect of blowing. In Fig. 4.42 and Fig. 4.44, the frequency marked in the right of the figure is the dominant frequency of each case. For Cases B1 and B4, the maximum values of the envelopes of N were larger than that of the baseline in the entire computational domain, especially with a more remarkable destabilization effect in Case B1. As is shown in Fig. 4.41, blowing excited the dominant frequencies (120 ~ 150 kHz) in Case B1 and much lower-frequency disturbances in Case B4. The destabilization effect of blowing in Case B4 is consistent with Refs. 32 and 33, in which blowing were also located at the entire calculation domain. Although the N factor at the blowing strip ($x = 0.08 \sim 0.12$ m) in Case 2 was smaller than that of the baseline, the destabilization effect of blowing contributed to a larger maximum value of N downstream. By contrast, compared with the baseline, the dominant disturbance in the second mode (120 kHz) at the blowing strip was damped in Case B3, leading to a lower maximum value of N at the end of the plate. It may explain the stabilization performance found in the study of Li et al. [13], where they placed the blowing at the dominant second mode region. Overall, the locations of the dominant second mode were crucial for the effectiveness of blowing in maintaining HBL stability. The HBL would be destabilized if the blowing strip was installed upstream of the

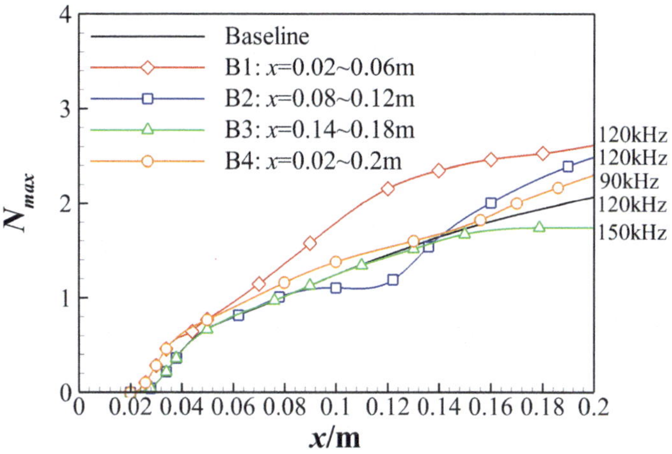

Fig. 4.42 Comparison of N-factor envelops in cases of only blowing at different locations

synchronization point of the dominant frequency of disturbance, and was otherwise (Case B3) generally stabilized.

We applied a metasurface as the medium of blowing to inject gas into the HBL. The sole use of the metasurface can suppress the second mode of high frequencies but excites the first mode of low frequencies [16]. The coupled performance of blowing and the metasurface in terms of inducing stability is shown in Fig. 4.43 and Fig. 4.44. Figure 4.43a and Fig. 4.43b show that the additional metasurface significantly reduced the growth rates of disturbances at higher frequencies (higher than ~ 180 kHz) at the leading edge, and suppressed the dominant disturbances at 120 ~ 150 kHz in the downstream, compared with Cases B1 and B2. This led to lower maximum values of N at the end of the plate (Figs. 4.44a and b). In Cases B3-Porous and B4-Porous (Figs. 4.44c and d), the overall stabilization performance deteriorated in the downstream region, compared with the counterparts in which only blowing was used (Cases B3 and B4) because the coating prematurely excited lower-frequency disturbances at the leading edge (Figs. 4.43c and d). And as shown in Fig. 4.43d, the additional metasurface diminished the second peaks of high-frequency (180, 210 kHz) disturbances in the Case B4 (Fig. 4.41d), which indicates that the metasurface had a stabilization effect on the Mack third mode. In summary, the stabilization performance in cases where only blowing was used changed when the porous blowing medium was considered. The metasurface generally helped suppress the dominant second mode excited by blowing, but could also reduce the overall effect of stabilization by increasing low-frequency disturbances growth rate of the first mode. This is consistent with the result reported in Ref. [44]. It can be concluded that the optimum strategy is to implement blowing and the metasurface only in the dominant region of the second mode to maximize the stability of the HBL.

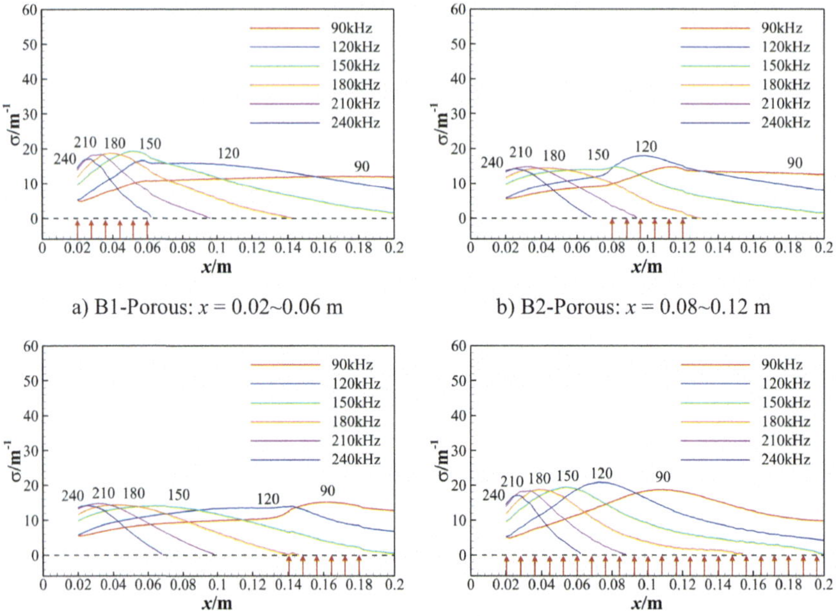

a) B1-Porous: $x = 0.02\sim0.06$ m

b) B2-Porous: $x = 0.08\sim0.12$ m

c) B3-Porous: $x = 0.14\sim0.18$ m

d) B4-Porous: $x = 0.02\sim0.2$ m

Fig. 4.43 Dimensional growth rates of unstable modes in cases of combination of blowing and acoustic metasurface

In the above analyses based on LST, we modeled the effect of blowing by using the average steady vertical velocity of flow while replacing the actual microstructures of the metasurface with the boundary condition for acoustic admittance at the wall. These simplifications ignore the complex interactions of the flow field with the microslits that might excite the unstable modes [16].As discussed above, compared with the baseline case, the disturbances were significantly destabilized in Cases B1/B1-Porous and were generally stabilized in Cases B3/B3-Porous. These cases were further investigated by using the DNS. Figure 4.45 shows the steady flow field in the region of blowing in Case B3-Porous. Alternating expansion and compression waves were induced at the edges of the microslits. Gas was injected from the bottom of the slit to the outer boundary layer, and the recirculation flows observed in the slits without blowing disappeared. Figure 4.46 compares the profiles of mean flow at different locations in Cases B1 and B1-Porous. The selected streamwise locations corresponded to the middle and downstream positions of the blowing strip. As shown in Fig. 4.46, the profiles of velocity and temperature above the metasurface were consistent with the corresponding profiles in the case where only blowing was used. Figure 4.47 shows the distributions of the thickness of the boundary layer as determined by $u^*(y)/u^*\infty = 0.99$. There were no visible discrepancies of the boundary layer thickness between the cases of blowing with and without involving

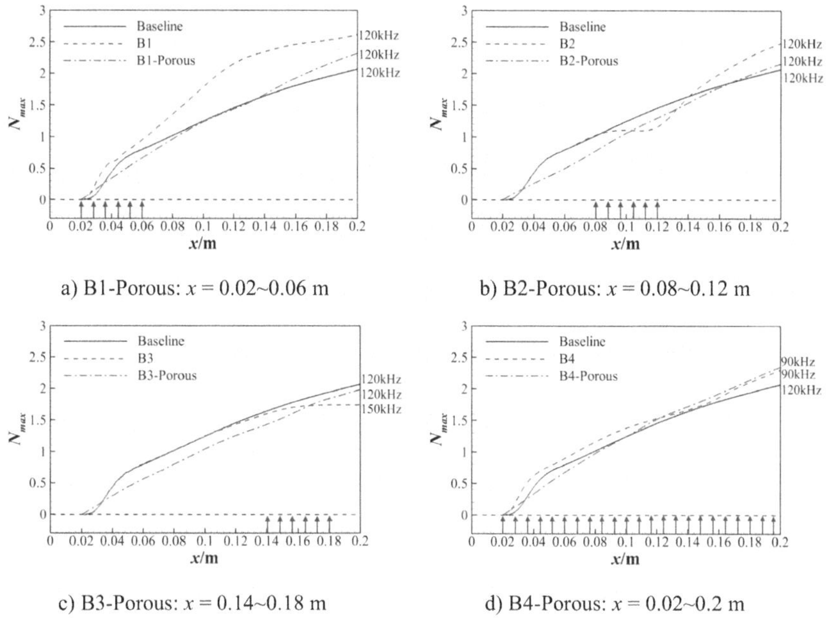

Fig. 4.44 Comparison of N-factor envelops in cases of combination of blowing and acoustic metasurface

the metasurface, which indicates that the acoustic metasurface had a minimal effect on the mean flow.

Figure 4.48 compares the contours of the instantaneous fluctuating pressure induced by a slot of the Gaussian pulse. As shown in Fig. 4.48a, typical two-cell structures were formed downstream of the smooth wall. Compared with the baseline in Fig. 4.48a, the intensity of the fluctuating pressure in Case B1 was higher (Fig. 4.48b) and that in Case B3 was lower (Fig. 4.48c). It also shows that the use of only blowing influenced the intensity of the second mode without destroying the two-cell structures. When the metasurface was used as well, the upper cells tended to join the lower part inside the slit when moving through the coating region. In Case B1-Porous (Fig. 4.48d), the intensity of the fluctuating pressure above the metasurface weakened compared with that in Case B1. By contrast, in Case B3-Porous (Fig. 4.48e), the fluctuating pressure slightly increased due to the metasurface. The amplitudes of the fluctuating pressure at the wall are compared in Fig. 4.49. The results of the DNS showed trends consistent with the predictions of the e^N method, which indicates that the roughness of the metasurface had little effect on the unstable disturbances in this HBL. Figure 4.50 shows the distributions of the fluctuating pressure amplitude at the wall in the frequency domain, calculated through the fast Fourier transform. Compared with the baseline (Fig. 4.50a), the blowing strip when placed upstream (Case B1, Fig. 4.50b) destabilized the broadband frequencies of disturbances such that the unstable modes extended to lower-frequency bands, which is

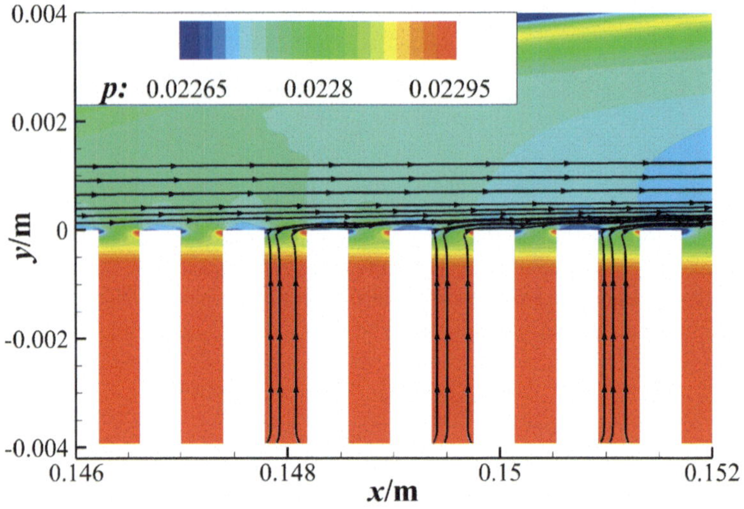

Fig. 4.45 Contours of local pressure around the microslits of the metasurface in Case B3-Porous (p is normalized by $\rho_\infty u_\infty{}^2$)

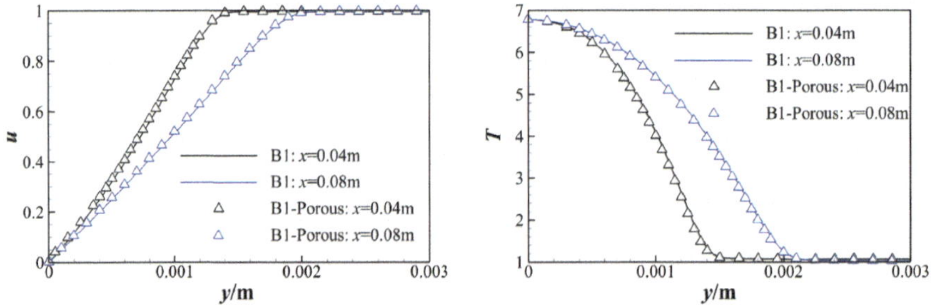

a) Streamwise velocity profile b) Temperature profiles

Fig. 4.46 Comparisons of profiles of the streamwise velocity and temperature between blowing-related Case B1 and blowing–acoustic metasurface Case B1-Porous, calculated by using the DNS (u and T are normalized by u_∞ and T_∞, respectively)

consistent with the LST-based analysis. In Case B3 (Fig. 4.50c), blowing significantly diminished the amplitudes of the dominant disturbances at frequencies 120 ~ 150 kHz, resulting in the stabilization of flow downstream. When the metasurface was added, the excited high-frequency disturbances in Case B1 were generally suppressed in Case B1-Porous (Fig. 4.50d), but the coating also increased the low-frequency first modes in the range 50 ~ 100 kHz (Case B3-Porous, Fig. 4.50e), and this slightly degraded the blowing stabilization performance in Case B3.

In this study, we investigated the stabilization effect of steady blowing and its medium on a Mach 6 flat-plate flow by using LST and the DNS. We emphasized the effect of the locations of blowing on the stabilization of flow, i.e., the blowing

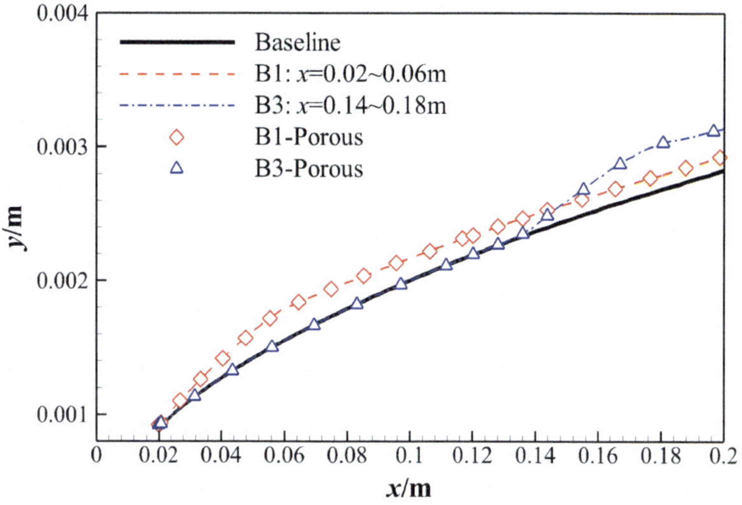

Fig. 4.47 Comparisons of distributions of the boundary layer thickness in cases of blowing at different locations

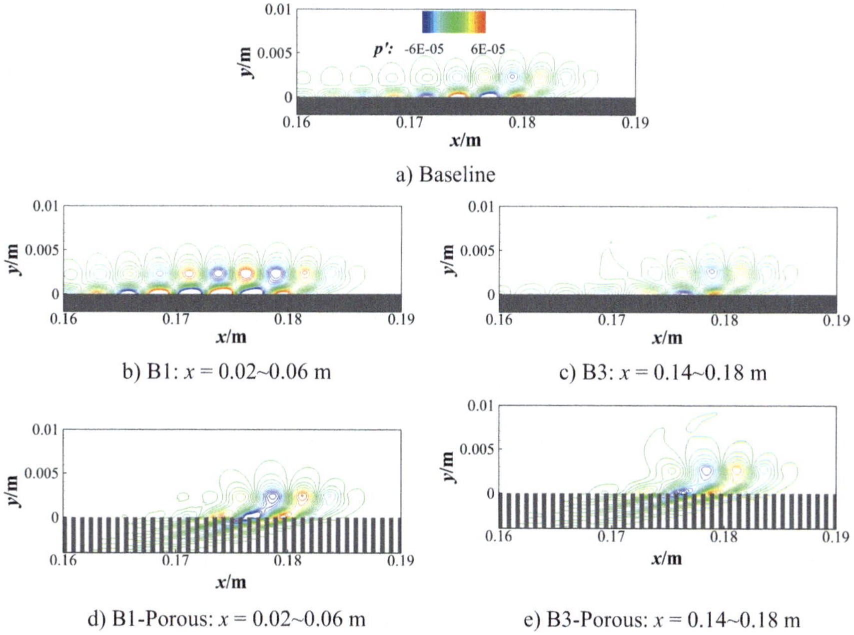

Fig. 4.48 Comparison of instantaneous fluctuating pressure fields (p' is normalized by $\rho_\infty u_\infty^2$)

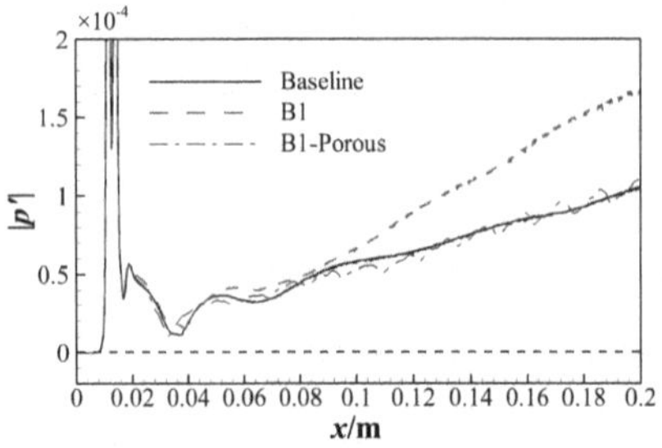

a) B1: $x = 0.02\sim0.06$ m

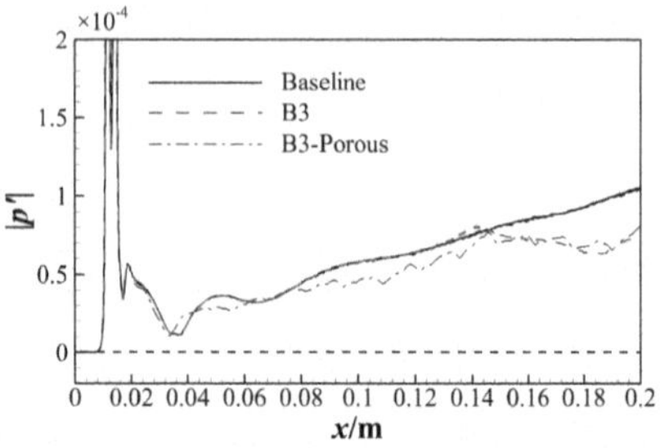

b) B3: $x = 0.14\sim0.18$ m

Fig. 4.49 Comparisons of the distributions of the amplitude of fluctuating pressure at the wall due to a Gaussian pulse in different cases ($|p'|$ is normalized by $\rho_\infty u_\infty^2$)

strip was placed upstream, middle, and downstream of flow as well as over the entire plate. A metasurface was used as the blowing medium to inject gas, and stabilization performance in the case of the combined use of blowing and the metasurface was examined.

When only blowing was applied without the metasurface, an LST-based analysis showed that the phase speed of mode F decreased prominently when approaching the blowing strip, and this directly led to the forward movement of the synchronization point. When the blowing strip was located upstream of the synchronization point, the disturbance was significantly excited, and vice versa. Therefore, the streamwise locations of the dominant second modes are crucial for blowing-related performance. The HBL is generally rendered unstable if the blowing strip is installed further upstream, and is stable when it is placed in the downstream region that sustains the dominant second modes.

The use of the porous blowing medium changed the stabilization performance in comparison with that when blowing alone was used. When the blowing occurred upstream, the metasurface generally suppressed high-frequency disturbances and helped stabilize the upstream HBL. Meanwhile, the coating prematurely excited lower-frequency disturbances to degrade the downstream and overall stabilization-related performance. The results here indicate that a possible optimum strategy to stabilize the HBL is to implement blowing and apply the acoustic metasurface only in the region in which the second mode is dominant.

4.4.1 Combination with Micro-suction

As discussed earlier, metasurface can suppress the second mode but excite the first mode, which means that a well-designed metasurface can only effectively suppress unstable perturbations within a certain range. One possible approach to address this limitation is to combine it with micro-suction. Boundary layer suction is a

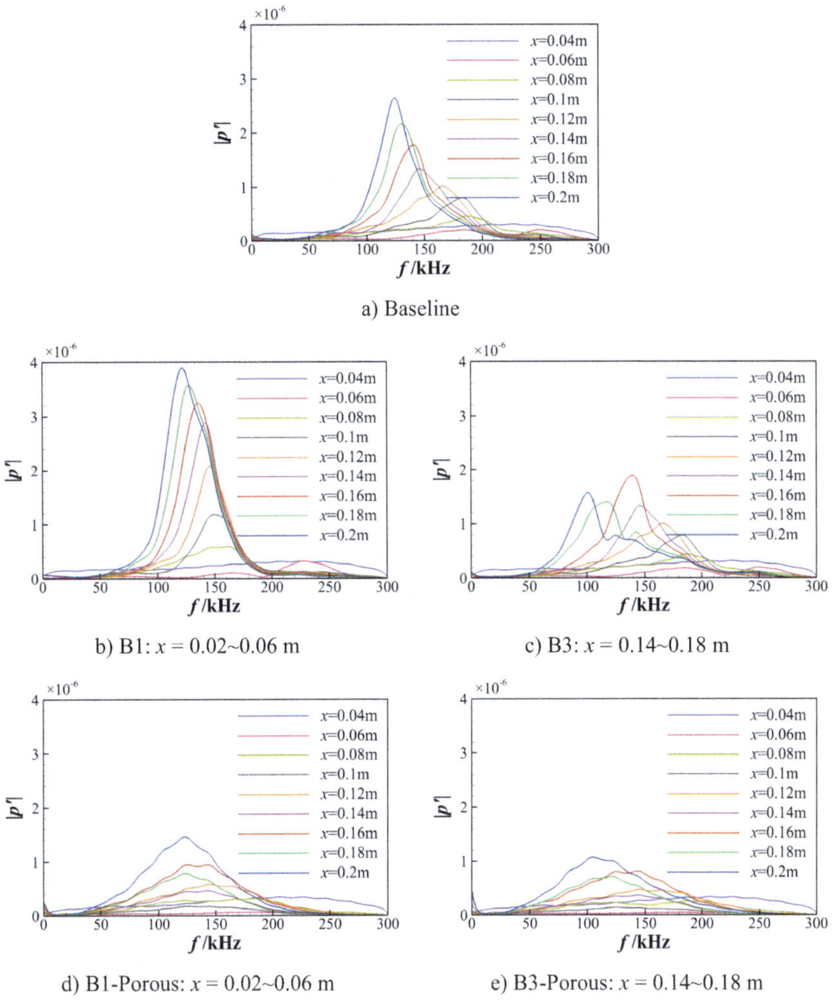

Fig. 4.50 Fourier analysis of fluctuating pressure at the wall

simple method that effectively reduces the boundary layer thickness and suppresses disturbance growth by removing low-speed fluid near the wall.

This study uses the "acoustic metasurfaces-suction" combination control method and focuses on the Ma 4.5 flat plate boundary layer. DNS and LST techniques are utilized to examine the impact of the combination control method on the unstable modes of high-speed boundary layers and investigate the viability of combined control.

The computational domain is shown in Fig. 4.51, which ranges from 144 to 224 mm, with the metasurfaces arranged from 164-184 mm. According to reference [42], the metasurfaces are set with a width of 1 mm and a depth of 1.5 mm. The

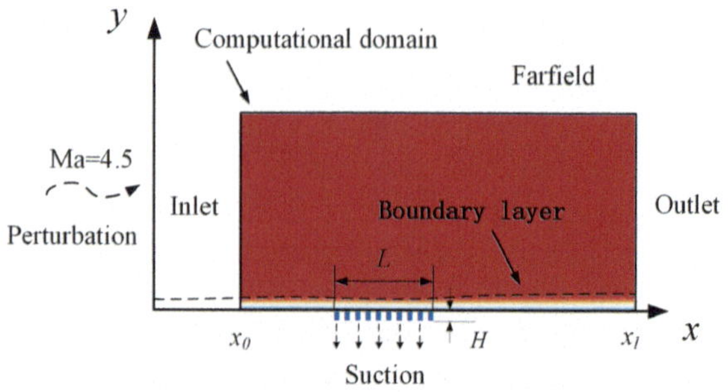

Fig. 4.51 Computational domain

inflow conditions are consistent with reference [43], with an inlet Mach number of 4.5, inflow temperature of 65.15 K, and a wall set as a no-slip isothermal wall with a temperature of 289.04 K. The computation uses a non-uniform grid, with grid refinement near the wall and in the channel region. Grid independence verification is performed, with different grid densities for DNS cases: Case1: 830 × 220, Case2: 623 × 165, Case3: 1245 × 330. In all three grid sets, the maximum height of the first grid layer does not exceed 0.005 mm, and there are at least 38 grid points per disturbance wavelength in the core region and at least 125 grid points in the boundary layer.

A two-dimensional neutral curve is shown in Fig. 4.52, and it is evident that both the instability first and second modes are present under the Ma4.5 flat-plate. The chosen mode for the second mode serves as the entrance unstable wave, with its growth rate and phase velocity illustrated in Fig. 4.53. The synchronization point of 150 kHz is located at 174 mm, hence, the metasurfaces arrangements designed in this study fall within the range of 164–184 mm. Within this range, the 150 kHz perturbation remains unstable and the first mode is set at 46 kHz, which is also observed to be in an unstable state within the computational domain.

The micro-suction is set at the bottom of the microslits, as shown in Fig. 4.54. Under the approximate assumption that the density of the extracted gas is the same as the density of the gas in the grooves, it can be derived that $\frac{\partial\rho}{\partial y} = 0$. Moreover, since the groove is very small, it can be approximately assumed that the gas temperature in the channel is the same as the wall temperature. Finally, based on the Poiseuille flow approximation, the suction function is given as: $(\rho^* v^*)' = q_0^* m\beta(x)$, where ρ^* represents density, v^* represents normal velocity, q_0^* represents the product of free inflow density and velocity, m is the amplitude, which is negative for inspiration, and the larger the absolute value, the stronger the suction effect. It can be regarded as the distribution function of inspiration control along the flow direction inside the groove, which can be defined as: $\beta(x) = (x - x_0)(2b - x + x_0)$. In this section, we choose 7 cases to investigate the influence of combination method on unstable modes, as shown in Table 4.5.

Figures 4.55 and 4.56 show the contour map and the profiles of the streamwise velocity at several positions for considered cases, respectively. Comparing Figs. 4.55a and 4.56b, it can be observed that the influence of applying microgrooves on the

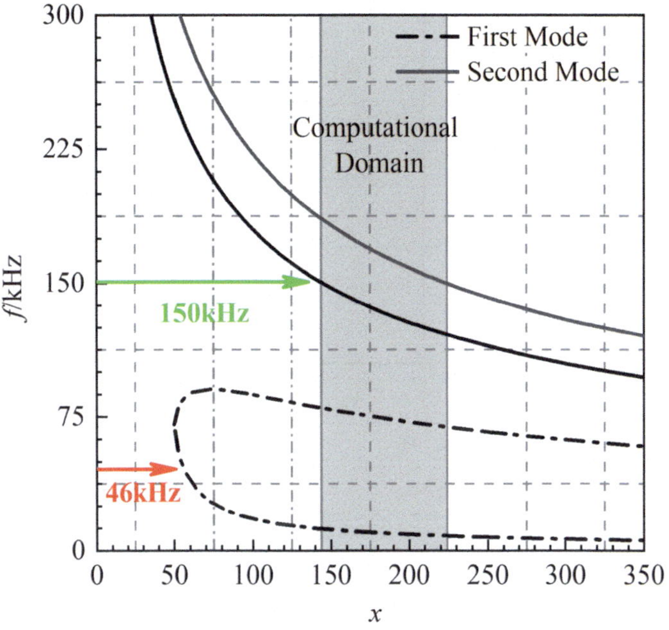

Fig. 4.52 Neutral curves of Ma 4.5 flat plate (two-dimensional)

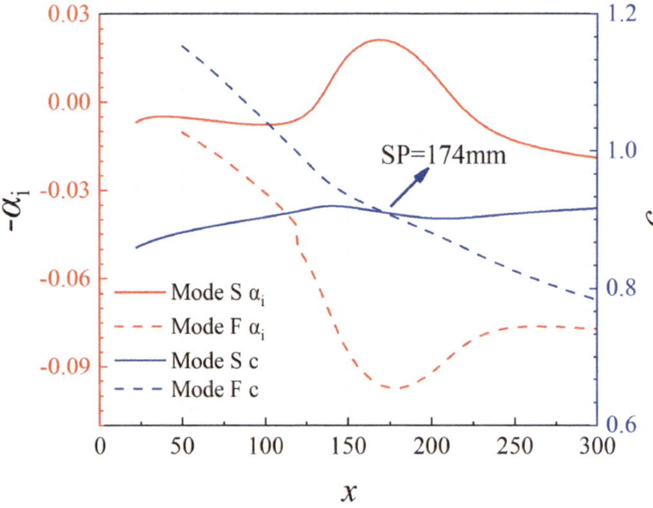

Fig. 4.53 Streamwise evolution of the growth rate and phase speed of the mode at 150 kHz

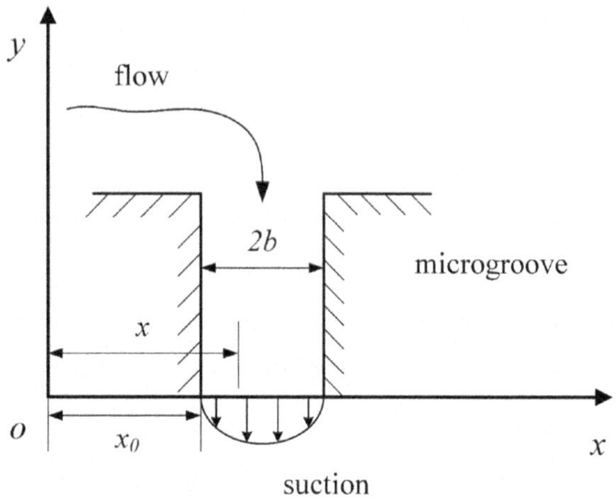

Fig. 4.54 Schematic diagram of suction in grooves

Table 4.5 Setups of calculation cases

Cases	Amplitude *m*	Region
P0	0	164 ~ 184
G0	0	164 ~ 184
G1	0.001	164 ~ 184 –
G2	0.002	164 ~ 184
G5	0.005	164 ~ 184
G10	0.01	164 ~ 184
G20	0.02	164 ~ 184

distribution of flow velocity is not significant, and the boundary layer represented by the dashed line (0.99 U_∞) remains unchanged. Moreover, it can be seen that with the increase in suction intensity, the flow velocity profile gradually becomes full, and the boundary layer in the controlling region becomes obviously thinner and the low-speed area is closer to the wall. As the redevelopment of boundary layer, the profile gradually close to the case P0. However, the boundary layer thickness still lower than that of case P0.

Figures 4.57 and 4.58 respectively show the velocity distribution in the normal direction and the cloud map of pressure near the wall surface. The surface microgrooves have a notable impact on the distribution of the wall normal velocity compared to a flat plate. Vortices, approximately the same width as the grooves, are formed in the channel. It also can be seen that there is an alternating distribution of expansion waves and compression waves. These wave structures have a similar scale to the groove width and the vortex structure is relatively symmetrical. When suction is applied, the main flow is drawn into the groove from the trailing edge, causing the vortex structure to move towards the leading edge. However, the position of the

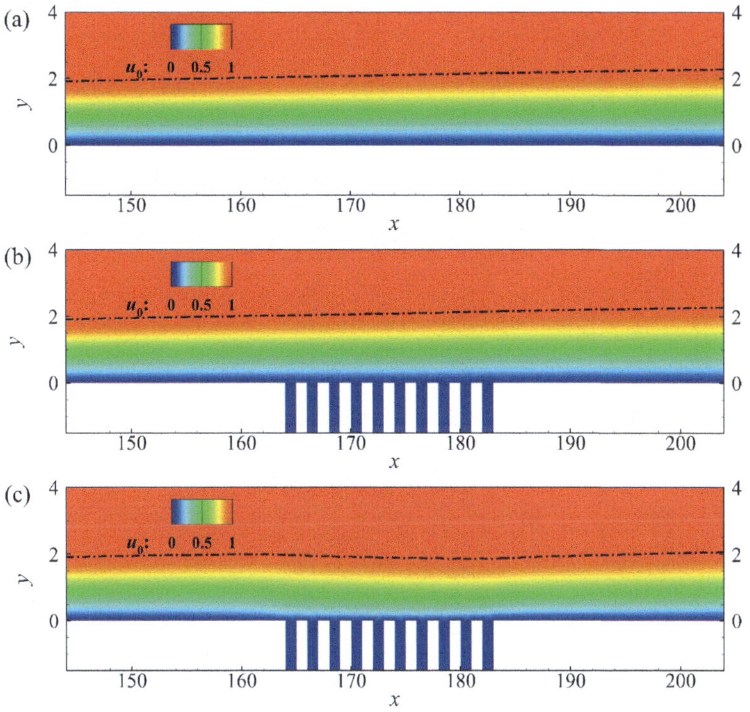

Fig. 4.55 Streamwise velocity contours of base flow: **a** P0; **b** G0; **c** G10

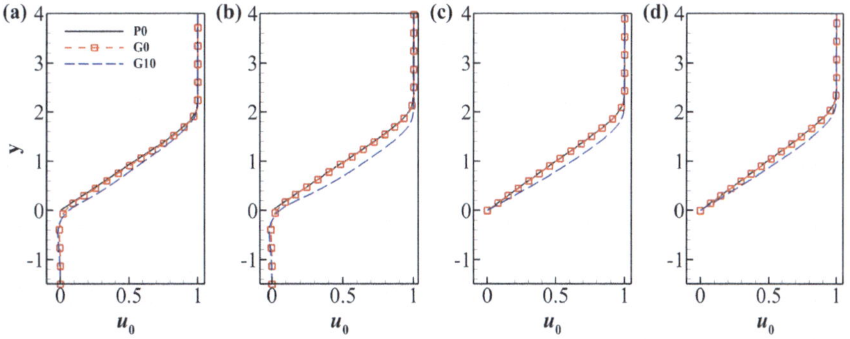

Fig. 4.56 Profiles of streamwise velocity at **a**164.5 mm; **b** 182.5 mm; **c** 190 mm; **d** 200 mm

compression/expansion wave remains relatively unchanged but an increase in its intensity shown in the pressure map.

In summary, the microgrooves have little effect on the streamwise velocity of the main flow, but it obviously affects the wall normal velocity distribution. The suction

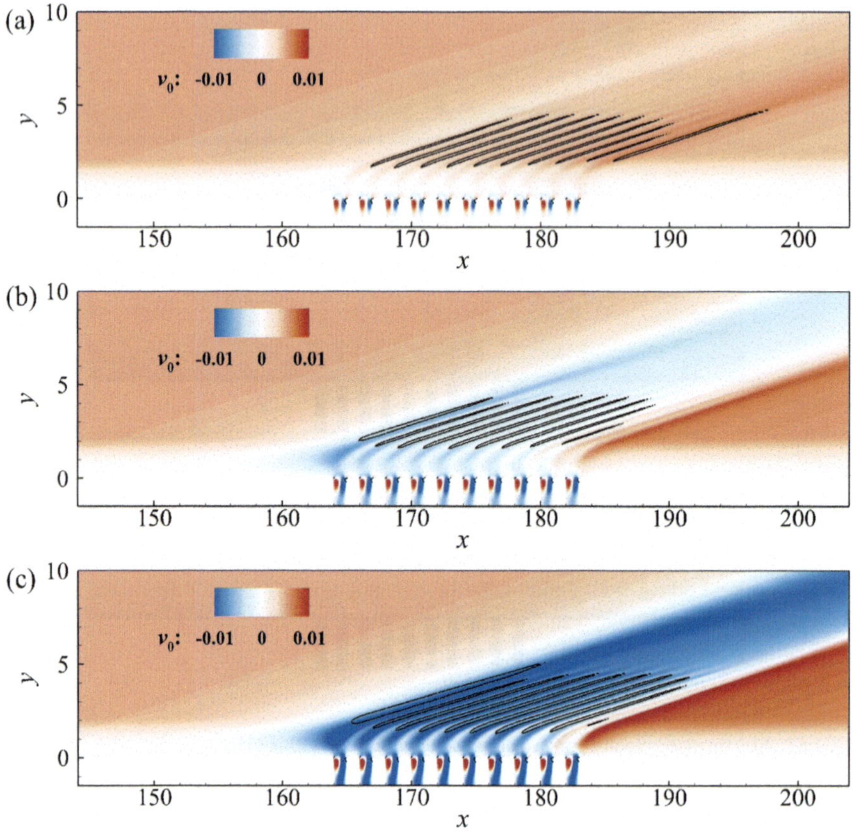

Fig. 4.57 Normal velocity contours of base flow (black lines represent the shock wave positions)

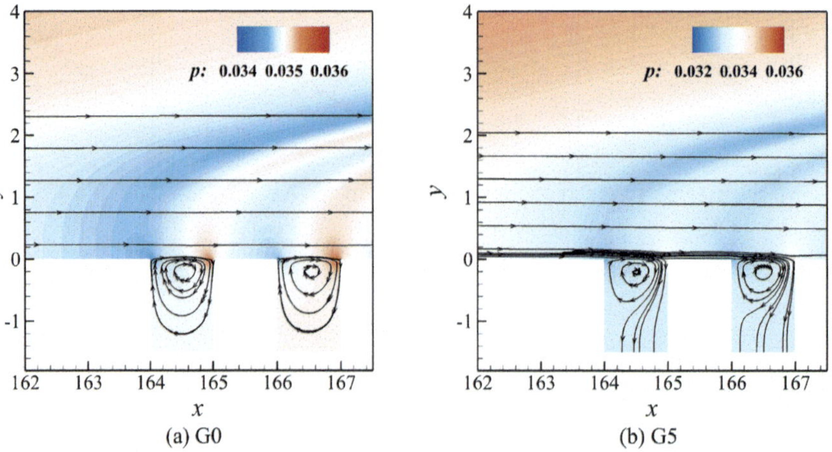

Fig. 4.58 Pressure contour around grooves for **a** G0; **b** G5 (black arrows represent streamlines)

modifies the velocity profile of the base flow direction and influences the boundary layer thickness of the downstream to some extent.

Figure 4.59 shows the map of pressure fluctuation at the same time under different conditions. It can be seen that the double-cell structures of the second mode gradually increases in the streamwise of the flat-plate (see Fig. 4.59a) and these structures scatter towards the boundary layer along the development of compression/expansion waves (see Fig. 4.59b). The attenuation of second mode in the downstream indicates that the microgrooves can suppress the 150 kHz perturbation. Moreover, Fig. 4.59cshows that the "microgrooves-suction" combination method has a more significant effect on the spatial structure of pressure fluctuation. The comparison of wall pressure fluctuation also indicates that the microgroove has no significant effect on the upstream flow field and the peak of pressure fluctuations decreases significantly, as shown in Fig. 4.60. When combined with suction control, the peak continues to decrease and shift further downstream. This is because suction thins the boundary layer, increases the frequency of the second mode wave, and shortens the wavelength.

When the frequency of incoming perturbation is not within the design range of the microgrooves, the microgrooves will excite the disturbance, which will lead to the deterioration of the control effect, as reported by Fedorov [33]. Therefore, we also set 130 kHz perturbation in the inlet to test the performance of microgrooves and the combination method, as shown in Fig. 4.61. As can be seen, the case G0 still inhibits the growth of the disturbance wave in the control region, but the disturbance grows faster downstream of the control region, and the pressure amplitude exceeds that of case P0 at $x = 200$ mm. However, the combination method achieves a good performance in suppressing perturbation, the growth rate of amplitude under case G1 decreases to a certain extent compared with case G0 and the growth rate

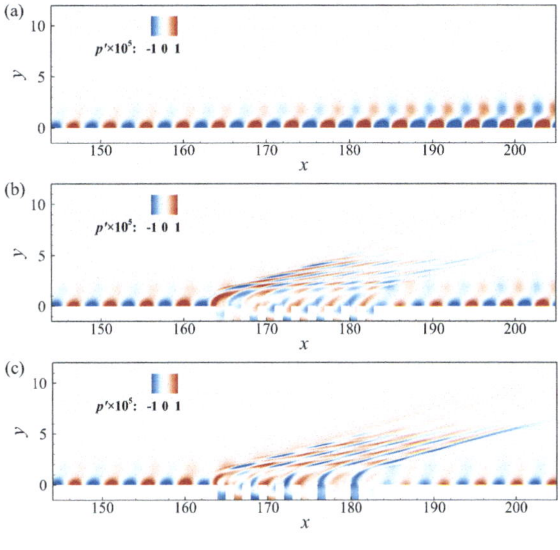

Fig. 4.59 Contours of instantaneous pressure fluctuations for **a** P0; **b** G0; **c** G10

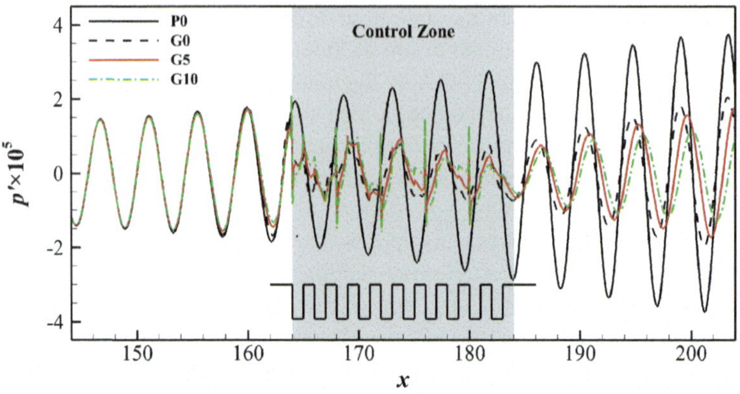

Fig. 4.60 Distribution of the wall pressure fluctuations (150 kHz)

even becomes negative near the exit in case G10. The results indicates that with the increase of suction intensity, combination method enhances correction effect on the boundary layer and suppress the 130 kHz disturbance.

Figure 4.62 depicts the contour of pressure fluctuation for considered cases, and it is evident that the structure of fluctuation is disrupted by the microgroove. Additionally, the amplitude of pressure fluctuation on the wall rises obviously in the downstream of microgrooves, which agrees well with Fedorov's findings, that is, microgrooves promote the development of first mode. Notably, the pressure fluctuation of case G10 is slightly lower than that of case G0, suggesting that suction method can be helpful to suppress first mode instability. Figure 4.63 shows the development of the first mode in the streamwise direction. When microgrooves are applied, the amplitude of pressure fluctuation changes sharply at the front and back of the microgrooves, and after passing through the control area the amplitude is higher than that in the flat-plate. However, the amplitude of case G10 is significantly lower than

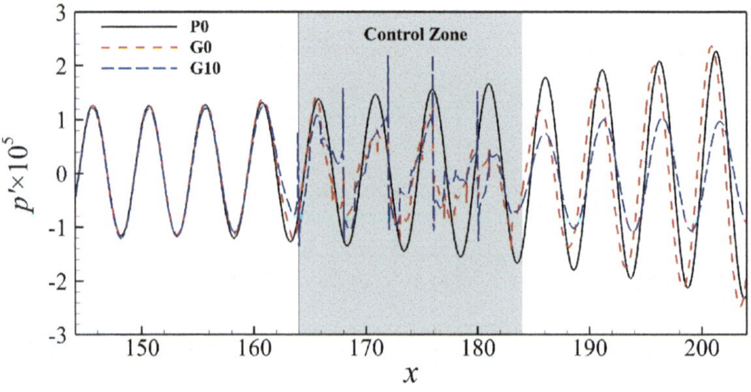

Fig. 4.61 Distribution of the wall pressure fluctuation (130 kHz)

that of case G0, although it is still higher than that of the case P0. This result shows that the combined control method can suppress the growth of first mode excited by the microgrooves. In addition, it is also found that the improvement of suppression effect is not obvious although the inspiratory intensity increases gradually, which indicates that there is an optimal value of the suction amplitude on suppressing the first mode.

Figure 4.64 shows the relationship between the growth rate and the dimensionless frequency in the spatial mode ($\beta = 0$) using LST technique. It can be seen that the microgrooves only have a weak inhibitory effect on the growth rate of the second mode, because of the localization effect of the microgrooves. The effect of suction on boundary layer correction is also reflected in the downstream, the thickness of the downstream boundary layer in the control region becomes thinner, as the amplitude of suction increases. LST shows that the growth rate of the second mode decreases significantly and unstable region moves to high frequency due to the effect of suction.

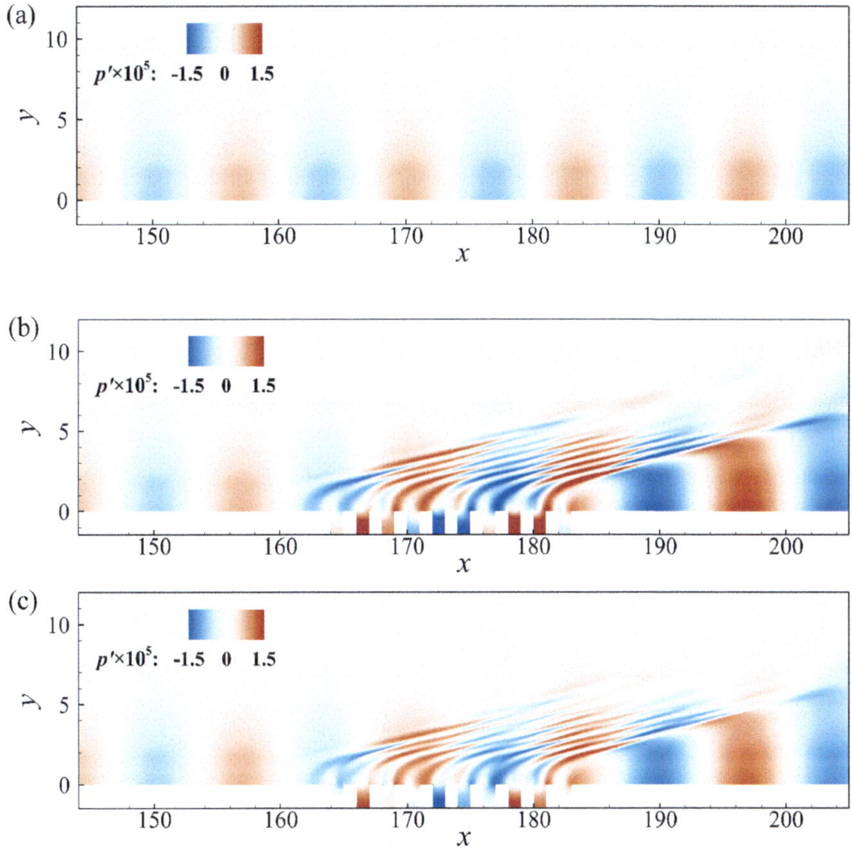

Fig. 4.62 Contours of instantaneous pressure fluctuation for **a** G0; **b** G0; **c** G10

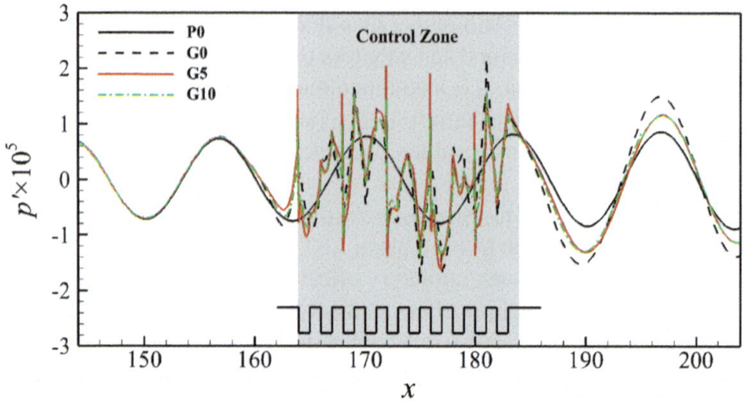

Fig. 4.63 Distribution of the wall pressure fluctuation (46 kHz)

Therefore, suction decrease the growth rate and rising the unstable frequency by influencing the thickness of the boundary layer, thereby achieving efficient suppression of the second mode. However, introducing suction has little impact on the growth rate of the first mode which is consistent with the results of DNS.

Figure 4.65 shows maps of spatial growth rate of unstable modes at $x = 190$ mm under considered cases. It can be clearly seen from the contour that the control effect of the combined control, with the increase of suction, the instability range of the second mode wave shrinks, the frequency band shifts upward, and the maximum growth rate decreases obviously. For the first mode, the maximum growth rate also

Fig. 4.64 Spatial growth rate of unstable modes near outlet

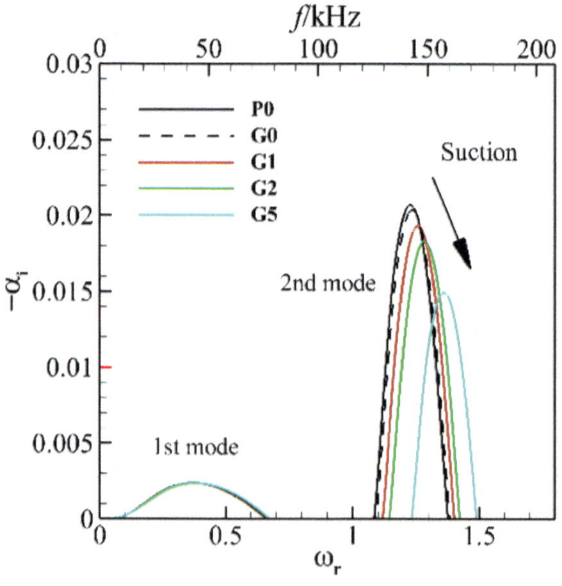

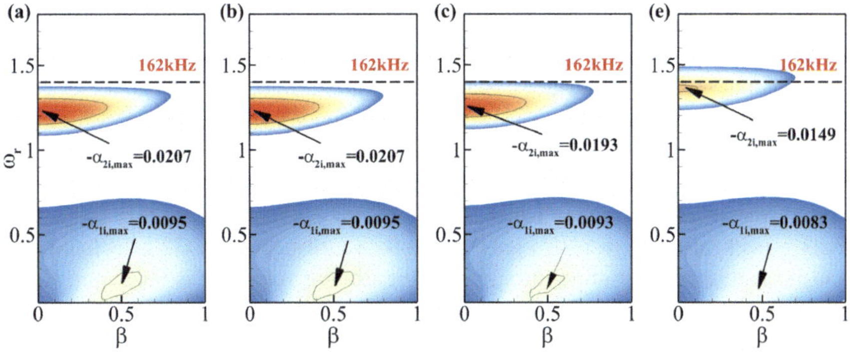

Fig. 4.65 Contours of spatial growth rate at different cases for unstable modes: **a** P0; **b** G0; **c** G1; **d** G5

decreases due to the suction, while the instability range has no obvious change. Compared with case G0, the combined control method (case G5) can inhibit the growth rate of the first and second modes of the least stable by 12.63% and 28.02%, respectively, which has a certain multi-mode control effect.

4.5 Machining Flaw Effect

In practice, machining flaw can exist on the edges of porous coatings as they are typically processed by laser cutting. The principle of laser cutting is that the laser beam output by the laser oscillator is focused through the focusing mirror to generate high density energy irradiation on the material, making it melt and evaporate for the cutting. When the laser is used to cut materials, slits which are wide at the top but narrow at the bottom, will be formed on the material surface. From the upper surface of the cutting seam to the lower surface of the plate, heating affected zones are formed on both sides of the cutting seam with the change of the cutting depth and the material laying angle. The material in the heat affected area is heated unevenly, resulting in the change of the geometry of the slit. The slit deformation scale is in the order of micron, which is similar to the scale of the rectangular acoustic metasurfaces. However, how this machining flaw affects the transition control of rectangular acoustic metasurfaces is still unclear, thus it is the subject of the present investigation.

In this study, we investigate the local shape mutation of rectangular acoustic metasurfaces caused by laser cutting, and quantify its effect on the second-mode disturbance for transition control. The frequency and synchronization interval of the most unstable second mode are obtained by the linear stability analysis. The amplitude suppression effect of the most unstable second-mode wave is studied by applying it to the upstream of the microgrooves in a series of direction numerical simulations (DNS).

In order to study the influence of the local deformation of rectangular acoustic metasurfaces caused by laser cutting on the transition control effect, three types of rectangular acoustic metasurface configurations are designed as shown in Fig. 4.66. G1 is the standard acoustic metasurface configuration, and the geometric parameters are set the same as those used in our previous studies [44, 45]. The width of the acoustic metasurface is $w = 0.4$ mm. The unit-cell period is $s = 1.6$ mm. The depth of the acoustic metasurface is $H = 1.392$ mm. The porosity ($\phi = 2w/s$) and cavity aspect ratio ($Ar = w/H$) are 0.5 and 0.287, respectively. G2 is a chamfered acoustic metasurface configuration, and the radius of rounding is 0.05 mm. G3 is the configuration with local bumps near the rectangular acoustic metasurface. The bump shape is obtained by fitting the acoustic metasurface shape after real laser cutting. In this study, three bulge heights are considered, i.e. $h = 0.03$ mm, 0.06 mm and 0.09 mm. The specific parameters are expressed in Formula (1) and the fitting coefficients are listed in Table 4.6.

$$y(x) = A\sin(ax) + B\cos(bx) + C \tag{4.10}$$

The freestream condition is the same as that used by Sandham and Lüdeke [46] and Wartemann et al. [47], i.e. Mach number M_∞=6.0, Reynolds number $Re_\infty = 1 \times 10^4$,

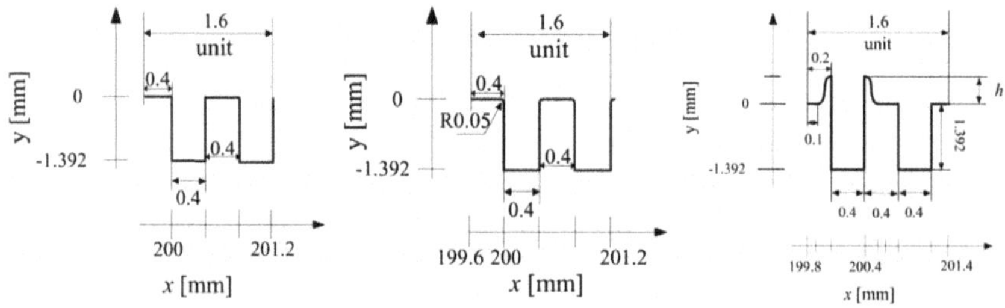

(a) G1 (b) G2 (c) G3

Fig. 4.66 Model and geometric parameters for the rectangular acoustic metasurfaces: **a** standard, G1; **b** chamfered, G2; **c** protruded, G3

Table 4.6 Geometric parameters of the protruded rectangular acoustic metasurfaces, G3

h	x	A	a	B	b	C
0.03	$x \in (199.9,200)$	-0.006645	34.61	-0.0134	34.61	0.01516
	$x \in (200.4,200.5)$	-0.01448	34.61	-0.03729	34.61	0.015116
0.06	$x \in (199.9,200)$	-0.02883	36.92	-6.977×10^{-5}	36.92	0.03046
	$x \in (200.4,200.5)$	0.01141	37.25	0.02625	37.25	0.03059
0.09	$x \in (199.9,200)$	0.04185	36.36	0.01122	36.36	0.04433
	$x \in (200.4,200.5)$	0.037	36.22	0.02266	36.22	0.04429

temperature $T_\infty = 216.65$ K. The working fluid is based on the ideal gas model, with the Prandtl number $P_r = 0.72$. The non-slip isothermal boundary condition is applied to the wall, with the wall temperature T_w=300 K.

Following our previous work [44], the computational domain is 85 mm in height and 54.7 mm in length (as shown inFig. 4.67), and the inlet is located at $x_s = 195.3$ mm, which corresponds to the starting point ($x - x_s = 0$ mm) of the computational domain. The wall-normal grid is stretched using a tangential function, and a total of 341 grid points are used to resolve the boundary layer. The streamwise grid is refined around the deformed location of the acoustic metasurface to ensure that the number of distributed grids is around 60 in each local bulge, as shown in Fig. 4.67 Simulation setup for boundary layer instability control using acoustic metasurfaces. Fig. 4.68 In order to avoid the wave reflection at the outlet, the streamwise grid is gradually stretched after the effective computational domain at $x - x_s = 21$ mm to dissipate the reflected waves. The inflow is taken as the superposition of the base flow and the second mode shown as below,

$$q(0, y, t) = q_0(0, y) + \varepsilon \cdot \mathrm{Re}\left[\hat{q}(y) \cdot e^{i(\omega t + \varphi)}\right], \tag{4.11}$$

where $q_0(0, y)$ is the base flow from the Blasius solution, and ε is the amplitude of the inlet disturbance, which is taken to be 0.0001 in the present study. $Re[\bullet]$ represents the real part.

LST is used to determine the streamwise installation layout of the rectangular acoustic metasurfaces. Figure 4.69 shows the LST calculated neutral curves and the N-value curves for unstable modes in the frequency range of $f^* \in [350, 470]$ kHz. As

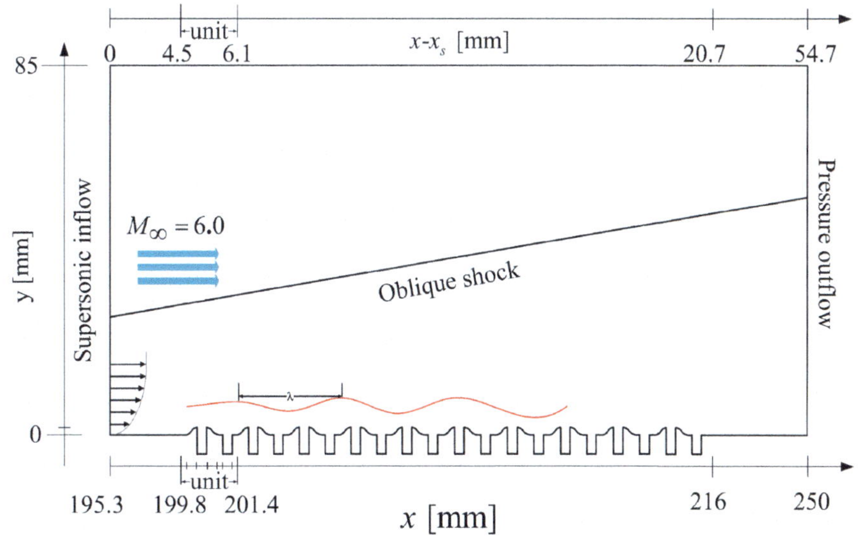

Fig. 4.67 Simulation setup for boundary layer instability control using acoustic metasurfaces

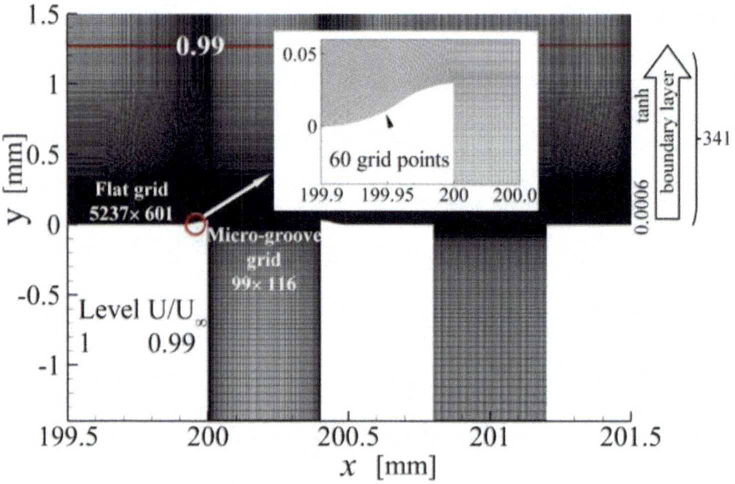

Fig. 4.68 Grid distribution in the computational domain

can be seen from Fig. 4.69b, the maximum N-value increases almost linearly for $x/L_0 > 180$, where $L_0 = 1$ mm is the reference length scale. Without loss of generality, we choose the computational domain of $x/L_0 \in [195.3, 250]$ (indicated by two vertical lines in Fig. 4.5b for detailed numerical simulations and analyses. In this domain, the frequency of the most unstable second-mode wave is about 400 kHz and the maximum N value is about 4. It is noticed that such a small N value is not large enough to trigger the nonlinear effect or transition in quiet environments, thus we only focus on the amplitude suppression of the second mode by the local deformation of rectangular acoustic metasurfaces. Previous studies [44] have found that, when porous coating is placed downstream of the synchronization location of Mode F and Mode S, the amplitude of the second mode can be effectively suppressed. However, we notice that the synchronization location between Mode S and Mode F is not at a fixed point under the present flow condition, but instead a region with a long length over 107 mm ($x/L_0 \in [154, 261]$, see Fig. 4.70). Therefore, in order to further reduce the computational cost, we arrange the rectangular acoustic metasurfaces within $x/L_0 \in [200, 216]$ of the synchronization interval, and investigate the influence of the porous coating on the amplitude suppression of the second mode.

Figure 4.71 shows the wall-normal velocity contours for the base flows of C1-C5. The local deformation of the acoustic metasurfaces has great influence on the wall-normal velocity above the coated surface. The leading and trailing edges of the rectangular acoustic metasurface form compression and expansion waves with the wavelength corresponding to the width of the acoustic metasurface. According to the streamline patterns of C1-C5 in Fig. 4.72, it is found that vortex structures with the size similar to the width of the acoustic metasurface are formed inside. Moreover, small separation bubbles are also formed at the rear edge of the acoustic metasurface of C4 and C5, and the size of the separation bubbles increases with the deformation

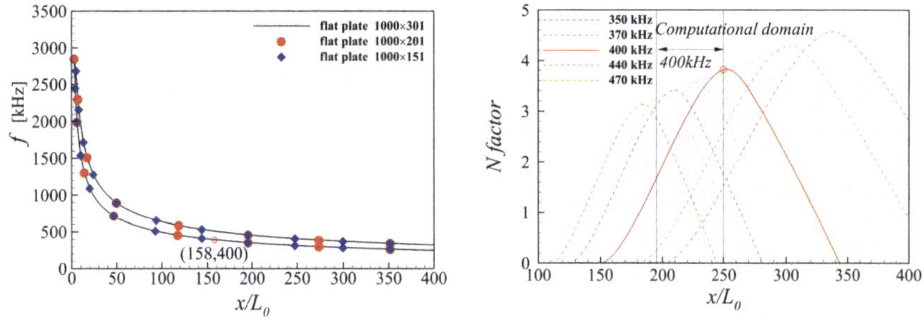

(a) **Neutral curve** (b) ***N*-value curve**

Fig. 4.69 Neutral curve and *N*-value curve for the stability of the flat-plate boundary layer

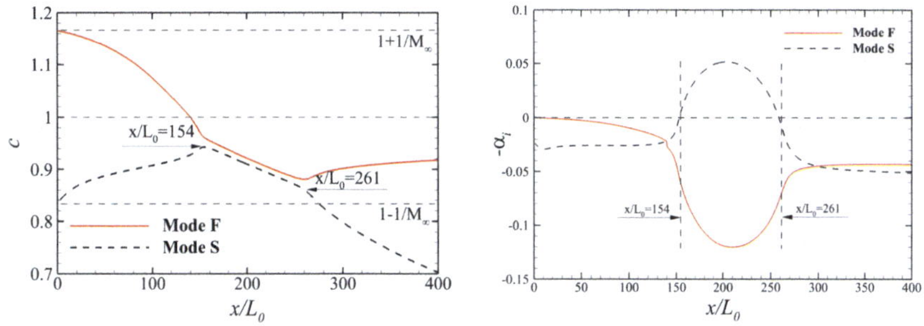

(a) Phase velocity (b) Growth rate

Fig. 4.70 Phase velocity and growth rate variations of the fast and slow modes at 400 kHz along the streamwise direction

height as indicated by the red circles in Fig. 4.72d and e. The streamwise velocity contours of the base flow near the acoustic metasurface are shown in Fig. 4.73. It can be seen that the streamwise velocity contour keeps good parallelism compared to the flat-plate case, and the local deformation of acoustic metasurfaces has little influence on the streamwise velocity above and downstream of them.

Figure 4.74 compares the variation of boundary layer thickness δ (99% of the freestream velocity) along the streamwise direction of the base flow for different acoustic metasurface configurations. The local deformation of acoustic metasurfaces clearly generates an oscillation in the boundary layer of the slotted area. Figure 4.75 displays the comparison of the streamwise mean velocity and temperature profiles at $x - x_s = 24$ mm. As can be seen that the profiles for different cases are in good agreement, which suggests that the local deformation of acoustic metasurfaces only has little influence on the boundary layer instability downstream.

According to the linear stability analysis, the frequency of the most unstable disturbance wave in the boundary layer of the investigated domain is about 400 kHz under the present freestream condition. In addition, Guo et al. [44] found that the

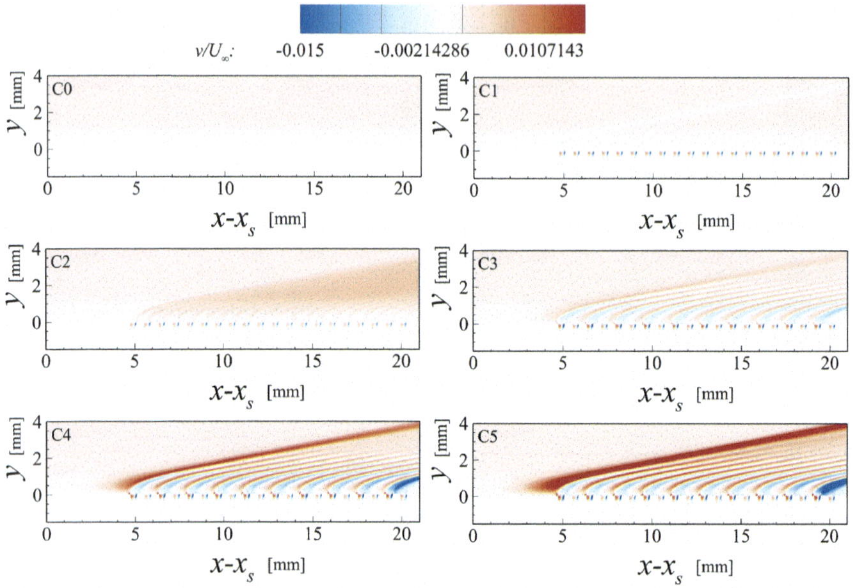

Fig. 4.71 Wall-normal velocity contours of the base flows for different acoustic metasurface configurations

evolutions of the second modes at different frequencies are linearly independent under small initial disturbance amplitude. Therefore, the inhibition of disturbance waves at the most unstable frequency becomes the key in delaying boundary layer transition. Figures 4.76 and 4.77 show the instantaneous fluctuating pressure $p\prime$ and kinetic energy E_k ($E_k = u\prime u\prime + v\prime v\prime$), respectively. It can be found that the second mode dissipates on the surface of the rectangular acoustic metasurface, and part of the disturbance penetrates into the cavity of the acoustic metasurface, while the rest is reflected out of the boundary layer, resulting in the reduction of the fluctuating pressure and kinetic energy. Figure 4.78a, b show that the amplitudes of the fluctuating pressure and the perturbation energy gradually decrease along the streamwise direction. By comparing C1-C5, we can further observe that local deformation of the rectangular acoustic metasurface enhances the dissipation of disturbances in the groove as well as the reflection effect of the boundary layer. The influence of the rectangular acoustic metasurfaces on the boundary layer is mainly reflected in the wall-normal velocity. The time–frequency analysis at the sampling point P_0 ($x - x_s = 20.7$ mm, $y = 0.1$ mm) of the rear edge of the slotting area confirms that the slotting of acoustic metasurfaces does not generate any new unstable mode (Fig. 4.79).

In order to further understand the suppression effect of the rectangular acoustic metasurfaces on the second mode, the normal amplitude integral U' and the amplitude amplification factor R are defined as below,

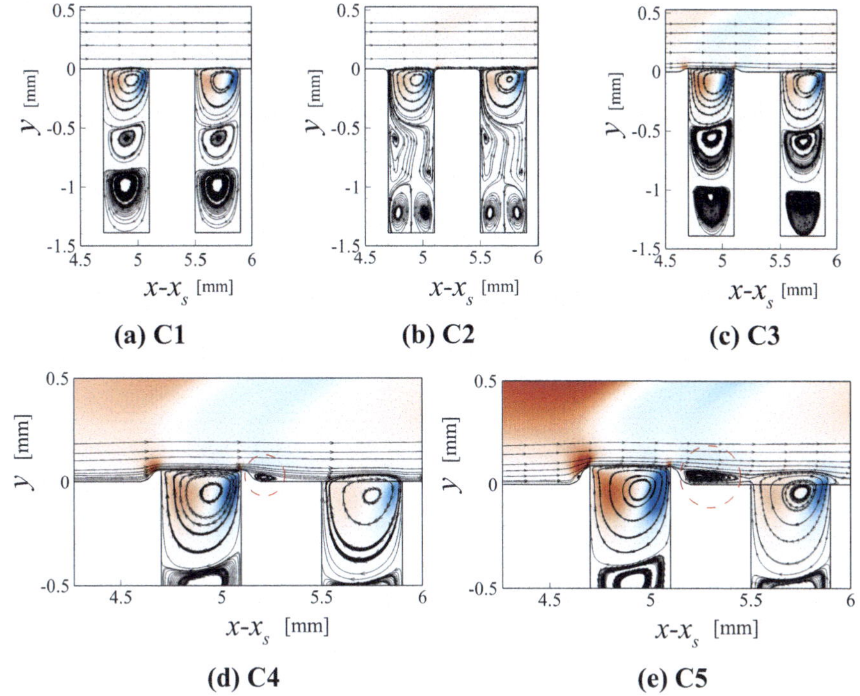

Fig. 4.72 Streamline patterns inside the cavity for different acoustic metasurface configurations

$$U'(x, f_0) = \int_0^{\infty} |\hat{u}(x, y, f_0)| dy, \tag{4.12}$$

$$R = \frac{\hat{u}_{max}(x - x_s = 20.7)}{\hat{u}_{max}(x - x_s = 0)} \tag{4.13}$$

and,

$$\hat{u}(x, y, f_0) = F\left[u'(x, y, t)\right], \tag{4.14}$$

$$\hat{u}_{max}(x, f_0) = max\left[|\hat{u}(x, y, f_0)|\right], \tag{4.15}$$

where $u'(x, y, t)$ is the fluctuating velocity obtained in the DNS, $\hat{u}_{max}$ is the maximum streamwise velocity disturbance amplitude, and f_0 is the frequency. $F[\bullet]$ represents the Fourier transform.

Figures 4.80a and b respectively show the amplitudes of the streamwise velocity disturbance $\hat{u}_{max}$ and U' along the streamwise direction for the second mode at $f^* = 400$ kHz. In the present study, both $\hat{u}_{max}$ and U' are normalized by the initial amplitude of the disturbance. Compared with C1, the configurations of C3-C5 present

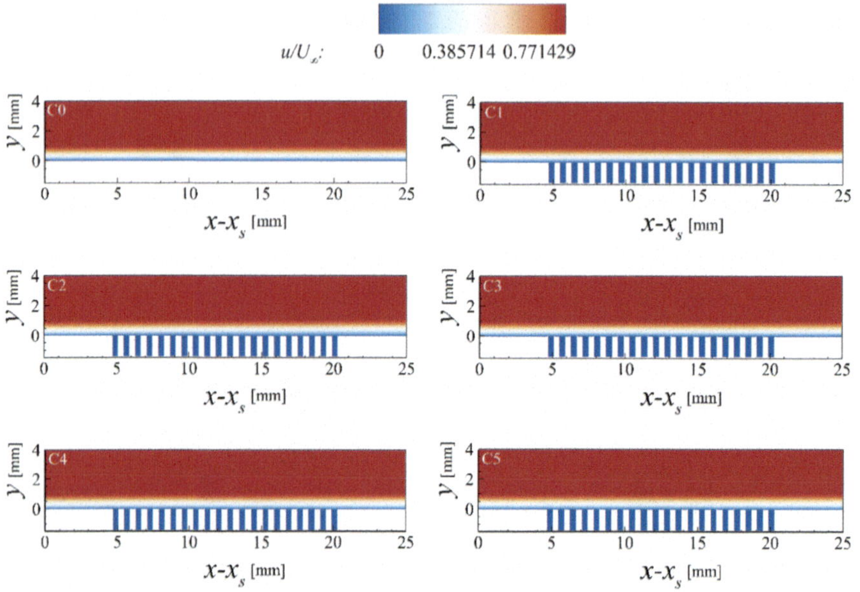

Fig. 4.73 Streamwise velocity contours of base flows for different acoustic metasurfaces configurations

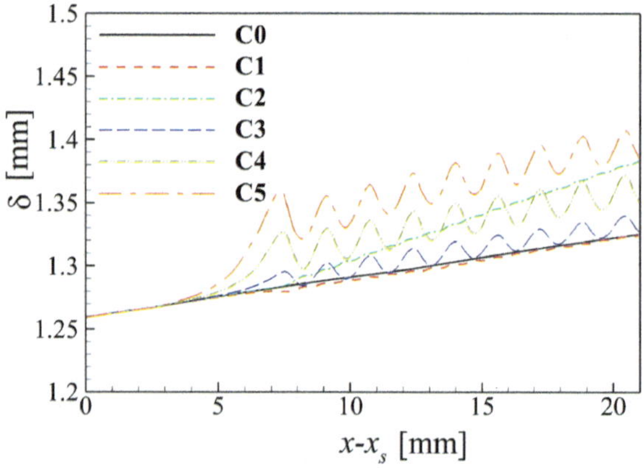

Fig. 4.74 Boundary layer thickness distribution along the streamwise direction for different acoustic metasurface configurations

dramatical changes of the velocity amplitude for different protrusion heights, and the second-mode wave is more suppressed for higher protrusions as the wave evolves downstream, resulting in a lower amplitude downstream the acoustic metasurfaces. The main reason is that the local deformation of acoustic metasurfaces enhances the wall shear effect and promotes the reflection and scattering of the second-mode wave on the wall, thus reducing the amplitude and energy of the disturbance. Multi-frequency second mode are also considered for C5 in appendix B, as the acoustic metasurfaces work in the linear stage of the mode evolution, the suppression effect

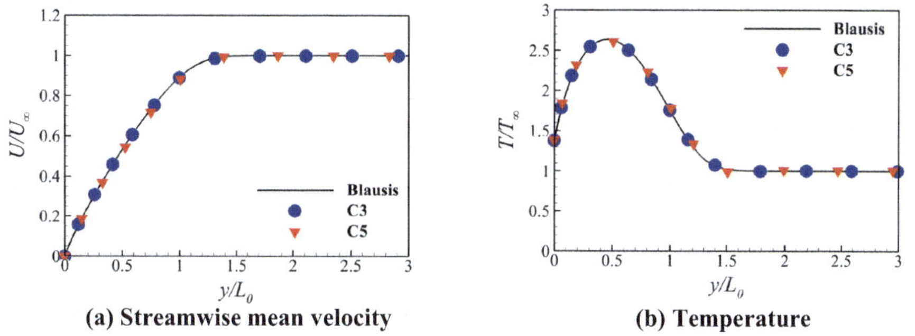

Fig. 4.75 Comparison of the streamwise mean velocity and temperature profiles at $x - x_s = 24$ mm

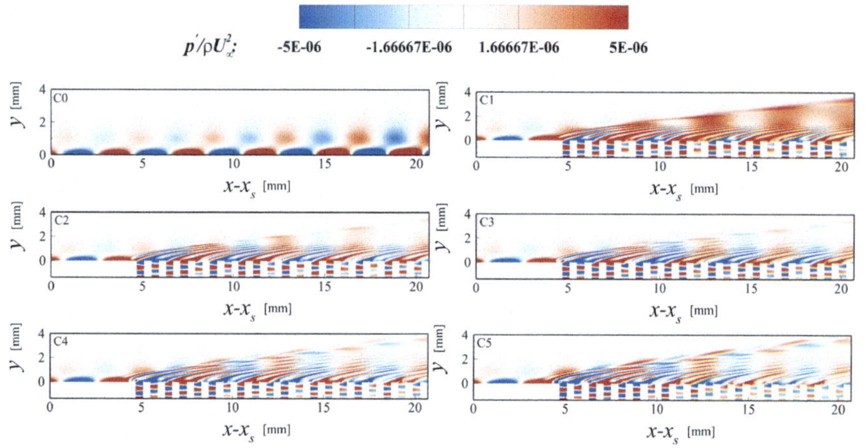

Fig. 4.76 Instantaneous fluctuating pressure fields for different acoustic metasurface configurations

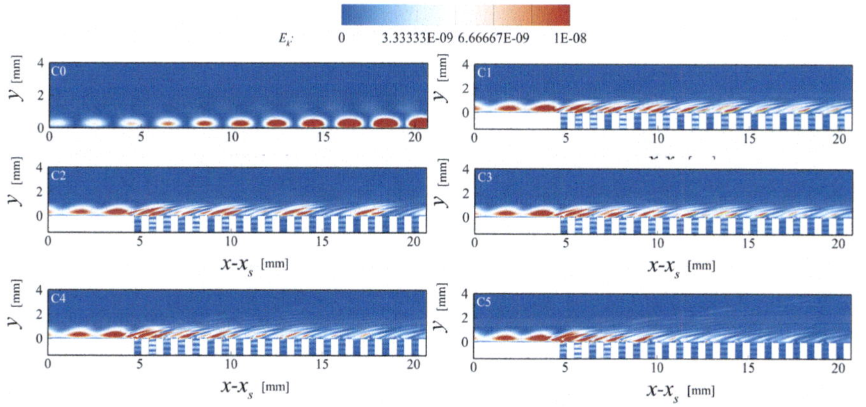

Fig. 4.77 Instantaneous fluctuating kinetic energy contours for different acoustic metasurface configurations

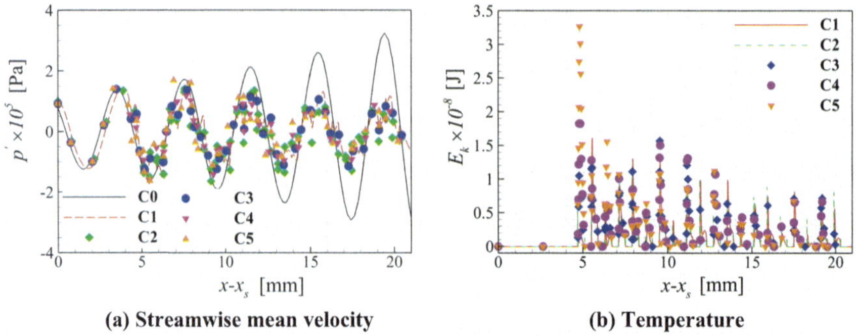

Fig. 4.78 Streamwise distributions of the fluctuating pressure and kinetic energy

Fig. 4.79 Fluctuating pressure spectrum at the sampling point P_0 ($x - x_s =$ 20.7 mm, $y = 0.1$ mm)

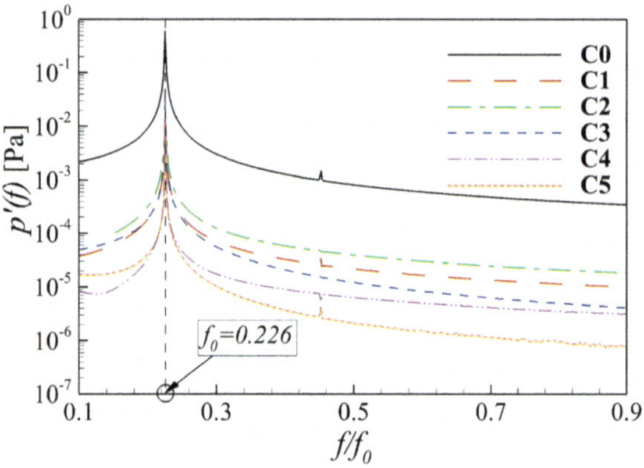

is the best for $f^* = 400$ kHz under the investigated configuration. Figure 4.81 quantitatively shows the amplification factors for different acoustic metasurface configurations. The amplification factor of C5 is decreased by about 44.7% compared to that of C1. Therefore, the small machining flaw of rectangular acoustic metasurfaces can actually have a favorable impact on the transition delay at least for the present studied protrusion scales. It is also noted that, for very large protrusion (typically around the local boundary layer thickness), it can act as local roughness and generate strong flow separation bubbles behind, which may eventually cause new instabilities in the flow leading to early transition. Fortunately, the machining flaw on the acoustic metasurfaces are in practice much smaller compared to the boundary layer thickness, thus we have deliberately limited our consideration of flaw protrusion at small scale only. Nevertheless, the present finding may give us a new choice to optimize the shape of rectangular acoustic metasurfaces for the boundary layer transition control.

In practice, the acoustic metasurfaces are typically installed in certain local regions of the wall rather than a full cover of the whole surface, such that it can reduce the maintenance cost. In order to further investigate the influence of the rectangular

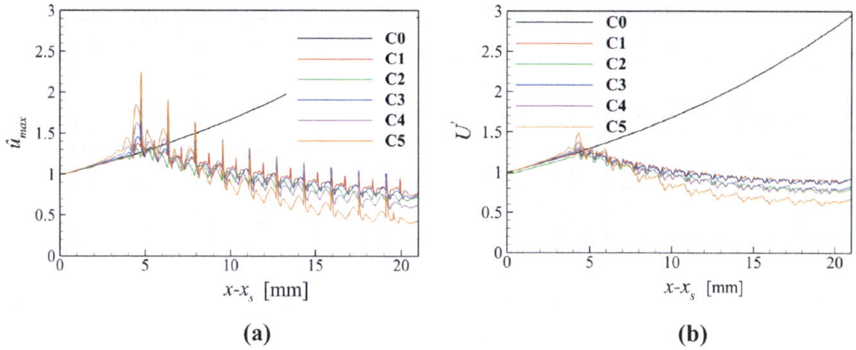

Fig. 4.80 Streamwise distributions of **a** the velocity disturbance amplitude and **b** the wall-normal integral of the velocity fluctuation amplitude

Fig. 4.81 Comparison of the amplification factors for different acoustic metasurface configurations

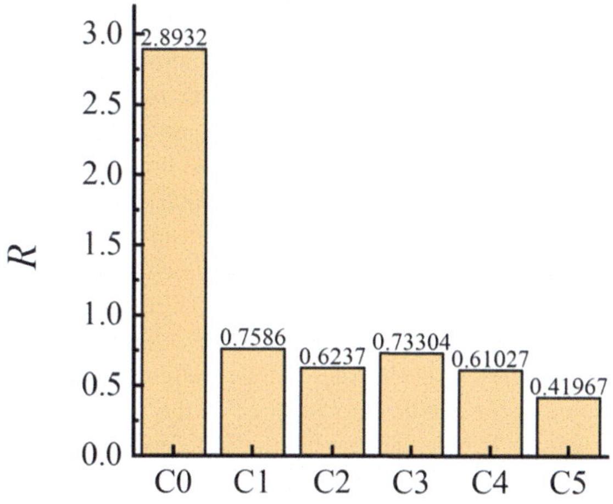

acoustic metasurface length on the second mode, a longer streamwise domain for C5 is adopted for the numerical simulation. Figure 4.82 compares the amplitude evolution of the second mode at 400 kHz between the two domains, with the short and long domains of 21 mm and 100 mm in length, respectively. In both cases, the acoustic metasurfaces are installed in the region $x - x_s \in [4.7, 20.7]$ mm. All the calculated amplitudes in the present study are normalized using the disturbance amplitude at the inlet. The results of LST and DNS perfectly agree with each other for the flat-plate boundary layer, which verifies the accuracy of the present DNS code. As discussed previously, the disturbance wave is dramatically suppressed on the acoustic metasurface coated surface. In the long domain, the disturbance wave starts to recover once it exits the rear of the acoustic metasurface region, but the amplitude remains lower than the flat-plate case (C0) for the whole downstream region. This

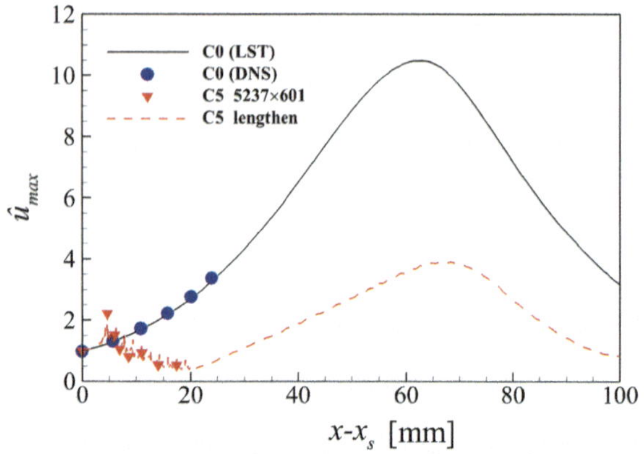

Fig. 4.82 Comparison of the velocity amplitude evolution of the second mode ($f^* = 400$ kHz) along the streamwise direction

demonstrates that the suppression of the second mode by acoustic metasurfaces only appears in the upstream region.

In this work, the influence of local deformation of transverse rectangular acoustic metasurfaces on the Mack second mode is studied in a Mach 6 hypersonic boundary layer using direct numerical simulations. It is found that the frequency of the most unstable second mode in the flat-plate boundary layer is 400 kHz under the given flow condition. The base flow calculation shows that the compression and expansion waves formed above each acoustic metasurface are enhanced under large deformation heights. The acoustic metasurfaces in the boundary layer cause shear effect which is more obvious for large deformations. The deformed acoustic metasurfaces distort the boundary layer in the wall-normal direction and leads to a wavy-like distribution of the boundary layer thickness in the slotted area. In the meantime, a small separation bubble is formed at the rear edge of the acoustic metasurface, and the size of the separation bubble increases with the deformation height of the acoustic metasurface.

The most unstable second mode is applied at the inlet of the computational domain to quantify the amplitude suppression effect due to the presence of acoustic metasurfaces. It is found that the suppression effect of the deformed micro grooves on the second mode is better than that of the regular grooves. In the investigated deformation range, the amplification factor of the second mode is negatively correlated with the deformation height of the micro grooves. Compared to the flat plate and the regular grooves, the amplitude suppression rate of C5 is about 85.6% and 44.7%, respectively. The present study indicates that small machining flaw can help suppress the second-mode waves, thus delay the hypersonic boundary layer transition.

4.6 Metasurfaces by Thermal Ablation of C/C Composites

4.6.1 Introduction of Thermal Ablation

Thermal protection composites mainly include silicon-based, carbon-based and carbonized composites. Liang et al. [48–51] have carried out a lot of research on the ablative behavior of silica-phenolic composites. Compared with silicon-based composites, carbon-based composites have higher chemical latent heat, so they are more and more widely used. Carbon fiber reinforced carbon matrix (C/C) braided composites with excellent thermo-dynamical and mechanical performances are widely used in advanced technologies, such as rocket engine nozzles, carbon brake discs, thermal protection systems of reentry vehicles, etc. [52–55]. As for the thermal protection materials, the thermochemical ablative behavior of C/C composites is an important issue in the application and the design of such materials. For instance, the performances of the C/C composites would dramatically degrade even if the temperature was as low as 500 °C in an oxidizing environment [56]. And thermal protection materials for the vehicles used in Earth's orbit or Mars orbit are exposed to environment with oxygen or carbon dioxide under temperature from 600 to 2000 °C and even higher [57]. A high interfacial mass transfer of thermochemical ablation induces a surface recession, and the surface recession is not uniform due to the ablation difference of fiber and matrix [58, 59]. As shown in Fig. 4.83, the metasurface with nonregular micropores are gradually formed during the ablation process.

The accurate predication of the ablation behavior is essential to characterize the required metasurface. This section is organized as follows. A diffusion/reaction coupling model considering the ablation difference of fiber and matrix is established. The thermochemical ablation is numerically solved on the basis of the volume-of-fluid method and the finite volume method using MATLAB codes. Then, the numerical results are validated solving the surface ablation compared to the analytical results. Finally, the behavior of 3D braid composites and 2.5D woven composites are characterized, and the effects of mesoscopic features and ablation calculation parameters on their ablation behavior are investigated.

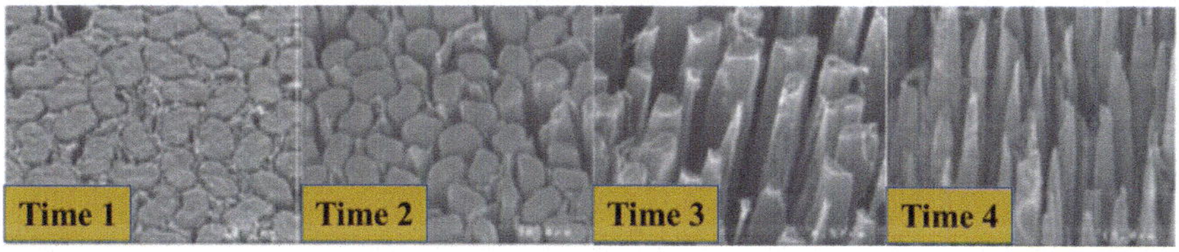

Fig. 4.83 Evolution of surface ablative morphology [52]

4.6.2 Thermochemical Ablation Model of C/C Composites

From low temperature to high temperature, the thermochemical ablation of C/C composites are oxidation, carbon nitrogen and sublimation, and the main ablative reaction is the oxidation. As for C/C composites, if the oxidizing gas concentration next to the solid–gas interface and the heterogeneous reaction rate at high temperature are known, the mass loss of the solid carbon phase would be computed. Then, the geometry change of the C/C composites can be known.

4.6.2.1 Mathematical Model

The thermochemical ablation process for C/C composites mainly includes the following reactions:

$$C_{(s)}+O_2 \rightarrow CO_2 \tag{4.16}$$

$$C_{(s)}+\frac{1}{2}O_2 \rightarrow CO \tag{4.17}$$

$$C_{(s)} + CO_2 \rightleftarrows 2CO \tag{4.18}$$

where the subscript s identifies the solid phase. Equations (4.16)–(4.18) express the heterogeneous reactions during the thermochemical ablation. It has been proven that the morphologies of the mesoscale structure during the ablation was controlled by the ratio of reaction/diffusion and the reactivity contrast between constituents in the previous studies [60, 61]. Thus, the oxidizing gas transfer can be simplified as a pure diffusion. In such cases, the geometry change of the carbon fiber reinforced composites and the diffusion of the gas during the oxidation process can be described by [60, 61]:

$$S(x, y, z, t) = 0 \tag{4.19}$$

$$\frac{\partial S}{\partial t} + \mathbf{v}_e \cdot \nabla S = 0 \tag{4.20}$$

$$\mathbf{v}_e = \Omega_e J_e \mathbf{n} \tag{4.21}$$

$$\mathbf{n} = \nabla S / \|\nabla S\| \tag{4.22}$$

$$J_e = k_e C \tag{4.23}$$

$$\frac{\partial C}{\partial t} + \nabla \cdot (-D\nabla C) = 0 \tag{4.24}$$

where S is the surface function of constituents within the composites, which has the first-order partial derivative. S is a non-zero value at points not lying on the interface, while S is 0 at the interface. The subscript e represents different phase carbon, i.e. the carbon fiber phase and the carbon matrix phase. $\mathbf{v}_e$ is the surface recession velocity of phase e, C is the molar concentration of the oxidizing gas, and D is the diffusion coefficient. Ω_e is the solid molar volume of phase, J_e is the molar rate of the oxidation at the interface, and $\mathbf{n}$ is the local normal vector at the interface.

Equations (4.19)–(4.24) describe the coupling of the gas diffusion and the surface recession during the ablation, while it is difficult to obtain the involved parameters accurately for solving the equations. However, it has been proven that the oxidation morphologies of C/C composites dominated by two dimensionless parameters Da and A [60]. The Damköhler numbers $Da = k_{yarn}l/D$ describing reaction/diffusion ratio and reactivity contrast between constituents within the braided composites $A = (k_{matrix}\Omega_{matrix})/(k_{yarn}\Omega_{yarn})$ are also used in the present study to simulate the different physico-chemical regime of the C/C composite ablation, where l is the characteristic length. Then, the reaction/diffusion ratio and reactivity contrast can be changed to study the oxidation behavior of C/C composites under different thermochemical ablative conditions.

4.6.2.2 Numerical Computation

The problem expressed by Eqs. (4.19)–(4.24) is similar to the gas liquid two-phase flow in computational fluid dynamics (CFD). Thus, the interface capturing method for solving the two-phase flow can be used. The volume-of-fluid (VOF) method is a common method for interface capturing. Similarly, different values for variable $\varepsilon = V_{solid}/V$ are corresponding to the gas ($\varepsilon = 0$), the solid ($\varepsilon = 1.0$) and the gas–solid interface ($\varepsilon = 0.5$). Then, Eqs. (4.19)–(4.24) can be rewritten as [62, 63]:

$$\frac{\partial C}{\partial t} + \nabla \cdot (J) = 0, \ \textit{in the fluid} \tag{4.25}$$

$$\frac{\partial \varepsilon_s}{\partial t} + v_s \nabla \cdot (-J) = 0, \ \textit{in the solid} \tag{4.26}$$

$$\begin{aligned} J &= -(1-\varepsilon_s)D\nabla C, \ \textit{in the fluid and silod} \\ J &= -kC|\nabla \varepsilon_s|^{-1}\nabla \varepsilon_s, \ \textit{at the interface} \end{aligned} \tag{4.27}$$

To solve the coupled partial differential Eqs. (4.25)–(4.27), the finite volume method is used. In a control volume, the Eqs. (4.25) and (4.26) can be expressed as [62, 63]:

$$\int_t^{t+\Delta t}\int_V (1-\varepsilon_s)\frac{\partial C}{\partial t}dV\,dt = \int_t^{t+\Delta t}\int_{V_f} \nabla\cdot(D\nabla C)dV\,dt - \int_t^{t+\Delta t}\int_{\sum_{f/s}} kCd\sum dt \tag{4.28}$$

$$\int_t^{t+\Delta t}\int_{V_s} \frac{\partial \varepsilon_s}{\partial t}dV\,dt = \int_t^{t+\Delta t} -v_s\int_{\sum_{f-s}} kCd\sum dt \tag{4.29}$$

where $\sum_{f/f}$ is the contact area of the interface between gas and gas in the control volume, $\sum_{f/s}$ is the contact area of the interface between solid and gas in the control volume, and V_f is the volume occupied by the gas in the control volume as shown in Fig. 4.84.

The gas concentration C and geometry change variable e are coupled interactively during ablation. A loosely coupled algorithm (LCA) is developed to decouple C and ε as shown in Fig. 4.85. In the time increment $t + \Delta t$, the geometry configurations of the constituents within the composites are reconstructed by using ε and C at time increment t. Then, the interfaces $\sum_{f/f}$ and $\sum_{f/s}$ are calculated from the reconstructed geometry configurations. The value of C can be updated corresponding to the time increment $t + \Delta t$.

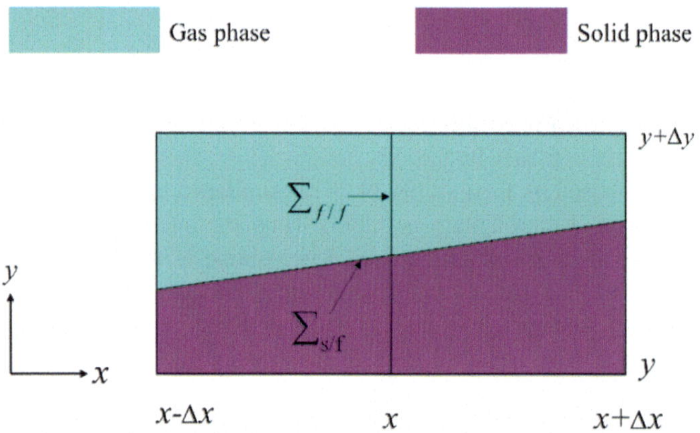

Fig. 4.84 The interface in control volumes (2D sketch)

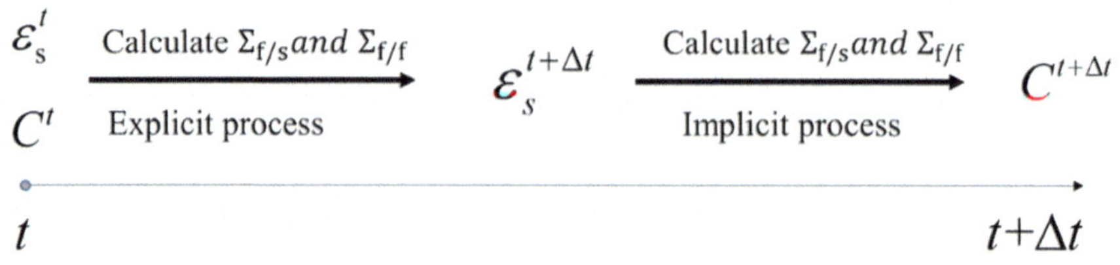

Fig. 4.85 The loosely coupled algorithm (LCA)

The piecewise linear interface calculation (PLIC) algorithm [64] is used for the geometry reconstruction during the simulation process. The central difference scheme is adopted for the spatial discretization, and the first order fully implicit scheme is used for the temporal discretization because the time accuracy of the loosely coupled algorithm does not exceed the first order. The numerical calculation solving Eqs. (4.28) and (4.29) is implemented by a mixed program using MATLAB and C languages.

4.6.2.3 Validation of Numerical Calculation

A one-dimensional diffusion/recession problem as shown in Fig. 4.86 is used to validate the loosely coupled algorithm and the discretization scheme mentioned above. The one-dimensional problem is not related to the complex geometry reconstruction of the interfaces, which also can be calculated using other solver (such as commercial software COMSOL). As shown in Fig. 4.86, A-B is gas phase zone, and B-C is the solid phase zone. The interface B will move to C due to the recession caused by the oxidation of the solid phase. At the initial stage, $t = 0$ s, $L_{\mathrm{AC}} = 1.0 \times 10^{-3}$ m, $L_{\mathrm{AB}} = 0.53 \times 10^{-3}$ m. Other parameters are listed as: $\Omega_{\mathrm{s}} = 6.67 \times 10^{-2}$ m^3/mol, $k_c = 2.0 \times 10^{-2}$ m/s, and $D = 2.0 \times 10^{-4}$ m^2/s. The boundary condition of the gas phase is $C_{\mathrm{A}} = 0.23$ mol/m^3. The numerical results obtained by LCA can be validated by that obtained COMSOL.

The total time of interface B moving from initial location to C is about 1.7 s. We choose the results of the location of gas–solid interface and the gas concentration distribution at $t = 0.5$ s and $t = 1.6$ s for comparison as illustrated in Fig. 4.87. The gas distribution and the location of the interface calculated from LCA are in good agreement with that from COMSOL.

Furthermore, a 3D model of fiber embedded in matrix is used to validate the 3D program. The geometry size of the computation region, the fiber diameter and the length of solid region along fiber direction are shown in Fig. 4.88. The Dirichlet boundary is applied on the upper surface of the model. The different reaction/diffusion ratios $Da = k_{yarn}R_f/D$ ($Da = 0.01$, 0.1, 1, 5, 10 and 100) are chosen

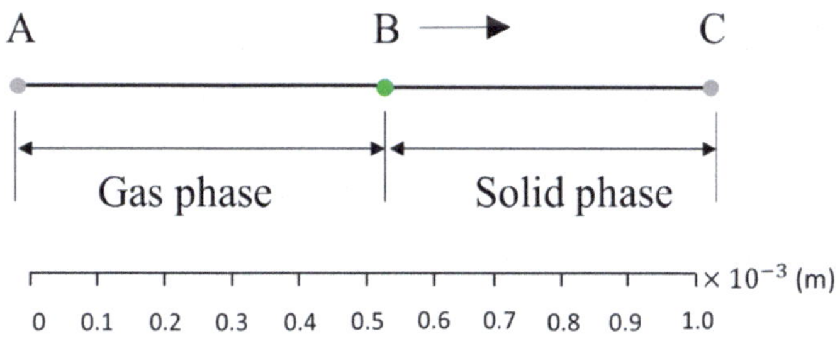

Fig. 4.86 The one-dimensional diffusion/recession problem

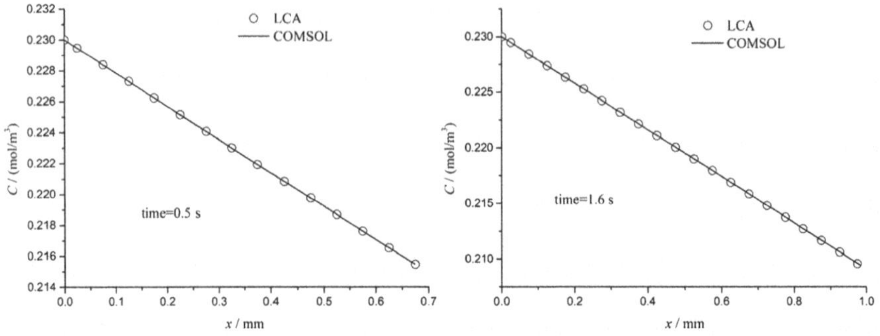

Fig. 4.87 Comparison of gas distribution obtained from LCA and COMSOL

to calculate the 3D morphologies of the carbon fiber under the thermochemical ablation. In the model, $R_f = 5$ μm, $C_0 = 2.713$ mol/m^3 and $D = 2 \times 10^{-4}$ m^2/s. The reactivity contrast between constituents A is equal to 5. As shown in Fig. 4.89, it can be found that the numerical morphologies from our program are in good agreement with the analytical results when Da equals to 1, 5, 10 and 100. The obvious discrepancies for Da equaling to 0.01 and 0.1 can be attributed to the assumption of vertical gas concentration gradient in the analytical model. When the exposure height of fiber is large, the assumption of vertical gas concentration will bring the bigger error for the morphologies from analytical results. On the other hand, the fiber in the analytical model has a semi-infinite length, while the simulation model has a finite length. Such difference between the numerical model and the analytical model will also lead to different results, especially when the height of exposed yarn in stable ablative morphology is relatively high (i.e. $Da = 0.01, 0.1$).

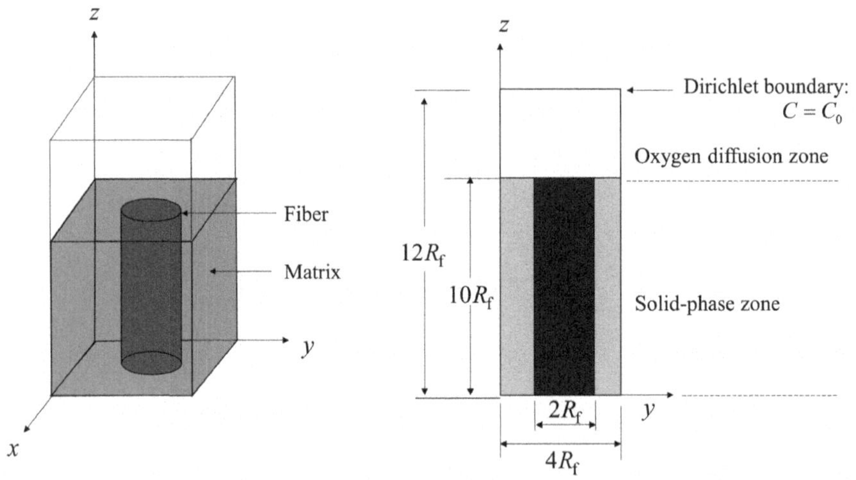

Fig. 4.88 Schematic diagram of simulation model

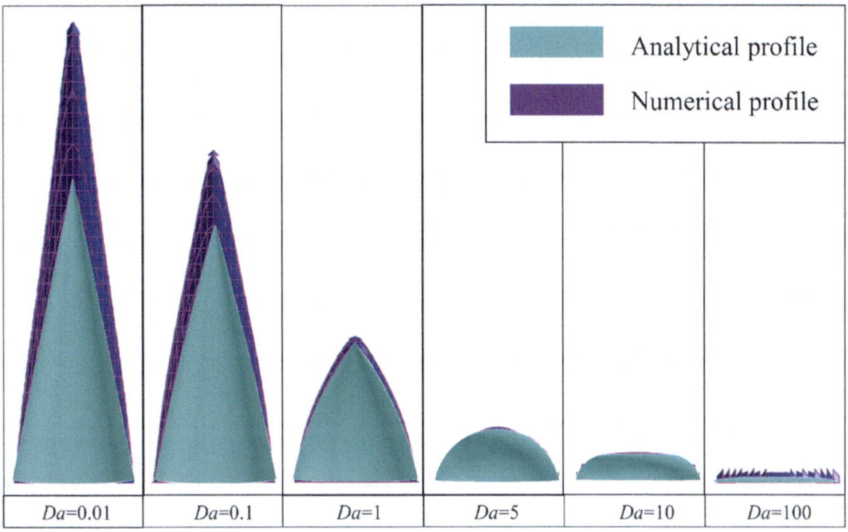

Fig. 4.89 Comparison of the morphologies from analytical results and simulation results

The ablative behavior of fiber under 898 K has been studied experimentally [65]. The fiber ablative morphologies are shown in Fig. 4.90, which are in good agreement with our analytical results as well. It should be noted that the geometry sizes of ablated fiber are difficult to be measured in the experiment.

4.6.3 Thermochemical Ablative Behavior of 3D C/C Braided Composites

4.6.3.1 Geometry Model and Computation Parameters of 3D C/C Braided Composites

The 3D C/C braided composites in the present study are composed of distributed four-directional braid yarns and carbon matrix. The distributed interior braid yarns within a representative volume cell (RVC) of the braided composites are shown in Fig. 4.91. A braid yarn is composed of 12 K (1 K represents one thousand fibers) T700 carbon fibers. In the numerical model, a braid yarn can be treated as homogenous solid on mesoscopic scale. The cross section of the yarn is an approximate ellipse with long axis a and short axis b. The height and width of the RVC are represented by h and W. Interior braid angle of the 3D braided composites is γ. The relationship of geometry parameters for the RVC are as follows: $b = 0.38$ mm, $a = \sqrt{3}b\cos\gamma$, $h = 8b/\tan\gamma$, and $W = 4\sqrt{2}b$ [60]. The braid yarn volume fraction V_{yarn} is related with the interior braid angle as listed in Table 4.7. The height of RVC of the 3D braided composites is related with the interior braid angle.

Fig. 4.90 Fiber ablative morphologies at 898 K [52]

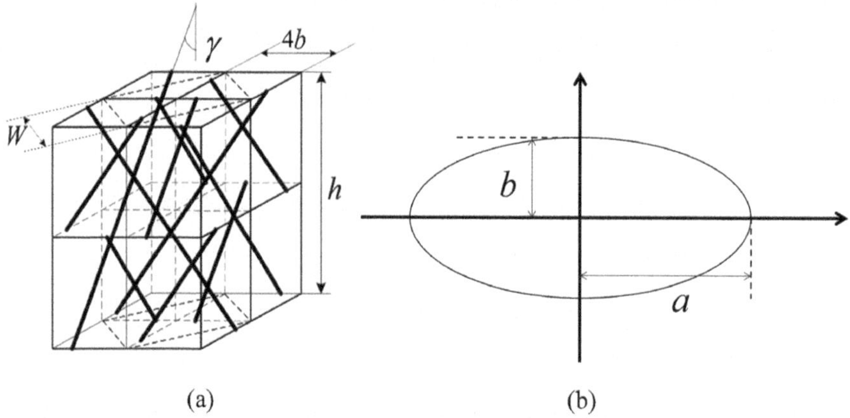

Fig. 4.91 Geometry model of 3D four-directional C/C braided composites, **a** interior cell, **b** cross-section of braid yarn

Table 4.7 Braided angle and braid yarn volume fraction

Parameter	Value				
Braided angle γ	20°	30°	40°	50°	60°
Domain yarn volume fraction V_{yarn}	67.68%	67.39%	67.89%	67.63%	68.27%

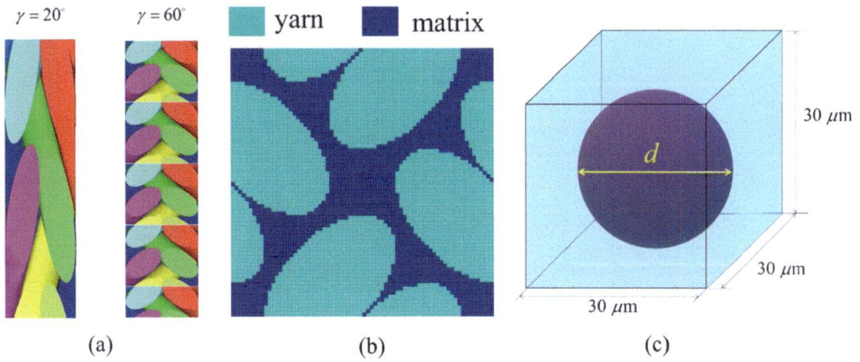

Fig. 4.92 Numerical geometry models of 3D C/C braided composites

The height of RVC increases with the decrease of interior braid angle. Thus, the height of computation model for the braided composites with different interior braid angles adopts the height of RVC with interior braid angle $\gamma = 20°$ as shown in Fig. 4.92. The size of grid element is $30 \times 30 \times 30\ \mu m^3$.

The mesoscale pores diameter (d) in C/C composites conforms can be divided into three regions, i.e. $d < 4\ \mu m$, $4\ \mu m < d < 20\ \mu m$, and $d > 20\ \mu m$. And if the diameter is smaller, the number of the pores are larger. In order to obtain the real distribution law of the pores, the pores diameter (d) distribution in the composite is measured by the mercury intrusion method as shown in Fig. 4.93a, in which B01, B02, B11 and B12 are the numbers of different simples. In the numerical model, the pores are randomly scattered into the carbon matrix to obtain the similar distribution laws excerpted in the experimental results. A spherical pore with diameter d is located in a cubic element as shown in Fig. 4.92c. The average diameter $\overline{d} = \left(\sum_{i=N} d_i\right)/N$ can be used to calculate different porosity $\varepsilon = \left(\sum_{i=1}^{N}\left(\pi d^3/6\right)\right)/\left(\sum_{i=1}^{N} 30^3\right)$, where N is the total number of the carbon matrix element. The effective density of the carbon matrix element and the molar volume are calculated by $\rho = \left(1 - \left(\pi d^3/6\right)/30^3\right)\rho_0$ and $v = v_0/\left(1 - \left(\pi d^3/6\right)/30^3\right)$, respectively, where ρ_0 and v_0 are the density and molar volume of matrix without pores. The distribution of pores in the numerical models are shown in Fig. 4.93b.

The oxidation behavior of 3D C/C braided composites can be simulated qualitatively by changing the relation of diffusion rate and reaction rate. The larger reaction rate is corresponding to the higher temperature situation for C/C composites. Da from 0.7 to 7.0 is used to simulate the reaction under temperature from 1850 to 3000 K with a characteristic length $l = 5 \times 10^{-3}$ m. In this study, the values of Da are 0.1, 1.0, 10.0, and the reactivity ratio between the braid yarns and the matrix A is equals to 1/2.

The material parameters for 3D braided composites used in the simulation are listed in Table 4.7. A dimensionless time $\tau = l/\left(C_0 k_{yarn}\Omega_{yarn}\right)$ is introduced. Because the initial height of the gas phase domain does not influence the oxidation

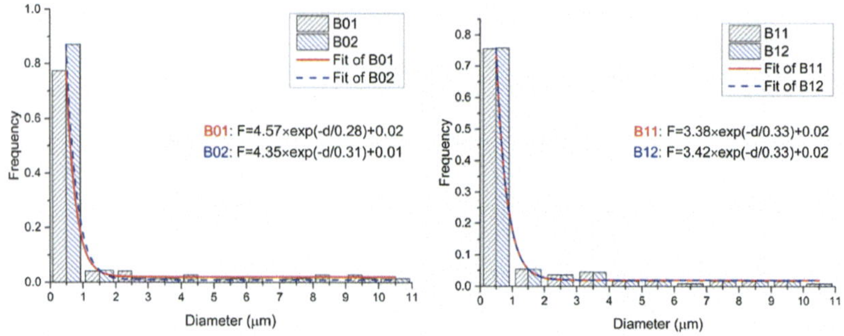

(a) Pores diameter distribution measured by the mercury intrusion method.

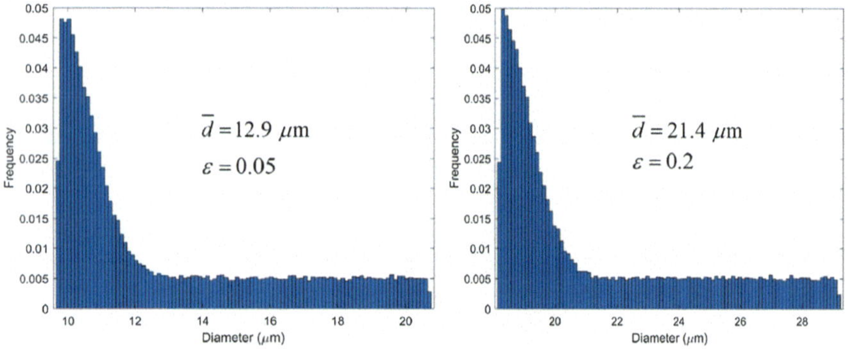

(b) Pores diameter distribution in numerical models.

Fig. 4.93 Pore diameter distribution laws

morphologies, a uniform height of 1 mm is used in the simulation. The oxidizing gas concentration C_0 at the boundary is 2.713 mol/m^3.

4.6.3.2 Effects of Braid Angle on Thermochemical Ablative Behavior

The oxidation morphologies with $Da = 0.1$, 1 and 10 at different time are shown in Figs. 4.94, 4.95, 4.96 and 4.97. During the oxidation process, the oxidation rate of the matrix is higher than that of the braid yarns, which makes the recession of the matrix larger than that of the braid yarns. The braid yarns gradually emerge from the matrix. When $Da = 0.1$ the gas diffusion rate is extremely faster than the rate of the reaction, the oxidizing gas is enough for both the braid yarns and the matrix. Under such conditions, the thermochemical ablative behaviors between the matrix and the braid yarn distinguish apparently because they have different reactivity ratios. The braid yarns next to the gas diffusion boundary are exposed to the oxidizing gas in a longer time than that close to the bottom, and appear the bamboo shoot shape as seen in Fig. 4.95. With the increase of Da, both the height of "bamboo shoot" and the height difference between the braid yarns and the matrix decrease as shown

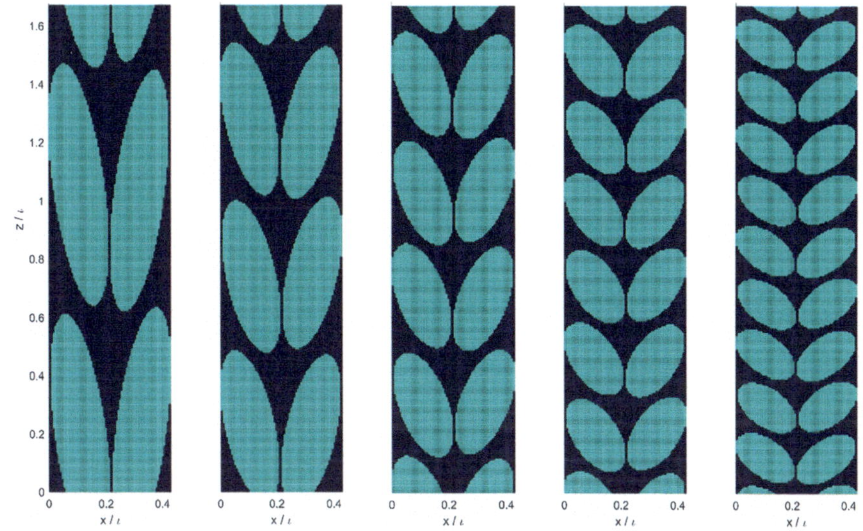

Fig. 4.94 Initial morphologies of 3D C/C braided composites with different braid angle at $t/\tau = 0$

in Fig. 4.96. When $Da = 10$, the gas diffusion rate is much lower than that of the reaction. Thus, the gas consumption is much lower than its supplement at the gas–solid interface. Under such conditions, the recession rates of both the braid yarns and the matrix will be constrained by the gas diffusion rates. Both the braid yarns and the matrix are lack of the oxidizing gas. The thermochemical ablative process enters the diffusion-controlled state. Both braid yarns and the matrix will recess synchronously as shown in Fig. 4.97. It was illustrated that the smaller value of *Da* would lead the larger morphology differences, and the morphology differences would disappear under a sufficiently large value of *Da*.

4.6.3.3 Relationship of Ablation Recession Rate Between Macro- and Mesoscale Results

During the ablation process, the braid yarns within the braided composites lose the protection by the carbon matrix, and the bare braid yarns will be destroyed by the external force. In order to evaluate the influence of the braid angle on the exposed height of the braid yarns, the recession surface locations of the matrix and the height difference between the braid yarns and the matrix are investigated. The recession surface locations are defined as the locations where the residual mass is 10% of the original mass of the elements in the layer of the direction z, i.e. $h_{matrix}(10\%) = 0.1 \times m_{matrix}/m_{matrix}^0$ and $h_{yarn}(10\%) = 0.1 \times m_{yarn}/m_{yarn}^0$, where m_{matrix} and m_{yarn} are the mass of matrix and braid yarns at the same layer of numerical elements, and the superscript 0 means the time at $t = 0$. Then, the height difference between the braid yarns and the matrix is defined as $\Delta h = h_{yarn}(10\%) - h_{matrix}(10\%)$.

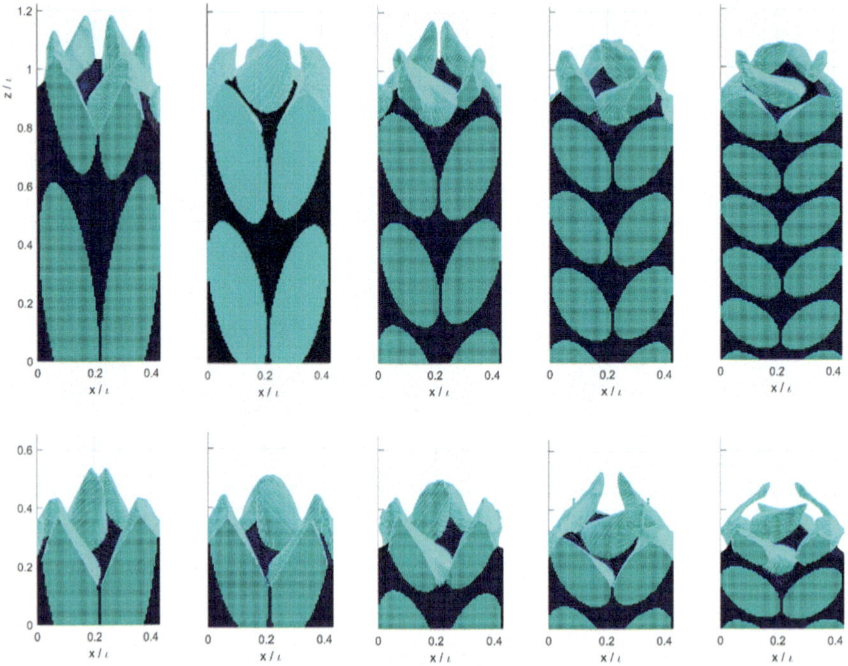

Fig. 4.95 Oxidation morphologies ($Da = 0.1$), at $t/\tau = 0.29$ and 0.55

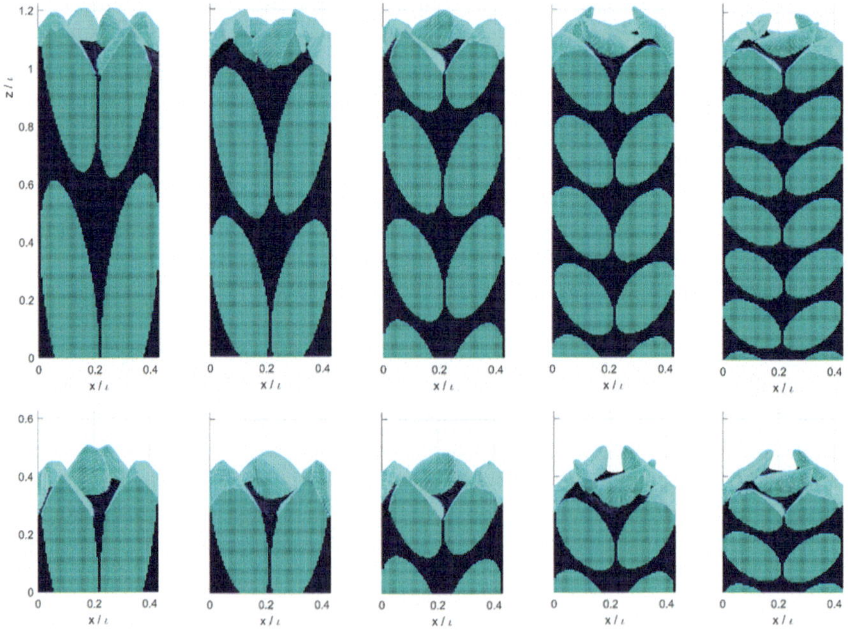

Fig. 4.96 Oxidation morphologies ($Da = 1$), at $t/\tau = 0.51$ and 1.52

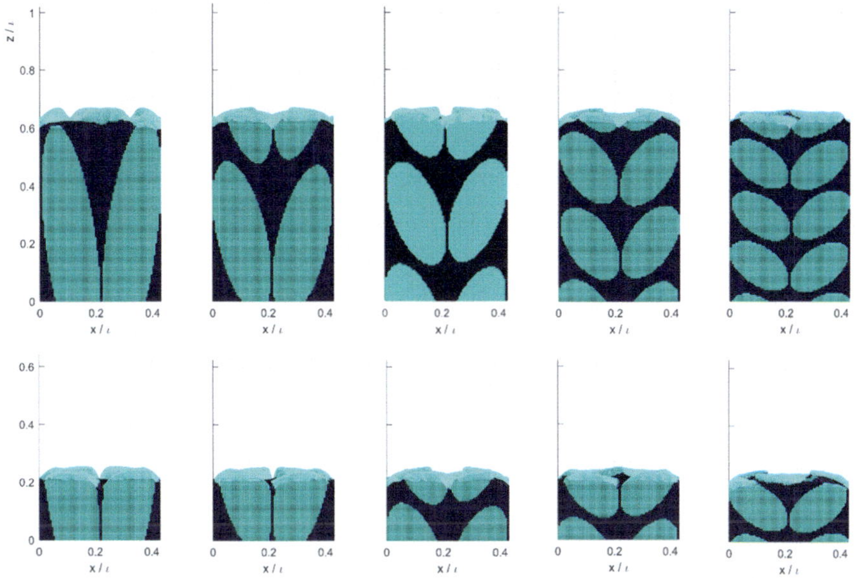

Fig. 4.97 Oxidation morphologies ($Da = 10$), at $t/\tau = 7.29$ and 12.99

The recession surface locations of matrix are shown in Fig. 4.98. When $Da = 0.1$ and $Da = 1$, the locations of the matrix surface with different braid angles show great discrepancies at the same time, and the discrepancies exhibit periodic. The periodic difference is owing to the periodic characteristics of the braided structures in Z direction. When $Da = 10$, the differences disappear because the ablation recession rates of the matrix and the braid yarns are same under the diffusion-controlled state. At this point, the braided structure will no longer affect the oxidation behavior. The height difference between the braid yarns and the matrix are shown in Fig. 4.99. The height difference also exhibits periodic, and the average height difference decreases with the increase of the braid angle. It should be noted that the braid yarn volume fraction changes from 67.39 to 68.27% for the different braid angles as listed in Table 4.7. The different braid yarn volume fraction may affect the height difference as well, however, as shown in Fig. 4.99, the average height difference decreases monotonously with the increase of the braid angle. The large height difference under small value of Da also illustrates quantitatively the ablation discrepancies of the predicted results in the macro- and meso-scales. We define the ablation recession distance as the distance of the matrix plane moving from the initial position, then the ablation mass in the macro- and mesoscale are calculated by:

$$m_{macro} = \overline{\rho} h_1 = \left[(L_A - L_B) - L_B \rho_y \right] h_1 \tag{4.30}$$

$$m_{meso} = L_A h_2 \rho_m - \frac{1}{4} L_B^2 \rho_y \tan\theta \tag{4.31}$$

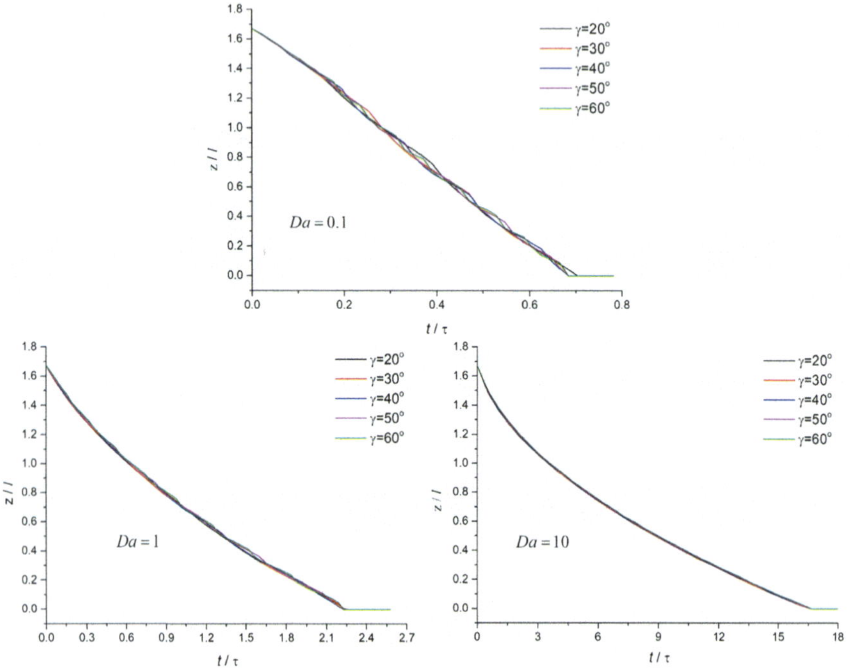

Fig. 4.98 Recession surface location of matrix

where ρ_m and ρ_y are the density of matrix and yarn, L_A and L_B are width of the composites and the yarn, h_1 and h_2 are the recession distances of the macroscopic and mesoscopic model (Fig. 4.100). Considering $\rho_m = \rho_y = \rho$ and $h_1 = h_2 = h$, then

$$m_{meso} = \rho\left[L_A h - \frac{1}{4}L_B^2 \tan\theta\right] < \rho L_A h = m_{macro} \tag{4.32}$$

Equation (4.32) illustrates that the mass obtained from macroscopic model is larger than that from mesoscopic model on the premise of the same ablation recession distance. It is also illustrated that if the calculated ablation mass is same, the recession distance from the macroscopic model is underestimated than that from the mesoscopic model. From Eq. (4.32), it can be found that m_{meso} decreases monotonously with the increase of h, illustrating that the sharper the braid yarns (i.e. smaller *Da*), the more underestimated the recession distances are. Hence, when using the macroscopic model to predict the recession distance of 3D C/C braided composite, the ablation differences caused by the asynchronous recession of the braid yarns and the matrix should be corrected, especially under a relative low temperature.

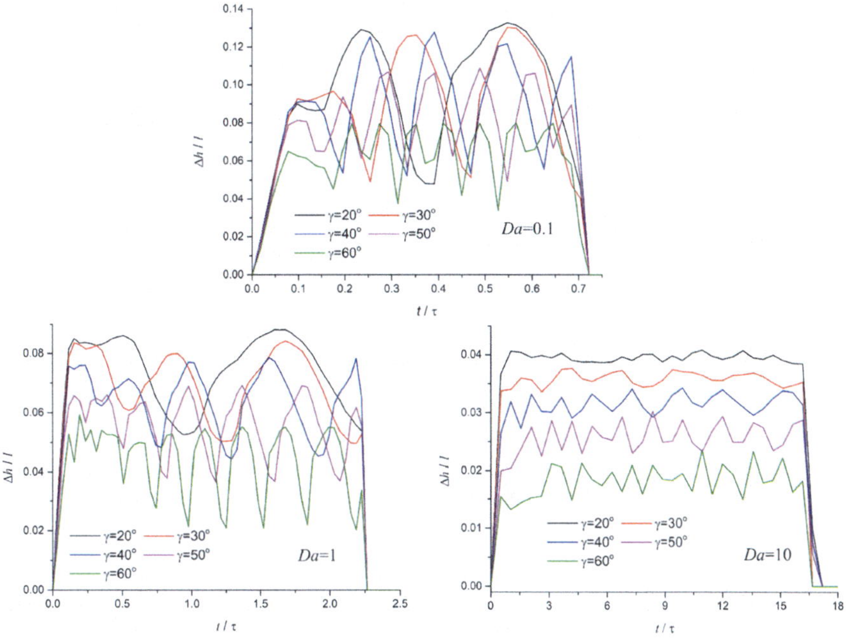

Fig. 4.99 Height difference between braid yarns and matrix during oxidation

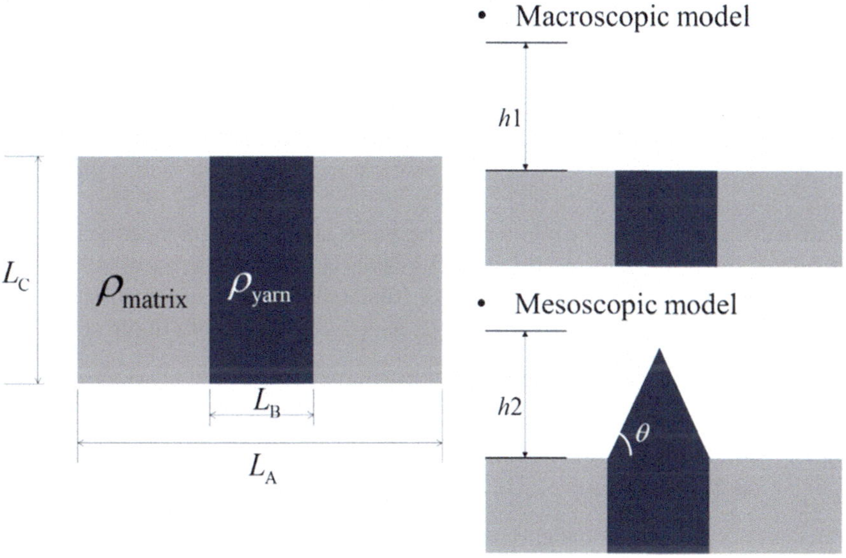

Fig. 4.100 Sketch of macroscopic and mesoscopic ablation model

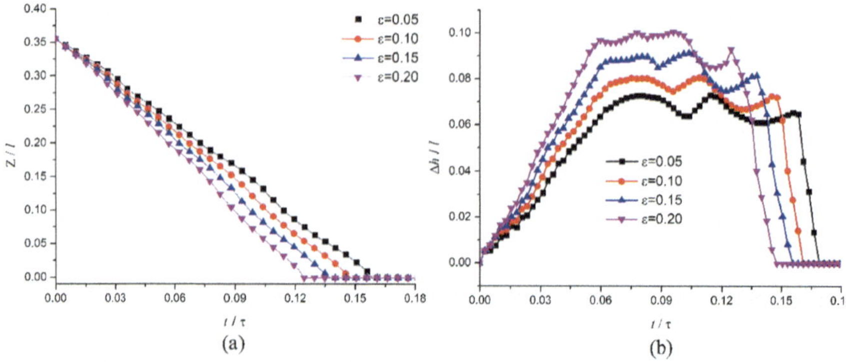

Fig. 4.101 Influence of *matrix porosity* on oxidation behavior ($Da = 0.1$), **a** recession surface location of matrix, and **b** height difference between braid yarns and matrix

4.6.3.4 Effect of Matrix Porosity on Thermochemical Ablative Behavior

The porosity of C/C composites is relevant with the manufacturing process, which can be more than 20%. To investigate the role of porosity, four kinds of porosity 5, 10, 15 and 20% are considered to study the ablation behavior of the 3D C/C braided composite with braid angle 60°. As shown in Figs. 4.101a and 4.102a, the higher porosity in the matrix, the faster the recession rate of matrix is. The height difference between the braid yarns and the matrix will increase with the increase of the porosity (see Figs. 4.101b and 4.102b). Comparing the results in Figs. 4.101 and 4.102, it can be found that the height difference caused by the porosity decreases with the increase of *Da*. The oxidation depth decreases with the increase of *Da*. With the increase of *Da*, the difference in oxidation behavior for different porosities will not appear. It is illustrated that the pores in the matrix will induce the severe oxidization for the braid yarns and matrix (see Fig. 4.103). Thus, as for the design of ablative C/C composites, the porosity of the matrix should be minimized as much as possible in the manufacturing for obtaining better antioxidant performances. Some phase-change materials, which can change from solid phase to liquid phase at high temperatures, can be added into the matrix to reduce the porosity of the surface in contact with oxygen at high temperatures.

4.6.3.5 Effect of Reactivity Contrast Between Matrix and Braid Yarn on Thermochemical Ablative Behavior

The reactivity contrast between matrix and braid yarn will change with different manufacturing process. To study the effect of reactivity contrast on the oxidation ablation, A changes from 2 to 5 by increasing k_m, and Da equals to 0.1. As shown in Fig. 4.104 and 4.105, it can be found that a higher reaction rate of the matrix makes the matrix recede faster, which also makes more braid yarns exposed to the

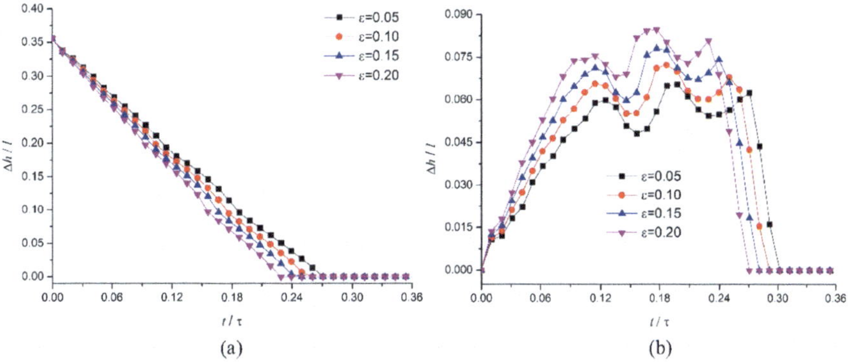

Fig. 4.102 Influence of *matrix porosity* on oxidation behavior ($Da = 1$), **a** recession surface location of matrix, and **b** height difference between braid yarns and matrix

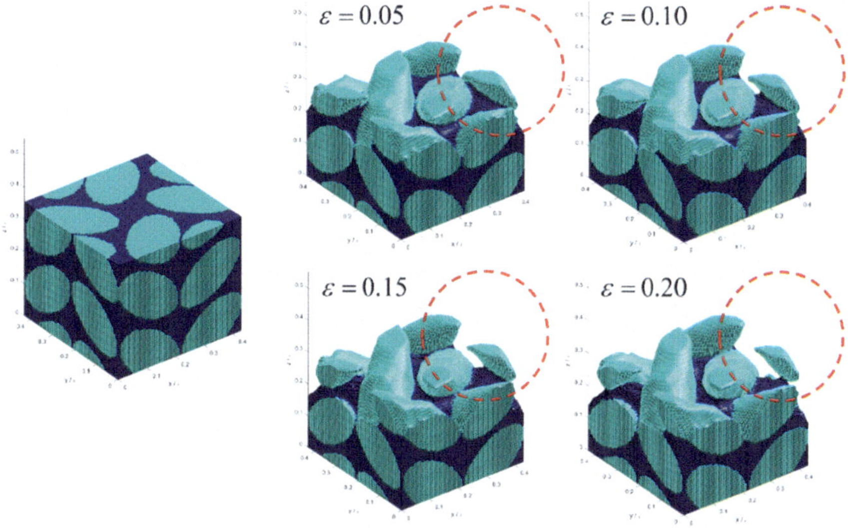

Fig. 4.103 Influence of matrix porosity on oxidation morphologies ($Da = 0.1$)

oxidizing gas, leading a larger height difference between the braid yarns and the matrix. Thus, the value of A should be as closer to 1 as possible to obtain a better ablation performance. The effective way is to enhance the oxidation resistance of the matrix, for example using the SiC as the matrix material, or improving the heat treatment temperature of C/C composites in the manufacturing process. It is useless to improve the antioxidant properties of the braid yarns without improving that of the matrix.

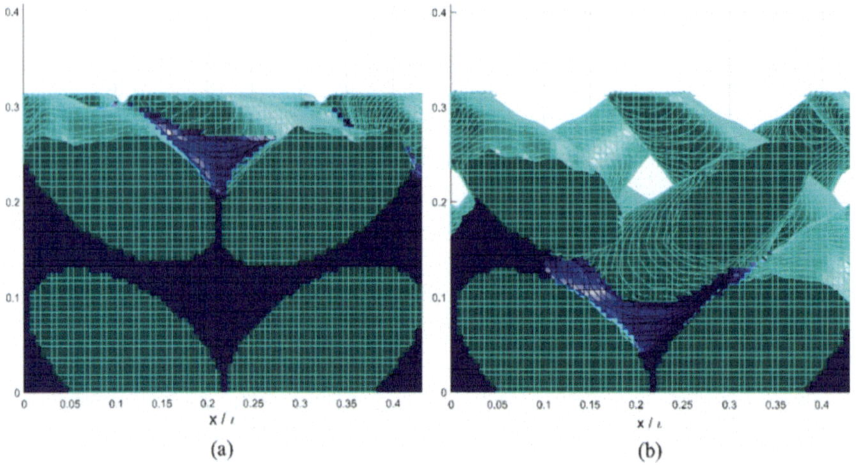

Fig. 4.104 Influence of A on oxidation morphologies ($Da = 0.1$): **a** oxidation morphology ($A = 2$), **b** oxidation morphology ($A = 5$)

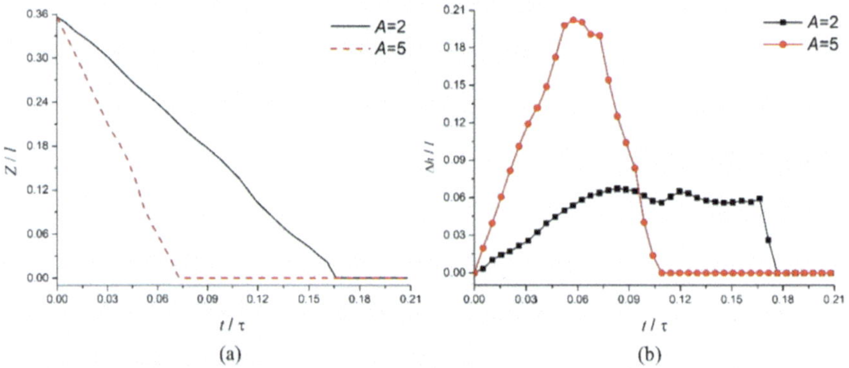

Fig. 4.105 Influence of A on oxidation behavior ($Da = 0.1$): **a** recession surface location of matrix, **b** height difference between braid yarns and matrix

4.6.4 Thermochemical Ablative Behavior of 2.5D C/C Woven Composites

4.6.4.1 Effect of Weaving Structures on Thermochemical Ablative Behavior

The ablation experiments showed that the recession rates and mass loss of different 2.5D woven structures were significantly different. It is known that 2.5D woven structures can be divided into three main types: orthogonal, angle interlock and layer-to-layer interlock structures. Their differences in structures are the number of weft yarn layers crossed by warp yarns and weaving directions. The RVCs (Represent volume cells) of three main types are shown in Fig. 4.106, and the x, y, z directions respectively represent warp/stuffer, weft and thickness directions of composites [66]. There are eight alternate warp and stuffer yarns in the y direction, and the number of the warp and stuffer yarn layers are respectively one and three. The yarns can be treated as the homogenous material on the mesoscale. Moreover, the sizes of the RVCs are all $4 \times 3.2 \times 0.99$ mm^3, and their weaving parameters are listed in Table 4.8.

Different from the macroscopic ablation in which the uniform recession rate is utilized to characterize the thermal ablation performance of C/C composites, the mesoscopic ablation needs to consider the reactivity difference between the fiber and matrix. The reaction rate of the matrix is faster than that of the fiber, the height of the matrix is lower than that of the fiber, so the height difference between the fiber and matrix is formed. Thus, the height of the matrix and the height difference between the fiber and matrix can respectively reflect the recession rate and roughness for the

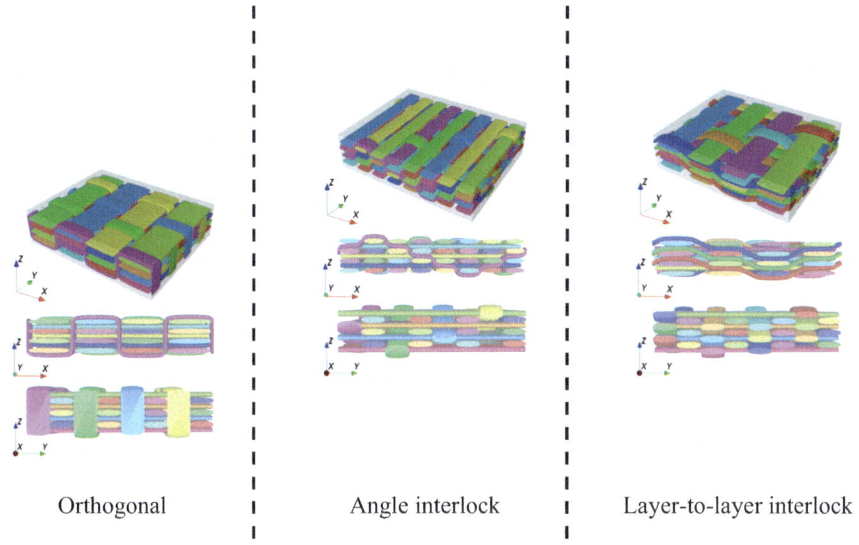

Fig. 4.106 Schematics of three main types of 2.5D woven structures

Table 4.8 Weaving parameters of three main types of 2.5D woven structures

2.5D woven structures	V_f	Weft yarn ($w \times h$)	Warp yarn ($w \times h$)	Stuffer yarn ($w \times h$)
Orthogonal	50.99%	0.8×0.1 mm^2	0.4×0.1 mm^2	0.4×0.1 mm^2
Angle interlock	46.89%	0.4×0.1 mm^2	0.4×0.1 mm^2	0.4×0.1 mm^2
Layer-to-layer interlock	51.91%	0.7×0.1 mm^2	0.4×0.1 mm^2	0.4×0.1 mm^2

where V_f is the fiber volume fraction, and w and h are the width and height of the ellipse section

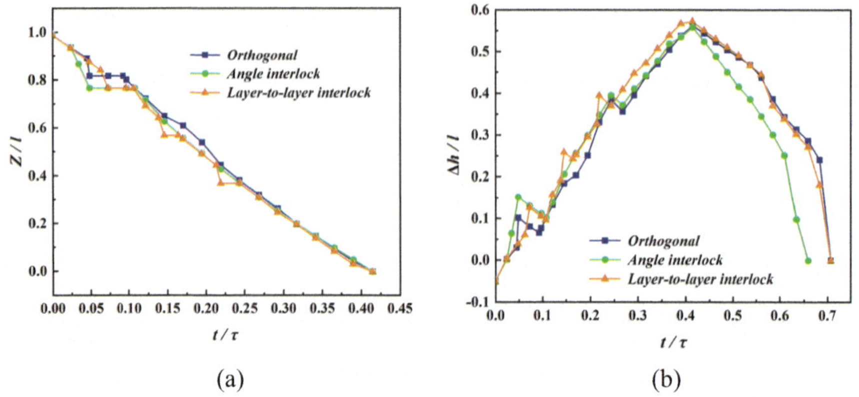

Fig. 4.107 Thermal ablation performances of three main types of 2.5D woven structures: **a** the variation of Z/l; **b** the variation of $\Delta h/l$

characterization of the thermal ablation performance of C/C composites. The larger the recession rate and roughness are, the worse the thermal ablation is. The large recession rate means the early consumption of thermal protection materials, and the large roughness induces an enhancement of heat and mass transfer and contributes to the laminar-to-turbulent transition in the dynamic boundary layer.

The thermal ablation behaviors of three main types of 2.5D woven structures are numerically solved with the following values: $A = 2$, $Da = 0.25$, $\Omega_m = \Omega_f = 6 \times 10^{-6}$ m^3/mol, $\tau = 1.23 \times 10^4$ s, and $l = 0.001$ m, and the variations of Z/l (the relative height of the matrix surface) and $\Delta h/l$ (the relative height difference between fiber and matrix surfaces) are shown in Fig. 4.107. It can be seen from Fig. 4.107a that the recession rates of the matrix of these three woven structures are almost the same. In Fig. 4.107b, when the relative time $t/\tau < 0.415$, the roughness of these three woven structures are nearly the same. When $t/\tau > 0.415$, the roughness of the angle interlock structure is smaller than those of the orthogonal and layer-to-layer interlock structures due to the smaller fiber volume fraction of the angle interlock structure. The surface ablation morphologies of these three woven structures with $t/\tau = 0$, 0.073, 0.147, 0.269, and 0.44 are shown in Fig. 4.108. It can be seen that porous structures appear on the surface of 2.5D woven composites due to heterogeneous ablation. In summary, the thermal ablation performances of these three woven structures are very close.

4.6.4.2 Effects of Computation Parameters on Thermal Ablation Behavior

The RVC of LA is numerically solved with different A, the other computation parameters are kept constant, and the porosity of the matrix is considered. The thermal ablation morphologies of LA are shown in Fig. 4.109a. When $A = 1.25$, the surface

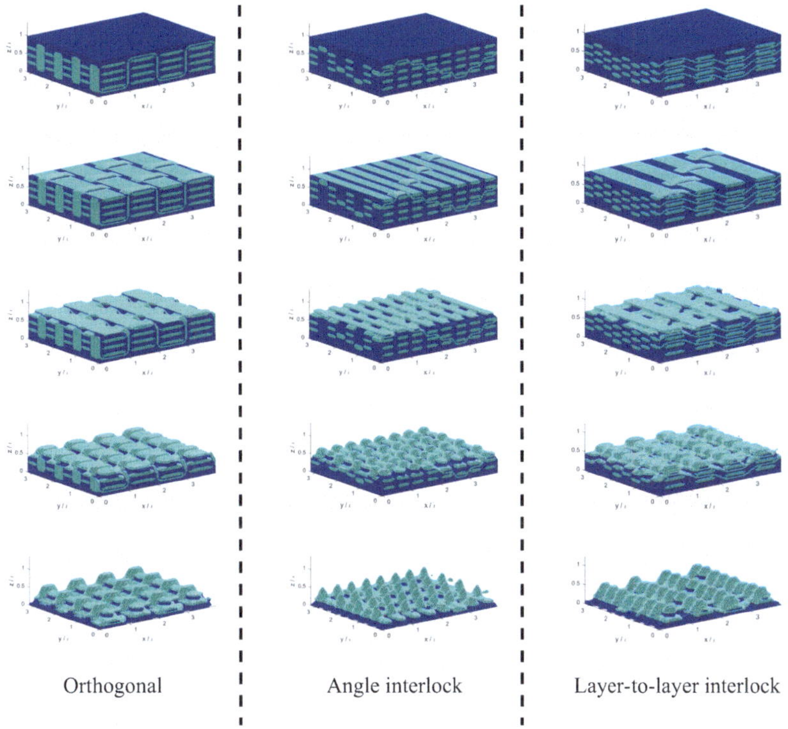

Fig. 4.108 Surface ablation morphologies of three main types of 2.5D woven structures with t/τ = 0, 0.073, 0.147, 0.269, and 0.44

of LA is not flat because *A* does not equal one due to the effect of φ. Moreover, the variations of *Z/l* and *Δh/l* at different *A* are shown in Figs. 4.110a and 4.107b. With the rise of t/τ, the matrix recession rate decreases, and the roughness firstly increases then decreases because the ablation rate of the matrix is greater than that of the fiber so that the matrix is consumed firstly, and then the fiber area exposed to the oxygen increases. Besides, with the rise of *A*, the matrix recession rate and roughness all increase. Thus, to improve the ablation performance of C/C composites, it is necessary to enhance the oxidation resistance of the matrix to reduce the reactivity difference between the fiber and matrix in the manufacture. It also can be seen from Figs. 4.107b and 4.110a that there are several fluctuations of the curves at t/τ = 0.103 ~ 0.138 s, 0.238 ~ 0.251 s, and 0.357 ~ 0.402 s because the matrix is discontinuous along direction *z* in the same *x* and *y*. Thus, each-layered yarn decelerates the recession of the matrix along direction *z*, which means that the fiber and matrix can protect each other from quick ablation in this mesostructure.

The thermal ablation morphologies of LA numerically solved with different *Da* are shown in Fig. 4.109b, and the variations of *Z/l* and *Δh/l* with different *Da* are shown in Figs. 4.110c and 4.107d. With the rise of *Da*, the matrix recession rate and the roughness all decrease. These are because when the ablation is biased towards

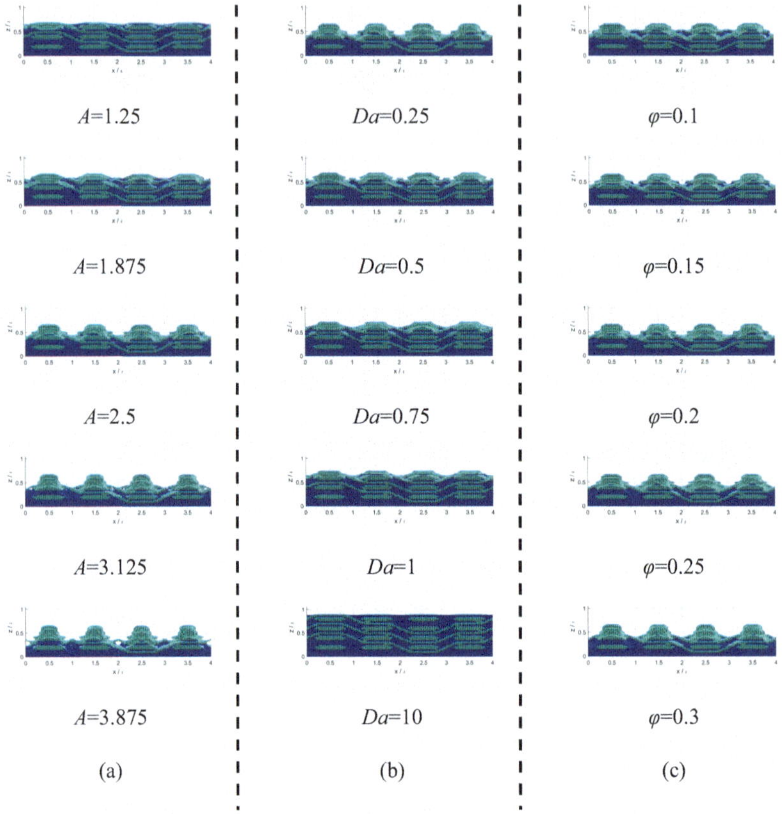

Fig. 4.109 Surface ablation morphologies of LA at $t/\tau = 0.317$: **a** with different A; **b** with different Da; **c** with different φ

the diffusion control, the surface recession decelerates, and the roughness narrows down.

The thermal ablation morphologies of LA numerically solved with different φ are shown in Fig. 4.109c, and the variations of Z/l and $\Delta h/l$ with different φ are shown in Fig. 4.110e and f. With the rise of φ, the matrix recession rate and the roughness all increase. Therefore, in the design and manufacture, the porosity of the matrix should be kept as low as possible to improve the thermal ablation performance.

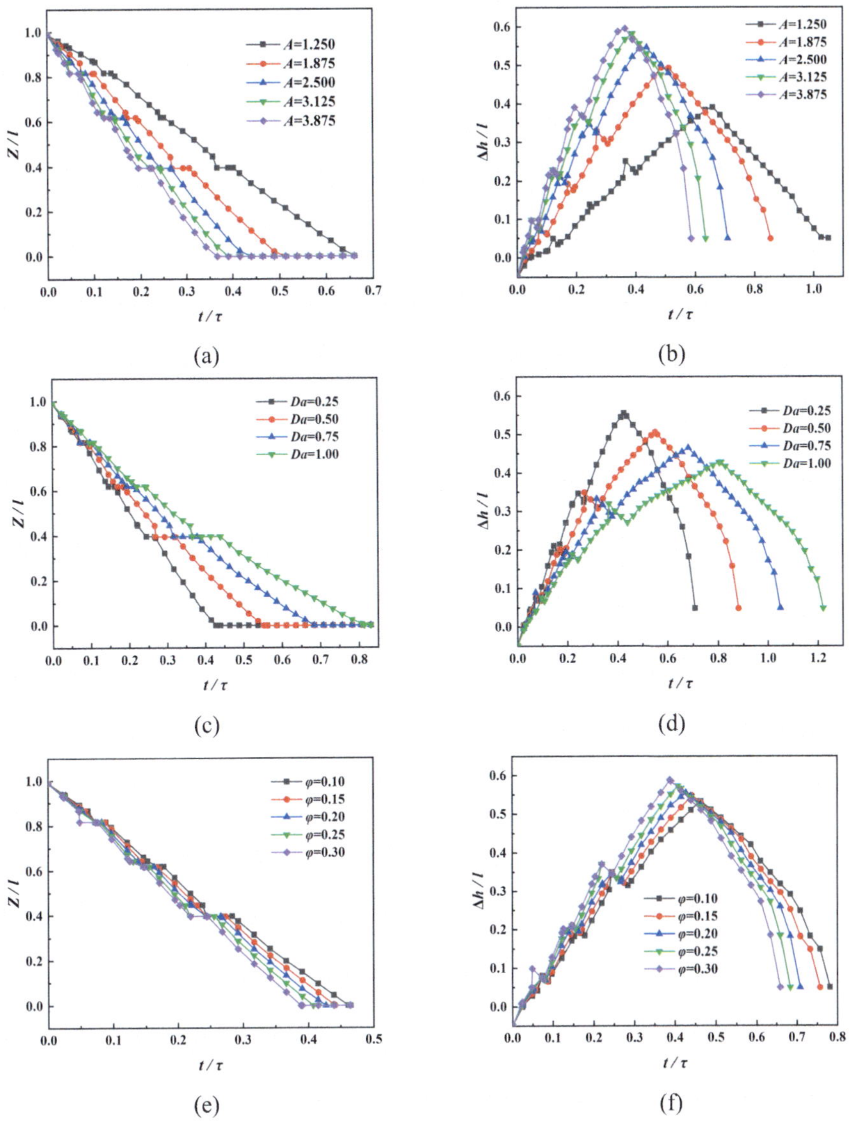

Fig. 4.110 Ablation performance of LA with different A, Da, and φ: **a** the variation of Z/l with different A; **b** the variation of $\Delta h/l$ with different A; **c** the variation of Z/l with different Da; **d** the variation of $\Delta h/l$ with different Da; **e** the variation of Z/l with different φ; **f** the variation of $\Delta h/l$ with different φ

4.7 Conclusion

By carefully designing the micropores or cavities, the metasurfaces could depict different acoustic characteristics. It can change the absorption, reflection direction, and acoustic phase. It seems numerous acoustic characteristics can affect the flow instabilities, and that worth further investigations. Most hypersonic vehicles have 3D configurations wherein the crossflow mode may dominate the transition, and thus, the wave control ability of the metasurface needs to be further investigated. Joint active control technologies may be required and improved to achieve a more effective and robust transition control capability. But in a near future, a realistic metasurface caused by the thermal ablation of C/C composites is also a possible way to be applied in engineering.

References

1. Brès GA, Colonius T, Fedorov AV (2010) Acoustic properties of metasurfaces for hypersonic boundary-layer control. AIAA J 48(2):267–274. https://doi.org/10.2514/1.40811
2. Brès GA, Inkman M, Colonius T, Fedorov AV (2013) Second-mode attenuation and cancellation by metasurfaces in a high-speed boundary layer. J Fluid Mech 726:312–337. https://doi.org/10.1017/jfm.2013.206
3. Bountin D, Chimitov T, Maslov A, Novikov A, Egorov I, Fedorov A, Utyuzhnikov S (2013) Stabilization of a hypersonic boundary layer using a wavy surface. AIAA J 51(5):1203–1210. https://doi.org/10.2514/1.J052044
4. Zhao R, Wen CY, Tian XD, Long TH, Yuan W (2018) Numerical simulation of local wall heating and cooling effect on the stability of a hypersonic boundary layer. Int J Heat Mass Tran 121:986
5. Li Y, Assouar BM (2016) Acoustic metasurface-based perfect absorber with deep subwavelength thickness. Appl Phys Lett 108:063502
6. Bok E, Park JJ, Choi H, Han CK, Wright OB, Lee SH (2018) Metasurface for water-to-air sound transmission. Phys Rev Lett 120:044302
7. Quan L, Zhong X, Liu X, Gong X, Johnson PA (2014) Effective impedance boundary optimization and its contribution to dipole radiation and radiation pattern control. Nat Commun 5:3188
8. Schwan L, Geslain A, Romero-Garcia V, Groby JP (2017) Complex dispersion relation of surface acoustic waves at a lossy metasurface. Appl Phys Lett 110:051902
9. Liu T, Liang S, Chen F, Zhu J (2018) Inherent losses induced absorptive acoustic rainbow trapping with a gradient metasurface. J Appl Phys 123:091702
10. Yu N, Genevet P, Kats MA, Aieta F, Tetienne J-P, Capasso F, Gaburro Z (2011) Light propagation with phase discontinuities: generalized laws of reflection and refraction. Science 334(6054):333–337. https://doi.org/10.1126/science.1210713
11. Asadchy VS, Albooyeh M, Tcvetkova SN, Diacuteaz-Rubio A, Ra'di Y, Tretyakov SA (2016) Perfect control of reflection and refraction using spatially dispersive metasurfaces. Phys Rev B 94(7):075142. https://doi.org/10.1103/PhysRevB.94.075142
12. Zhao R, Wen CY, Long TH, Tian XD, Zhou L, Wu Y (2019) Spatial direct numerical simulation of the hypersonic boundary-layer stabilization using metasurfaces. AIAA J 57(11):5061–5065. https://doi.org/10.2514/1.J058467
13. Fedorov AV, Shiplyuk AN, Maslov AA, Burov E, Malmuth ND (2003) Stabilization of a hypersonic boundary layer using an ultrasonically absorptive coating. J Fluid Mech 479:99–124

14. Stephen SO, Michael V (2013) Effects of porous walls on hypersonic boundary layers over a sharp cone. AIAA J 51:1234–1244
15. Tian X, Zhao R, Long T, Wen C (2019) Reverse design of ultrasonic absorptive coating for the stabilization of Mack modes. AIAA J 57(6):2264–2269
16. Zhao R, Liu X, Wen C, Wang X (2022) Broadband design of acoustic metasurfaces for the stabilization of a mach 4 boundary layer flow. Adv Aerodyn 4(15)
17. Tian X, Liu T, Wang T, Zhu J, Wen C (2022) Double-layer acoustic metasurface for the suppression of the Mack second mode in hypersonic boundary-layer flow. Phys Fluids 34:074105
18. Kimmel R (2003) Aspects of hypersonic boundary layer transition control. In: 41st Aerospace sciences meeting and exhibit, Reno, Nevada, 6–9 Jan 2003
19. Carpenter PW, Porter LJ (2001) Effects of passive porous walls on boundary-layer instability. AIAA J 39(4):597–604
20. Wang X, Zhong X (2011) Phase angle of metasurface admittance and its effect on boundary-layer stabilization. In: AIAA Paper 2011, p 3080
21. Zhao R, Yan H, Xi K, Wen CY (2020) Research on acoustic metasurfaces for the suppression of the first mode. Aeronaut Sci Technol 31:104–112 (Chinese)
22. Gui Y, Wang W, Zhao R, Zhao J, Wu J (2022) Hypersonic boundary-layer instability characteristics on sharp cone with metasurface. AIAA J 27(6):1–9
23. Lukashevich SV, Morozov SO, Shiplyuk AN (2016) Experimental study of the effect of a passive metasurface on disturbances in a hypersonic boundary layer 2: effect of the metasurface location. J Appl Mech Tech Phys 57(5):873–878
24. Li F, Choudhari MM, Chang CL, White J (2013) Effects of injection on the instability of boundary layers over hypersonic configurations. Phys Fluids 25(10):104107
25. Miró FM, Fabio P (2013) Effect of uneven wall blowing on hypersonic boundary-layer stability and transition. Phys Fluids 25:104107
26. Mortensen C, Zhong X (2016) Real-gas and surface-ablation effects on hypersonic boundary-layer instability over a blunt cone. AIAA J 54(3):980–998
27. Holden MS, Mundy E, Maclean M (2011) Heat transfer measurements to examine surface roughness and blowing effects in hypersonic flows. In: AIAA Paper, p 760
28. Schneider S (2008) Hypersonic boundary-layer transition with ablation and blowing. In: AIAA Paper, p 3730
29. Johnson HB, Gronvall JE, Chandler GV (2009) Reacting hypersonic boundary layer stability with blowing and suction. In: AIAA Paper, p 938
30. Ghaffari S, Marxen O, Iaccarino G, Shaqfeh ESG (2010) Numerical simulations of hypersonic boundary-layer instability with wall blowing. In: AIAA Paper, p 706
31. Pagella A, Rist U, Wagner S (2002) Numerical investigations of small-amplitude disturbances in a boundary layer with impinging shock wave at Ma = 4.8. Phys Fluids 14(7):2088–2100
32. Marvin JG, Akin CM (1970) Combined effects of mass addition and nose bluntness on boundary-layer transition. AIAA J 8(5):857–863
33. Fedorov AV, Malmuth N (2001) Stabilization of hypersonic boundary layers by metasurfaces. AIAA J 39:605–610
34. Brès GA, Colonius T, Fedorov AV (2010) Acoustic properties of metasurfaces for hypersonic boundary-layer control. AIAA J 48(2):267–274
35. Wang X, Zhong X (2012) The stabilization of a hypersonic boundary layer using local sections of metasurface. Phys Fluids 24:034105
36. Lukashevich SV, Morozov SO, Shiplyuk AN (2016) Experimental study of the effect of a passive metasurface on disturbances in a hypersonic boundary layer 2: effect of the metasurface location. J Appl Mech Tech Phys 57(5):873–878
37. Zhao R, Wen C, Zhou Y, Tu G, Lei J (2022) Review of acoustic metasurfaces for hypersonic boundary layer stabilization. Prog Aerosp Sci 130:100808
38. Zhu W, Shi M, Zhu Y, Lee C (2020) Experimental study of hypersonic boundary layer transition on a permeable wall of a flared cone. Phys Fluids 32:011701

39. Zhu W, Chen X, Zhu Y, Lee C (2020) Nonlinear interactions in the hypersonic boundary layer on the permeable wall. Phys Fluids 32:104110
40. Fedorov AV, Soudakov VG (2014) Theoretical-numerical analysis of boundary-layer stability with combined injection and acoustic absorptive coating. Moscow Inst Phys Technol Dolgoprudny (Russia)
41. Miró FM, Dehairs P, Pinna F et al (2019) Effect of wall blowing on hypersonic boundary-layer transition. AIAA J 57(4):1567–1578
42. Malmuth N, Fedorov A, Shalaev V, Cole J (1998) Problems in high speed flow prediction relevant to control. In: 2nd AIAA theoretical fluid mechanics meeting
43. Fedorov AV, Malmuth ND, Rasheed A et al (2001) Stabilization of hypersonic boundary layers by porous coatings. AIAA J 39(4):605–610
44. Guo QL, Tu GH, Chen JQ, Yuan XX, Wan BB (2020) Control of hypersonic boundary layer instability by transverse rectangular micro-cavities. J Aeros Power 35(1):135–143
45. Guo Q, Li C, Tu G, Chen J, Wan B, Liu Y (2021) Investigation on the depth effects of the acoustic metasurfaces on the suppression of the second modes in the hypersonic boundary layer. J Phys: Conf Ser 1786(1)
46. Sandham ND, Lüdeke H (2009) Numerical study of mach 6 boundary-layer stabilization by means of a porous surface. AIAA J 47(9):2243–2252
47. Wartemann V, Lüdeke H, Sandham ND (2012) Numerical investigation of hypersonic boundary-layer stabilization by porous surfaces. AIAA J 50(6):1281–1290
48. Shi S, Liang J, Yi F, Fang G (2013) Modeling of one-dimensional thermal response of silica-phenolic composites with volume ablation. J Compos Mater 47(18):2219–2235
49. Shi S, Liang J (2014) Modeling of coupled temperature-displacement-diffusion problem for silica-phenolic composite under high temperature. Polym Polym Compos 22(2):85–95
50. Shi SB, Liang J, Fang GD, Wang LL (2011) Evaluation of the effective properties of high silica/phenolic composite under volumetric ablation condition. In: International conference on heterogeneous material mechanics
51. Shi S, Liang J, Lin G, Fang G (2013) High temperature thermomechanical behavior of silica-phenolic composite exposed to heat flux environments. Combust Sci Technol 87:204–209
52. Vignoles GL, Aspa Y, Quintard M (2010) Modelling of carbon–carbon composite ablation in rocket nozzles. Compos Sci Technol 70(9):1303–1311
53. Kuo KK, Keswani ST (2007) A comprehensive theoretical model for carbon-carbon composite nozzle recession. Combust Sci Technol 42(3–4)
54. Wu S, Yi M, Ge Y, Ran L, Peng K (2017) Effect of carbon fiber reinforcement on the tribological performance and behavior of aircraft carbon brake discs. Carbon, 117(Complete):279–292
55. Chen L, Jin H, Xu Z, Li J, Guo Q, Shan M et al (2015) Role of a gradient interface layer in interfacial enhancement of carbon fiber/epoxy hierarchical composites. J Mater Sci 50(1):112–121
56. Fitzer E, Manocha LM (1998) Carbon reinforcements and carbon/carbon composites. In: Processing of carbon/carbon composites (Chapter 4), pp 97–147. https://doi.org/10.1007/978-3-642-58745-0
57. Dimitrienko YI (2016) [Solid mechanics and its applications] Thermo mechanics of composite structures under high temperatures, vol 224. Mechanics of composite thin-walled shells under high temperatures (Chap 12), pp 309–337. https://doi.org/10.1007/978-94-017-7494-9
58. Lachaud J, Vignoles GL, Goyhénèche J-M, Epherre JF (2011) Ablation in carbon/carbon composites: microscopic observations and 3D numerical simulation of surface roughness evolution. Wiley
59. Yang J, Jing Z, Ge J, Zhang B, Li W, Zhang Q, Shang T, Liang J (2024) Mesoscopic modeling of mechanical erosion for the ablation-erosion behavior of 3D C/C composites. Acta Mech Sin 40:423277
60. Lachaud J, Aspa Y, Vignoles GL (2008) Analytical modeling of the steady state ablation of a 3d c/c composite. Int J Heat Mass Transf 51(9–10), 2614–2627
61. Lachaud J, Vignoles GL (2009) A brownian motion technique to simulate gasification and its application to c/c composite ablation. Comput Mater Sci 44(4):1034–1041

62. Aspa Y, Quintard M, Plazanet F, Descamps C, Vignoles GL (2008) Ablation of carbon/carbon composites: direct numerical simulation and effective behavior. Wiley
63. Li W, Fang G, Li W, Liang J, Li M (2019) Role of mesoscopic features on thermochemical ablative behavior of 3d c/c braided composites. Int J Heat Mass Transf 144:118602
64. Gueyffier D, Li J, Nadim A, Scardovelli R, Zaleski S (1999) Volume-of-fluid interface tracking with smoothed surface stress methods for three-dimensional flows. J Comput Phys, 152(2), 423–456
65. Lachaud J (2006) Physico-chemical modeling of carbon-based composite ablation. L'UNIVERSITE BORDEAUX I
66. Yang J, Li W, Ge J, Li W, Li Y, Liu S, Liang J (2023) A mesoscopic model of mechanical erosion for the characterization of ablation behavior of C/C woven composites. Int J Heat Mass Transf 206:123962

Chapter 5
Stabilization Mechanisms

5.1 Introduction

The theoretical study by LST shows the existence of unstable modes in the hypersonic boundary layer by mathematically giving the growth rate of these modes. However, it is still a mystery how these unstable modes are amplified downstream from a physical perspective. There have been attempts to provide a physical interpretation of the growth of small disturbances in the flow field based on energy approaches. Reynolds [1] proposed a two-dimensional (2-D) kinetic energy equation for incompressible parallel shear flows. The total disturbance kinetic energy growth rate is related to a production term due to the Reynolds stress and a dissipation term due to the viscous effect. The small 2-D disturbance in the incompressible boundary layer grows if and only if the production term outweighs the dissipation term. Kuehl [2] derived a Lagrangian acoustic energy equation by ignoring the viscous effect for the compressible boundary layer. With the Lagrangian acoustic energy equation, Kuehl showed that the second mode is related to a forced, resonating, thermo-acoustic standing wave trapped in a thermo-acoustic impedance well formed by the sonic line and the solid wall. The thermo-acoustic Reynolds stress draws energy from the mean flow, which is the energy source for the amplification of the second mode. Tian and Wen [3] performed the stability analysis for the second mode in a Mach 6.0 boundary layer based on the fluctuating internal energy. The second mode has the largest growth rate when the wall-normal transport of energy is in phase to the change in fluctuating internal energy in the vicinity of the critical layer, similar to the mechanism of the Rijke tube that 'depends upon the phase of the vibration at which the transfer of heat takes place'[4]. With the energy approach, the results of Kuehl [2] and Tian [3] provide some insight into the physical mechanism of the second mode from different perspectives. However, their results do not explicitly present the relation between the growth rate of the second mode and energy sources.

R. Zhao et al., *Stabilization of Hypersonic Boundary-Layer: Acoustic Metasurfaces*,
https://doi.org/10.1007/978-981-97-8623-7_5

Most previous energy analyses ignored the earlier evolvement of the second mode instability wave through the receptivity process. It is well known that infinitesimal freestream disturbances can be decomposed into different modes, including the vortical, acoustic, and entropic components, and each of them is governed by an individual equation [5]. However, when the three types of freestream disturbances enter the boundary layer, they couple with each other and the base flow in quite a complicated manner. Therefore, although the evolution of the second mode is closely linked with the slow acoustic waves and others [6], the energy analysis considering the decomposed components is difficult to perform. Recently, Unnikrishnan and Gaitonde [7] utilized the momentum potential theory (MPT) to decompose the second mode into the vortical, acoustic, and entropic components and found that they interact with each other and the base flow in a well-described process. The highlight is that the MPT technique provides a straightforward way to reveal the coupling of the three components without the leading-edge receptivity analysis. In detail, the momentum potential theory (MPT) proposed by Doak [8] obtained an energy budget equation for disturbances in a time-stationary mean flow field. The fluctuation momentum density is decomposed into vortical, acoustic, and thermal components. The transport of the total energy flux is governed by the superimposition of the source terms due to each MPT component and an additional source term. The vortical component displays a series of rope-shaped recirculation-cell patterns in the boundary layer, while both acoustic and thermal components display 'trapped' structures. Furthermore, Unnikrishnan and Gaitonde examined source terms in the boundary layer with different wall temperatures based on the MPT's energy budget equation and claimed that the thermal source is the primary source for the growth of the second mode. The finding of Unnikrishnan and Gaitonde is inspiring. However, their conclusion that the modal growth of the second mode is caused by the thermal source is mainly based on numerical observations of the term magnitude itself. The work lacks a direct link between the growth rate and source terms in Doak's energy budget equation, yielding a debatable question. In this chapter, the mechanisms of the amplification of the unstable modes are investigated by Doak's MPT and Thermal-Acoustic Coupling Theory.

5.2 Momentum Potential Theory

5.2.1 MPT Method

5.2.1.1 MPT Decomposition

It is generally acknowledged that Mack's second mode and higher modes belong to the family of trapped acoustic waves in the waveguide formed by the sonic line and the solid wall. However, the term "acoustic wave" adopted here is not well defined. Kovásznauy [5] showed that random fluctuations in the flow field with uniform mean velocity and temperature can be decomposed into vorticity, entropy, and acoustic modes. Under the small disturbance assumption, the interaction between different modes is negligible and the modes obey three independent governing equations. If the intensity of fluctuation is large that the linear assumption is invalid, quadratic interaction terms between different modes appear in the governing equation as source terms. However, Kovásznay's approach is not suitable for extension to the boundary layer flows, where the gradient of mean velocity and mean temperature are strong. In the flow field with a strong gradient, complicated interaction terms exist between different modes even fluctuations are small.

The MPT approach proposed by Doak [8] overcomes the restriction in Kovásznay's approach for a general flow which is time stationary. In the time stationary flow, any physic quantity ϕ can be expressed as the sum of the time-averaged quantity $\overline{\phi}$ and the fluctuation quantity $\phi\prime$. The time-averaged term $\overline{\phi}$ is independent on time t. In Doak's MPT approach, the momentum density, $\boldsymbol{m} = \rho\boldsymbol{u}$, is split into its vortical, acoustic, and thermal components. Considering the continuity equation,

$$\frac{\partial\rho}{\partial t} + \nabla\cdot\boldsymbol{m} = 0 \tag{5.1}$$

Then, the vector field $\boldsymbol{m}$ is decomposed into the rotational part $\boldsymbol{m}_B$ and irrotational part $-\nabla\varphi$ through the Helmholtz decomposition.

$$\boldsymbol{m} = \boldsymbol{m}_B - \nabla\varphi, \nabla\cdot\boldsymbol{m}_B = 0 \tag{5.2}$$

Here, the scalar field φ is the momentum potential.

The time average of Eq. (5.1) is

$$\nabla\cdot\overline{\boldsymbol{m}} = 0 \tag{5.3}$$

The vector field $\overline{\boldsymbol{m}} = \overline{\boldsymbol{m}}_B$ is solenoidal. Thus, the fluctuation momentum density $\boldsymbol{m}'$ is expressed as

$$\boldsymbol{m}' = \boldsymbol{m}'_B - \nabla\varphi,\ \nabla\cdot\boldsymbol{m}'_B = 0 \tag{5.4}$$

Subtract Eq. (5.1) by Eq. (5.3), we have the fluctuation continuity equation

$$\frac{\partial\rho'}{\partial t} + \nabla\cdot\boldsymbol{m}' = 0 \tag{5.5}$$

Furthermore, substitute Eq. (5.4) into Eq. (5.5), we obtain the Poisson equation

$$\frac{\partial\rho'}{\partial t} = \nabla^2\varphi \tag{5.6}$$

For a general flow in the thermal equilibrium state, we can express the density ρ as the function of the pressure p and the entropy S.

$$\rho = \rho(p, S) \tag{5.7}$$

With the small disturbance assumption, we have

$$\rho' = \left(\frac{\partial \rho}{\partial p}\right)_S p' + \left(\frac{\partial \rho}{\partial S}\right)_p S' \tag{5.8}$$

Here,

$$\left(\frac{\partial \rho}{\partial p}\right)_S = \frac{1}{\overline{a}^2} = \frac{1}{\gamma R \overline{T}}, \quad \left(\frac{\partial \rho}{\partial S}\right)_p = -\frac{(\gamma - 1)\overline{\rho}}{\gamma R} \tag{5.9}$$

Hence, Eq. (5.6) can be written as

$$\frac{1}{\overline{a}^2}\frac{\partial p'}{\partial t} - \frac{(\gamma - 1)\overline{\rho}}{\gamma R}\frac{\partial S'}{\partial t} = \nabla^2 \varphi \tag{5.10}$$

Because the momentum potential φ is linear in Eq. (5.10), φ can be regarded as the linear superposition of uniquely defined acoustic component φ_A and thermal component φ_T.

$$\varphi = \varphi_A + \varphi_T, \quad \frac{1}{\overline{a}^2}\frac{\partial p'}{\partial t} = \nabla^2 \varphi_A, \quad -\frac{(\gamma - 1)\overline{\rho}}{\gamma R}\frac{\partial S'}{\partial t} = \nabla^2 \varphi_T \tag{5.11}$$

As a consequence, the fluctuation momentum density $\boldsymbol{m}\prime$ can be written as the sum of the vortical component $\boldsymbol{m}'_B$, the acoustic component $\boldsymbol{m}'_A$, and the thermal component $\boldsymbol{m}'_T$.

$$\boldsymbol{m}' = \boldsymbol{m}'_B + \boldsymbol{m}'_A + \boldsymbol{m}'_T, \ \nabla \cdot \boldsymbol{m}'_B = 0, \ \boldsymbol{m}'_A = -\nabla \varphi_A, \ \boldsymbol{m}'_T = -\nabla \varphi_T \tag{5.12}$$

Vortical component $\boldsymbol{m}'_B$ is solenoidal, isobaric, and isentropic. Acoustic component $\boldsymbol{m}'_A$ represents the irrotational and isentropic part of $\boldsymbol{m}'$, which is related to the fluctuation pressure p' in the flow field. Thermal component $\boldsymbol{m}'_T$ represents the irrotational and isobaric part of $\boldsymbol{m}'$, which is related to the fluctuation entropy S' in the flow field.

The definition of momentum potential φ can be extended into the multi-chemical-component flow with chemical reaction by expressing the fluctuation density ρ' in the form

$$\rho = \rho(p, S, \alpha, \beta, \cdots) \tag{5.13}$$

$\alpha, \beta, \cdots$ are chemical-component concentrations or other suitable independent thermodynamic variables. Consequently, Eq. (5.8) can be rewritten as

$$\rho' = \left(\frac{\partial \rho}{\partial p}\right)_{S,\alpha,\beta,\cdots} p' + \left(\frac{\partial \rho}{\partial S}\right)_{p,\alpha,\beta,\cdots} S' + \left(\frac{\partial \rho}{\partial \alpha}\right)_{p,S,\beta,\cdots} \alpha' + \left(\frac{\partial \rho}{\partial \beta}\right)_{p,S,\alpha,\cdots} \beta' + \cdots \tag{5.14}$$

and the definitions (5.11) can be extended to

$$\varphi = \varphi_A + \varphi_T + \varphi_\alpha + \varphi_\beta + \cdots$$

$$\left(\frac{\partial \rho}{\partial \alpha}\right)_{p,S,\beta,\cdots} \frac{\partial \alpha'}{\partial t} = \nabla^2 \varphi_\alpha, \ \left(\frac{\partial \rho}{\partial \beta}\right)_{p,S,\alpha,\cdots} \frac{\partial \beta'}{\partial t} = \nabla^2 \varphi_\beta, \cdots \tag{5.15}$$

Similarly, we can define components related to chemical components $\boldsymbol{m}'_\alpha, \boldsymbol{m}'_\beta, \cdots$ as

$$\boldsymbol{m}'_\alpha = -\nabla \varphi_\alpha, \ \boldsymbol{m}'_\beta = -\nabla \varphi_\beta, \cdots \tag{5.16}$$

5.2.1.2 MPT Energy Budget Equation

Besides the MPT decomposition which provides a unique definition of vortical, acoustic, and thermal components in a general time-stationary flow, the other important aspect of the MPT approach is the energy budget equation governing the fluctuation energy transported by different components. The derivation of the energy budget equation is reviewed as follows.

Firstly, considering the momentum conservation equation

$$\frac{\partial \boldsymbol{u}}{\partial t} + (\boldsymbol{u} \cdot \nabla)\boldsymbol{u} = -\frac{1}{\rho}\nabla p + \frac{1}{\rho}\nabla \cdot \overline{\overline{\boldsymbol{S}}} \tag{5.17}$$

Here, $\overline{\overline{\boldsymbol{S}}}$ is the viscous stress tensor.

Since

$$(\boldsymbol{u} \cdot \nabla)\boldsymbol{u} = \boldsymbol{\Omega} \times \boldsymbol{u} + \nabla\left(\frac{\boldsymbol{u}^2}{2}\right) \tag{5.18}$$

and

$$\nabla h = T\nabla S + \frac{1}{\rho}\nabla p \tag{5.19}$$

Equation (5.17) can be expressed as

$$\frac{\partial \boldsymbol{u}}{\partial t} + \nabla\left(\frac{\boldsymbol{u}^2}{2} + h\right) = -\boldsymbol{\Omega} \times \boldsymbol{u} + T\nabla S + \frac{1}{\rho}\nabla \cdot \overline{\overline{S}} \tag{5.20}$$

Here, $\boldsymbol{\Omega} = \nabla \times \boldsymbol{u}$ is the vorticity vector and $-\boldsymbol{\Omega} \times \boldsymbol{u}$ is the Coriolis acceleration. h is the enthalpy per unit mass.

For the convenience of discussion, Eq. (5.20) is rewritten as

$$\frac{\partial \boldsymbol{u}}{\partial t} + \nabla H = \boldsymbol{\alpha} \tag{5.21}$$

Here, $H = \frac{u^2}{2} + h$ is the total enthalpy per unit mass and $\boldsymbol{\alpha} = -\boldsymbol{\Omega} \times \boldsymbol{u} + T\nabla S + \frac{1}{\rho}\nabla \cdot \overline{\overline{S}}$ is the so-called "acceleration" vector.

Thus, the fluctuation momentum conservation equation is

$$\frac{\partial \boldsymbol{u}'}{\partial t} + \nabla H' = \boldsymbol{\alpha}' \tag{5.22}$$

(1) Vortical energy budget equation

Taking the scalar production of Eq. (5.22) and the fluctuation momentum density $\boldsymbol{m}'$

$$\boldsymbol{m}' \cdot \frac{\partial \boldsymbol{u}'}{\partial t} + \boldsymbol{m}' \cdot \nabla H' = \boldsymbol{m}' \cdot \boldsymbol{\alpha}' \tag{5.23}$$

From Eq. (5.23), we have

$$\frac{\partial}{\partial t}(\boldsymbol{m}' \cdot \boldsymbol{u}') + \nabla \cdot (\boldsymbol{m}'H') - \boldsymbol{u}' \cdot \frac{\partial \boldsymbol{m}'}{\partial t} - (\nabla \cdot \boldsymbol{m}')H' = \boldsymbol{m}' \cdot \boldsymbol{\alpha}' \tag{5.24}$$

Noting that

$$\boldsymbol{m}' = \overline{\rho}\boldsymbol{u}' + \overline{\boldsymbol{u}}\rho' \tag{5.25}$$

With the continuity Eq. (5.5) and Eq. (5.25),

$$\frac{\partial}{\partial t}\left(\boldsymbol{m}' \cdot \boldsymbol{u}' - \frac{1}{2}\overline{\rho}\boldsymbol{u}'^2\right) + \nabla \cdot (\boldsymbol{m}'H') + h'\frac{\partial \rho'}{\partial t} = \boldsymbol{m}' \cdot \boldsymbol{\alpha}' \tag{5.26}$$

The time-average of Eq. (5.26) is

$$\nabla \cdot \left(\overline{\boldsymbol{m}'H'}\right) + \overline{h'\frac{\partial \rho'}{\partial t}} = \overline{\boldsymbol{m}' \cdot \boldsymbol{\alpha}'} \tag{5.27}$$

It is easy to obtain that

$$\overline{h' \frac{\partial \rho'}{\partial t}} = -\overline{\frac{p'}{R} \frac{\partial S'}{\partial t}} \tag{5.28}$$

Finally, we get the energy budget equation

$$\nabla \cdot \left(\overline{\boldsymbol{m}' H'} \right) = \overline{\boldsymbol{m}' \cdot \boldsymbol{\alpha}'} + \overline{\frac{p'}{R} \frac{\partial S'}{\partial t}} \tag{5.29}$$

Because $\boldsymbol{m}'$ is the linear superposition of the vortical component $\boldsymbol{m}'_B$, the acoustic component $\boldsymbol{m}'_A$, and the thermal component $\boldsymbol{m}'_T$, the energy budget Eq. (5.29) can be expressed as

$$\nabla \cdot \left(\overline{\boldsymbol{m}'_B H'} + \overline{\boldsymbol{m}'_A H'} + \overline{\boldsymbol{m}'_T H'} \right) = \overline{\boldsymbol{m}'_B \cdot \boldsymbol{\alpha}'} + \overline{\boldsymbol{m}'_A \cdot \boldsymbol{\alpha}'} + \overline{\boldsymbol{m}'_T \cdot \boldsymbol{\alpha}'} + \overline{\frac{p'}{R} \frac{\partial S'}{\partial t}} \tag{5.30}$$

On the left-hand side of Eq. (5.30), $\overline{\boldsymbol{m}'_B H'}$, $\overline{\boldsymbol{m}'_A H'}$, and $\overline{\boldsymbol{m}'_T H'}$ are fluctuation energy fluxes carried by $\boldsymbol{m}'_B$, $\boldsymbol{m}'_A$, and $\boldsymbol{m}'_T$, respectively. The first three terms on the right-hand side indicate energy sources due to the interaction of the vector $\boldsymbol{\zeta}'$ with different MPT components. In addition to the source terms due to the MPT components, the source term, $\overline{\frac{p'}{R} \frac{\partial S'}{\partial t}}$, is the interaction between fluctuation pressure and entropy. It is noted that fluctuation pressure and entropy are related to the acoustic and thermal components, respectively. Thus, the source, $\overline{\frac{p'}{R} \frac{\partial S'}{\partial t}}$, represents the thermo-acoustic coupling effect in the disturbances flow field. For convenience of discussion, the vortical source $-\overline{\boldsymbol{m}'_B \cdot \boldsymbol{\zeta}'}$, acoustic source $-\overline{\boldsymbol{m}'_A \cdot \boldsymbol{\zeta}'}$, and thermal source $-\overline{\boldsymbol{m}'_T \cdot \boldsymbol{\zeta}'}$ are designated as P_B, P_A, and P_T, respectively. The source due to the thermo-acoustic coupling effect is designated as P_{ta}.

The energy budget Eq. (5.30) proposed by Doak obtains an energy balance between energy fluxes and source terms. The existence of multiple energy fluxes, as well as source terms, makes it hard to clarify how the energy flux due to each component is affected by different source terms. Hence, the independent energy budget equation for each component is pursued to remove the ambiguity in the Eq. (5.30).

First of all, the fluctuation velocity $\boldsymbol{u}'$ is decomposed into different parts corresponding to different components of $\boldsymbol{m}'$. From Eq. (5.25),

$$\boldsymbol{u}' = \frac{\boldsymbol{m}'}{\overline{\rho}} - \frac{\overline{\boldsymbol{u}}}{\overline{\rho}} \rho' = \frac{\boldsymbol{m}'}{\overline{\rho}} - \frac{\overline{\boldsymbol{u}}}{\overline{\rho}} \left(\frac{p'}{\gamma R \overline{T}} - \frac{(\gamma - 1)\overline{\rho} S'}{\gamma R} \right) \tag{5.31}$$

Based on Eq. (5.31), we define

$$\begin{cases} \boldsymbol{u'}_B = \frac{\boldsymbol{m'}_B}{\overline{\rho}} \\ \boldsymbol{u'}_A = \frac{\boldsymbol{m'}_A}{\overline{\rho}} - \frac{\overline{\boldsymbol{u}}}{\gamma\overline{p}}p' \\ \boldsymbol{u'}_T = \frac{\boldsymbol{m'}_T}{\overline{\rho}} + \frac{(\gamma-1)\overline{\boldsymbol{u}}}{\gamma R}S' \end{cases}, \boldsymbol{u'} = \boldsymbol{u'}_B + \boldsymbol{u'}_A + \boldsymbol{u'}_T \tag{5.32}$$

Taking the scalar production of Eq. (5.22) and the fluctuation momentum density $\boldsymbol{m'}_B$

$$\boldsymbol{m'}_B \cdot \frac{\partial \boldsymbol{u'}}{\partial t} + \boldsymbol{m'}_B \cdot \nabla H' = \boldsymbol{m'}_B \cdot \boldsymbol{\alpha'} \tag{5.33}$$

Equation (5.33) can be rewritten as

$$\frac{\partial}{\partial t}\left(\boldsymbol{m'}_B \cdot \boldsymbol{u'}\right) - \boldsymbol{u'} \cdot \frac{\partial \boldsymbol{m'}_B}{\partial t} + \nabla \cdot \left(\boldsymbol{m'}_B H'\right) = \boldsymbol{m'}_B \cdot \boldsymbol{\alpha'} \tag{5.34}$$

The second term on the left-hand side of Eq. (5.34)

$$\boldsymbol{u'} \cdot \frac{\partial \boldsymbol{m'}_B}{\partial t} = \left(\boldsymbol{u'}_B + \boldsymbol{u'}_A + \boldsymbol{u'}_T\right) \cdot \overline{\rho}\frac{\partial \boldsymbol{u'}_B}{\partial t} \tag{5.35}$$

Instituting Eq. (5.35) into Eq. (5.34), we have

$$\frac{\partial}{\partial t}\left(\boldsymbol{m'}_B \cdot \boldsymbol{u'} - \frac{1}{2}\overline{\rho}\boldsymbol{u'}^2_B\right) + \nabla \cdot \left(\boldsymbol{m'}_B H'\right) = \boldsymbol{m'}_B \cdot \boldsymbol{\alpha'} + \overline{\rho}\boldsymbol{u'}_A \cdot \frac{\partial \boldsymbol{u}_{B'}}{\partial t} + \overline{\rho}\boldsymbol{u'}_T \cdot \frac{\partial \boldsymbol{u'}_B}{\partial t} \tag{5.36}$$

Taking the time-average of Eq. (5.22), we obtain

$$\nabla \cdot \left(\overline{\boldsymbol{m'}_B H'}\right) = \overline{\boldsymbol{m'}_B \cdot \boldsymbol{\alpha'}} + \overline{\overline{\rho}\boldsymbol{u'}_A \cdot \frac{\partial \boldsymbol{u'}_B}{\partial t}} + \overline{\overline{\rho}\boldsymbol{u'}_T \cdot \frac{\partial \boldsymbol{u'}_B}{\partial t}} \tag{5.37}$$

The vortical energy flux $\overline{\boldsymbol{m'}_B H'}$ is governed by the vortical source $\overline{\boldsymbol{m'}_B \cdot \boldsymbol{\alpha'}}$ and two additional source terms. Additional source terms $\overline{\overline{\rho}\boldsymbol{u'}_A \cdot \frac{\partial \boldsymbol{u'}_B}{\partial t}}$ and $\overline{\overline{\rho}\boldsymbol{u'}_T \cdot \frac{\partial \boldsymbol{u'}_B}{\partial t}}$ don't present in Doak's energy budget Eq. (5.30).

(2) Acoustic energy budget equation

Taking the scalar production of Eq. (5.22) and the fluctuation momentum density $\boldsymbol{m'}_A$

$$\boldsymbol{m'}_A \cdot \frac{\partial \boldsymbol{u'}}{\partial t} + \boldsymbol{m'}_A \cdot \nabla H' = \boldsymbol{m'}_A \cdot \boldsymbol{\alpha'} \tag{5.38}$$

Rewriting Eq. (5.38) as

$$\frac{\partial}{\partial t}\left(\boldsymbol{m'}_A \cdot \boldsymbol{u'}\right) - \boldsymbol{u'} \cdot \frac{\partial \boldsymbol{m'}_A}{\partial t} + \nabla \cdot \left(\boldsymbol{m'}_A H'\right) - \left(\nabla \cdot \boldsymbol{m'}_A\right)H' = \boldsymbol{m'}_A \cdot \boldsymbol{\alpha'} \tag{5.39}$$

In the left-hand side of Eq. (5.39), we have

$$\boldsymbol{u}' \cdot \frac{\partial \boldsymbol{m}'_A}{\partial t} = \boldsymbol{u}' \cdot \left(\overline{\rho}\frac{\partial \boldsymbol{u}_{A'}}{\partial t} + \frac{\overline{\boldsymbol{u}}}{\overline{a}^2}\frac{\partial p'}{\partial t}\right) \tag{5.40}$$

$$\left(\nabla \cdot \boldsymbol{m}'_A\right)H' = -\frac{1}{\overline{a}^2}\frac{\partial p'}{\partial t}\left(h' + \overline{\boldsymbol{u}} \cdot \boldsymbol{u}'\right) \tag{5.41}$$

Hence, Eq. (5.39) is rewritten as

$$\frac{\partial}{\partial t}\left(\boldsymbol{m}'_A \cdot \boldsymbol{u}'\right) + \nabla \cdot \left(\boldsymbol{m}'_A H'\right) = \boldsymbol{m}'_A \cdot \boldsymbol{\alpha}' + \overline{\rho}\boldsymbol{u}'_B \cdot \frac{\partial \boldsymbol{u}'_A}{\partial t} + \overline{\rho}\boldsymbol{u}'_T \cdot \frac{\partial \boldsymbol{u}'_A}{\partial t} - \frac{h'}{\overline{a}^2}\frac{\partial p'}{\partial t} \tag{5.42}$$

The time-average of Eq. (5.42) is

$$\nabla \cdot \left(\overline{\boldsymbol{m}'_A H'}\right) = \overline{\boldsymbol{m}'_A \cdot \boldsymbol{\alpha}'} + \overline{\overline{\rho}\boldsymbol{u}'_B \cdot \frac{\partial \boldsymbol{u}'_A}{\partial t}} + \overline{\overline{\rho}\boldsymbol{u}'_T \cdot \frac{\partial \boldsymbol{u}'_A}{\partial t}} - \overline{\frac{h'}{\overline{a}^2}\frac{\partial p'}{\partial t}} \tag{5.43}$$

The fourth term on the right-hand side of Eq. (5.43) satisfies

$$\overline{\frac{h'}{\overline{a}^2}\frac{\partial p'}{\partial t}} = \overline{\frac{S'}{\gamma R}\frac{\partial p'}{\partial t}} = -\overline{\frac{p'}{\gamma R}\frac{\partial S'}{\partial t}} \tag{5.44}$$

Finally, we have

$$\nabla \cdot \left(\overline{\boldsymbol{m}'_A H'}\right) = \overline{\boldsymbol{m}'_A \cdot \boldsymbol{\alpha}'} + \overline{\overline{\rho}\boldsymbol{u}'_B \cdot \frac{\partial \boldsymbol{u}'_A}{\partial t}} + \overline{\overline{\rho}\boldsymbol{u}'_T \cdot \frac{\partial \boldsymbol{u}'_A}{\partial t}} + \overline{\frac{p'}{\gamma R}\frac{\partial S'}{\partial t}} \tag{5.45}$$

The acoustic energy flux $\overline{\boldsymbol{m}'_A H'}$ is governed by the acoustic source $\overline{\boldsymbol{m}'_A \cdot \boldsymbol{\alpha}'}$, $\overline{\overline{\rho}\boldsymbol{u}'_B \cdot \frac{\partial \boldsymbol{u}'_A}{\partial t}}$, $\overline{\overline{\rho}\boldsymbol{u}_{T'} \cdot \frac{\partial \boldsymbol{u}_{A'}}{\partial t}}$, and $\overline{\frac{p'}{\gamma R}\frac{\partial S'}{\partial t}}$ that is a part of the thermo-acoustic coupling term $\overline{\frac{p'}{R}\frac{\partial S'}{\partial t}}$ in Eq. (5.30). Source terms $\overline{\overline{\rho}\boldsymbol{u}'_B \cdot \frac{\partial \boldsymbol{u}'_A}{\partial t}}$ and $\overline{\overline{\rho}\boldsymbol{u}'_T \cdot \frac{\partial \boldsymbol{u}'_A}{\partial t}}$ also don't present in Doak's energy budget Eq. (5.30).

(3) Thermal energy budget equation

Taking the scalar production of Eq. (5.22) and the fluctuation momentum density $\boldsymbol{m}'_T$

$$\boldsymbol{m}'_T \cdot \frac{\partial \boldsymbol{u}'}{\partial t} + \boldsymbol{m}'_T \cdot \nabla H' = \boldsymbol{m}'_T \cdot \boldsymbol{\alpha}' \tag{5.46}$$

Rewriting Eq. (5.46) as

$$\frac{\partial}{\partial t}\left(\boldsymbol{m}'_T \cdot \boldsymbol{u}'\right) - \boldsymbol{u}' \cdot \frac{\partial \boldsymbol{m}_{T'}}{\partial t} + \nabla \cdot \left(\boldsymbol{m}'_T H'\right) - \left(\nabla \cdot \boldsymbol{m}'_T\right)H' = \boldsymbol{m}'_A \cdot \boldsymbol{\alpha}' \tag{5.47}$$

Similarly, we have

$$\boldsymbol{u}' \cdot \frac{\partial \boldsymbol{m}'_T}{\partial t} = \boldsymbol{u}' \cdot \left[\bar{\rho} \frac{\partial \boldsymbol{u}'_T}{\partial t} - \frac{(\gamma - 1)\bar{\rho}\bar{\boldsymbol{u}}}{\gamma R} \frac{\partial S'}{\partial t} \right] \tag{5.48}$$

$$(\nabla \cdot \boldsymbol{m}'_T) H' = \frac{(\gamma - 1)\bar{\rho}}{\gamma R} (h' + \bar{\boldsymbol{u}} \cdot \boldsymbol{u}') \tag{5.49}$$

Hence, Eq. (5.47) is rewritten as

$$\frac{\partial}{\partial t}(\boldsymbol{m}'_T \cdot \boldsymbol{u}') + \nabla \cdot (\boldsymbol{m}'_T H') = \boldsymbol{m}'_T \cdot \boldsymbol{\alpha}' + \bar{\rho}\boldsymbol{u}'_A \cdot \frac{\partial \boldsymbol{u}'_T}{\partial t} + \bar{\rho}\boldsymbol{u}'_B \cdot \frac{\partial \boldsymbol{u}'_T}{\partial t} + \frac{(\gamma - 1)\bar{\rho}h'}{\gamma R} \frac{\partial S'}{\partial t} \tag{5.50}$$

The time-average of Eq. (5.50) is

$$\nabla \cdot \left(\overline{\boldsymbol{m}'_T H'}\right) = \overline{\boldsymbol{m}'_T \cdot \boldsymbol{\alpha}'} + \overline{\bar{\rho}\boldsymbol{u}'_A \cdot \frac{\partial \boldsymbol{u}'_T}{\partial t}} + \overline{\bar{\rho}\boldsymbol{u}'_B \cdot \frac{\partial \boldsymbol{u}'_T}{\partial t}} + \frac{(\gamma - 1)\bar{\rho}}{\gamma R} \overline{h' \frac{\partial S'}{\partial t}} \tag{5.51}$$

The fourth term on the right-hand side of Eq. (5.51) satisfies

$$\frac{(\gamma - 1)\bar{\rho}}{\gamma R} \overline{h' \frac{\partial S'}{\partial t}} = \overline{\frac{(\gamma - 1)p'}{\gamma R} \frac{\partial S'}{\partial t}} \tag{5.52}$$

Finally, we obtain

$$\nabla \cdot \left(\overline{\boldsymbol{m}'_T H'}\right) = \overline{\boldsymbol{m}'_T \cdot \boldsymbol{\alpha}'} + \overline{\bar{\rho}\boldsymbol{u}'_A \cdot \frac{\partial \boldsymbol{u}'_T}{\partial t}} + \overline{\bar{\rho}\boldsymbol{u}'_B \cdot \frac{\partial \boldsymbol{u}'_T}{\partial t}} + \overline{\frac{(\gamma - 1)p'}{\gamma R} \frac{\partial S'}{\partial t}} \tag{5.53}$$

The thermal energy flux $\overline{\boldsymbol{m}'_T H'}$ is governed by the thermal source $\overline{\boldsymbol{m}'_T \cdot \boldsymbol{\alpha}'}$, $\overline{\bar{\rho}\boldsymbol{u}'_A \cdot \frac{\partial \boldsymbol{u}'_T}{\partial t}}$, $\overline{\bar{\rho}\boldsymbol{u}'_B \cdot \frac{\partial \boldsymbol{u}'_T}{\partial t}}$, and $\overline{\frac{(\gamma-1)p'}{\gamma R} \frac{\partial S'}{\partial t}}$ that is the other part of the thermo-acoustic coupling term $\overline{\frac{p'}{R} \frac{\partial S'}{\partial t}}$ in Eq. (5.30). Source terms $\overline{\bar{\rho}\boldsymbol{u}'_A \cdot \frac{\partial \boldsymbol{u}'_T}{\partial t}}$ and $\overline{\bar{\rho}\boldsymbol{u}'_B \cdot \frac{\partial \boldsymbol{u}'_T}{\partial t}}$ also don't present in Doak's energy budget Eq. (5.30).

(4) Energy exchange terms

In three energy budget Eqs. (5.37), (5.45), and (5.53), which govern the energy fluxes carried by each MPT component independently, we find source terms that don't present in Doak's original MPT budget equation. Those new source terms satisfy

$$\overline{\bar{\rho}\boldsymbol{u}'_A \cdot \frac{\partial \boldsymbol{u}'_B}{\partial t}} + \overline{\bar{\rho}\boldsymbol{u}'_B \cdot \frac{\partial \boldsymbol{u}'_A}{\partial t}} = 0, \ \overline{\bar{\rho}\boldsymbol{u}'_A \cdot \frac{\partial \boldsymbol{u}'_T}{\partial t}} + \overline{\bar{\rho}\boldsymbol{u}'_T \cdot \frac{\partial \boldsymbol{u}'_A}{\partial t}} = 0,$$
$$\overline{\bar{\rho}\boldsymbol{u}'_B \cdot \frac{\partial \boldsymbol{u}'_T}{\partial t}} + \overline{\bar{\rho}\boldsymbol{u}'_T \cdot \frac{\partial \boldsymbol{u}'_B}{\partial t}} = 0 \tag{5.54}$$

Therefore, we conclude that the physical meaning of the source term $\overline{\overline{\rho}\boldsymbol{u}'_a \cdot \frac{\partial \boldsymbol{u}'_b}{\partial t}}$ is the energy transfer from a component to b component during the energy transport due to b component. This finding clarifies in which manner fluctuation energy exchanges between different components.

5.2.1.3 Summary

In this chapter, we first reviewed the MPT decomposition and the derivation of the energy budget equation in Doak's MPT approach. In the MPT decomposition, the fluctuation momentum density $\boldsymbol{m}' = (\rho \boldsymbol{u})'$ is spilled into the rotational and the irrotational part by the Helmholtz decomposition. The rotational part of $\boldsymbol{m}'$ is defined as the vortical component $\boldsymbol{m}'_B$ of disturbances. Furthermore, the irrotational part of $\boldsymbol{m}'$ is decomposed into the acoustic component $\boldsymbol{m}'_A$ and the thermal component $\boldsymbol{m}'_T$ based on the continuity equation and the thermodynamic state equation. The MPT decomposition provides a unique definition of the vortical, acoustic, and thermal components for a general time stationary flow, while the classic decomposition in Kovásznay's approach is only applicable for the uniform flow. Besides the MPT decomposition, Doak obtained a mean energy budget equation for the fluctuation energy flux $\overline{\boldsymbol{m}'H'}$ based on the continuity equation and the momentum conservation equation. The energy flux $\overline{\boldsymbol{m}'H'}$ and the source term $\overline{\boldsymbol{m}' \cdot \boldsymbol{\alpha}'}$ can be expressed as a linear superposition of energy fluxes and source terms due to each MPT component. Besides the source term caused by the interaction between the acceleration vector $\boldsymbol{\alpha}'$ and each MPT component, there exists another source term $\overline{\frac{p'}{R}\frac{\partial S'}{\partial t}}$ which is related to the fluctuation thermal diffusion process. Source terms due to each component and the thermo-acoustic coupling source term $\overline{\frac{p'}{R}\frac{\partial S'}{\partial t}}$ indicate the energy exchange between disturbances and mean flow during the fluctuation energy transport carried by $\boldsymbol{m}'$.

However, Doak's original budget equation has energy fluxes and source terms in one equation. The effect of source terms on each energy fluxes is confused. The ambiguity in Doak's original budget equation makes it difficult to provide a distinct physical interpretation for the development of disturbances in the flow field. To this end, we proposed three independent energy budget equations for each MPT component by introducing the decomposition of the fluctuation velocity $\boldsymbol{u}'$. In the energy budget equation for the energy flux carried by each MPT component, energy exchange terms $\overline{\overline{\rho}\boldsymbol{u}'_a \cdot \frac{\partial \boldsymbol{u}'_b}{\partial t}}$ from a component to b component are discovered for the first time. The effect of source terms on each energy flux is clear in these three independent energy budget equations. The vortical energy flux $\overline{\boldsymbol{m}'_B H'}$ is governed by the vortical source $\overline{\boldsymbol{m}'_B \cdot \boldsymbol{\alpha}'}$ and energy exchange terms, $\overline{\overline{\rho}\boldsymbol{u}'_A \cdot \frac{\partial \boldsymbol{u}'_B}{\partial t}}$ and $\overline{\overline{\rho}\boldsymbol{u}'_T \cdot \frac{\partial \boldsymbol{u}'_B}{\partial t}}$. The acoustic energy flux $\overline{\boldsymbol{m}_{A'} H'}$ is governed by the acoustic source $\overline{\boldsymbol{m}'_A \cdot \boldsymbol{\alpha}'}$, energy exchange terms $\overline{\overline{\rho}\boldsymbol{u}'_B \cdot \frac{\partial \boldsymbol{u}'_A}{\partial t}}$ and $\overline{\overline{\rho}\boldsymbol{u}'_T \cdot \frac{\partial \boldsymbol{u}'_A}{\partial t}}$, and the thermo-acoustic source $\overline{\frac{p'}{\gamma R}\frac{\partial S'}{\partial t}}$.

The thermal energy flux $\overline{\boldsymbol{m'}_T H'}$ is governed by the thermal source $\overline{\boldsymbol{m'}_T \cdot \boldsymbol{\alpha'}}$, energy exchange terms $\overline{\overline{\rho}\boldsymbol{u'}_A \cdot \frac{\partial \boldsymbol{u'}_T}{\partial t}}$ and $\overline{\overline{\rho}\boldsymbol{u'}_B \cdot \frac{\partial \boldsymbol{u'}_A}{\partial t}}$, and the thermo-acoustic source $\overline{\frac{(\gamma-1)p'}{\gamma R}\frac{\partial S'}{\partial t}}$.

5.2.2 *MPT Analysis of the Supersonic Mode*

5.2.2.1 Supersonic Mode

Bitter and Shepherd [9] studied Mack's second mode over a highly cooled wall. The wall-to-edge temperature ratio was $T_w/T_e < 1$, which corresponds to the state obtained in certain high-enthalpy wind tunnel experiments and real flight cases. They found that a special supersonic mode emerged at the edge of the BL and traveled outside it. Inside the BL over a highly cooled wall, another sonic line appeared, where $\overline{M}(y_{s2}) = \frac{\overline{u}(y_{s2})-c}{\overline{a}(y_{s2})} = 1$ above the first sonic line. The disturbance traveled subsonically between these sonic lines, but supersonically above the upper sonic line (along the edge of the BL) and below the lower sonic line. Notably, LST analysis [10–13] showed that the synchronization of the free-stream slow sound waves and Mack's second mode leads to the supersonic mode. The supersonic mode was oscillatory outside the BL. This supersonic mode broadens the unstable frequency band for disturbance in the flow field. Chuvakhov and Fedorov [14] called this supersonic mode the "spontaneous radiation of sound," which emphasizes the oscillatory disturbance of the supersonic mode outside the BL. Owing to the lower amplification rate of this supersonic mode compared with that of subsonic Mack's second mode, the supersonic mode is considered to be insignificant in the transition of the high-speed BL. However, Mortensen [15] recently reported that the rate of growth of the supersonic mode exceeds that of Mack's subsonic second mode in a Mach 20 flow over highly blunt cones. As Mack's second mode and supersonic mode both exhibits acoustic-wavelike behavior between the lower sonic line and the solid wall, it is reasonable that acoustic metasurface, which suppresses Mack's second mode by dissipating the acoustic wave energy, can also stabilize the supersonic mode because of its similar acoustic characteristics.

Note that the above description of the supersonic mode as a "spontaneous radiation of sound" has not been rigorously validated in realistic circumstances. The difficulty pertains to the lack of a universal definition of an "acoustic" or "sound" wave in a high-speed BL that exhibits a strong gradient in the mean flow field. Doak's MPT approach provides an elegant tool for decomposing a time-stationary fluctuation flow field into three components and overcomes the restriction of Kovásznauy's approach in inhomogeneous systems. Unnikrishnan and Gaitonde [16] performed MPT decomposition of instability waves in the Mach 6 flat plate boundary layer at different wall temperatures. They showed that oscillatory waves outside the highly cooled BL were mostly composed of vortical and acoustic fluctuations. The net outward acoustic flux of the supersonic modes increased with higher levels of wall cooling, as did the perturbation in the wall pressure. Their inspirational work revea67led that the MPT approach is promised in the analysis of supersonic modes. However, their energy consideration was based on Doak's original energy budget equation and thus cannot clarify the effect of sources terms on the "sound radiation" of supersonic modes. To this end, MPT decomposition and three independent energy budget equations derived above are applied to DNS data of supersonic modes with high-level wall cooling effect over smooth solid wall and acoustic metasurface to provide a physical interpretation of the sound radiation mechanism in the supersonic mode and the stabilization mechanism of acoustic metasurfaces to the supersonic mode.

5.2.2.2 Direct Numerical Simulation Results

Two numerical simulations are performed for Mach 6.0 BL flows on a solid wall and a acoustic metasurface with very low wall temperature. In both cases, the unit Reynolds number is 1×10^7 m^{-1}, and Mach number in the freestream M_∞ is 6.0. The freestream temperature and wall temperature are 300 K and 150 K, respectively. An instantaneous snapshot of p', $(\rho u)'$ and $(\rho v)'$ from the results of the DNS of the solid wall and the acoustic metasurface are shown in the left column and the right column of Fig. 5.1, respectively.

In the solid wall case, the fluctuations induced by the suction/blowing actuator are bifurcated into two regions, one aligning along the shock wave and the other

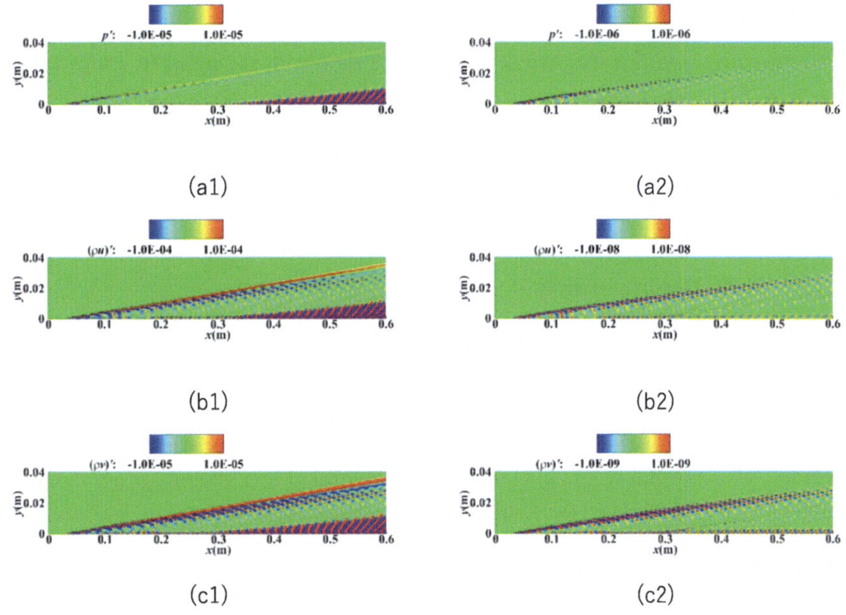

Fig. 5.1 Instantaneous snapshots of **a1**, **a2** p' (Pa), **b1**, **b2** $(\rho u)'$ (kg/m2/s), and **c1**, **c2** $(\rho v)'$ (kg/m2/s) in the case of the solid wall (left column) and the porous wall (right column)

traveling downstream within the boundary layer. An unstable supersonic mode occurs at $x \sim 0.34m$, and p', $(\rho u)\prime$, and $(\rho v)\prime$ are not evanescent above the BL. The so-called region of "spontaneous sound radiation" is enlarged as the perturbations propagated downstream, and it appears as though they were radiating outward at a fixed degree. In general, the peak values of fluctuations in the streamwise momentum (Fig. 5.1b1 and b2) are an order of magnitude higher than those of the wall-normal component (Fig. 5.1c1 and c2).

In contrast to the case of the solid wall, when the disturbances travel downstream through the acoustic metasurface ($x = 0.3 - 0.6$ m), there is no visible "spontaneous sound radiation" phenomenon (Fig. 5.1a2). The fluctuations within the BL are significantly suppressed, and this indicates that the acoustic metasurfaces could effectively suppress the amplification of Mack's second mode and the supersonic mode.

5.2.2.3 Linear Stability Theory Results

The spatial LST analysis is performed for the highly cooled wall case with the same flow condition and the same blow-suction frequency as in the DNS calculation. In this highly cooled wall case, Mack's second mode is the F mode instead of the S mode in the adiabatic wall case. The trajectory of the F mode in the $(\frac{c_{ph}}{u_\infty}, \sigma)$ plane are shown in Fig. 5.2. Three vertical dashed lines in Fig. 5.2 on the left (red), center (blue), and right (green) indicate the location $(1 - 1/M_\infty)$, 1, and $(1 + 1/M_\infty)$, which correspond to slow sound waves, vorticity/entropy waves, and fast sound waves in the freestream, respectively. Arrows in Fig. 5.2 represent the trend in $\frac{c_{ph}}{u_\infty}$ of the F mode as x increases. The F mode originates from fast sound waves is stable ($\sigma < 0$). Then, the F mode synchronizes with vorticity/entropy waves and becomes unstable ($\sigma > 0$) after cross vorticity/entropy waves. Finally, the F mode coalesces with slow sound waves and leads to the supersonic mode and a new stable mode.

5.2.2.4 MPT Decomposition

The different components are extracted by applying Doak's MPT approach to the instantaneous fluctuation field. The MPT decomposition is accomplished by solving two Poisson equations in Eq. (5.11). The Dirichlet boundary condition for momentum potential ($\varphi = 0$) is adopted at the boundaries of the 2D computation region. An instantaneous snapshot of the magnitudes of the vortical component ($\boldsymbol{m'}_B$), acoustic component ($\boldsymbol{m'}_A$), and thermal component ($\boldsymbol{m'}_T$) of the solid wall and the porous wall are depicted in the left column and the right column of Fig. 5.3, respectively.

In the solid wall case, the result captures vortical perturbations from small fluctuations induced at the leading shock wave by the blowing–suction actuator of the wall. The vortical and thermal perturbations have the largest and smallest magnitudes, respectively. Both $\boldsymbol{m'}_B$ and $\boldsymbol{m'}_A$ are present in the region of oscillation outside the BL, while the thermal component is not prominent in the "sound radiation."

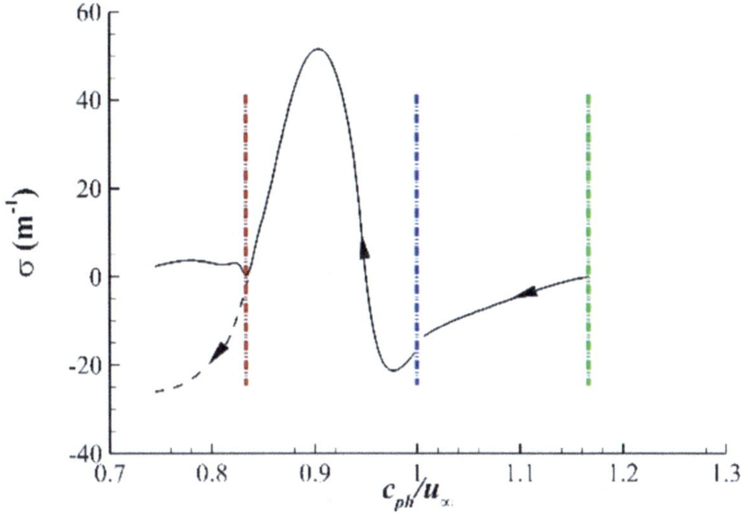

Fig. 5.2 The trajectory of the F mode in the ($\frac{c_{ph}}{u_\infty}$,σ) plane. The dashed line shows the new stable mode. Here, u_∞ is the free stream velocity

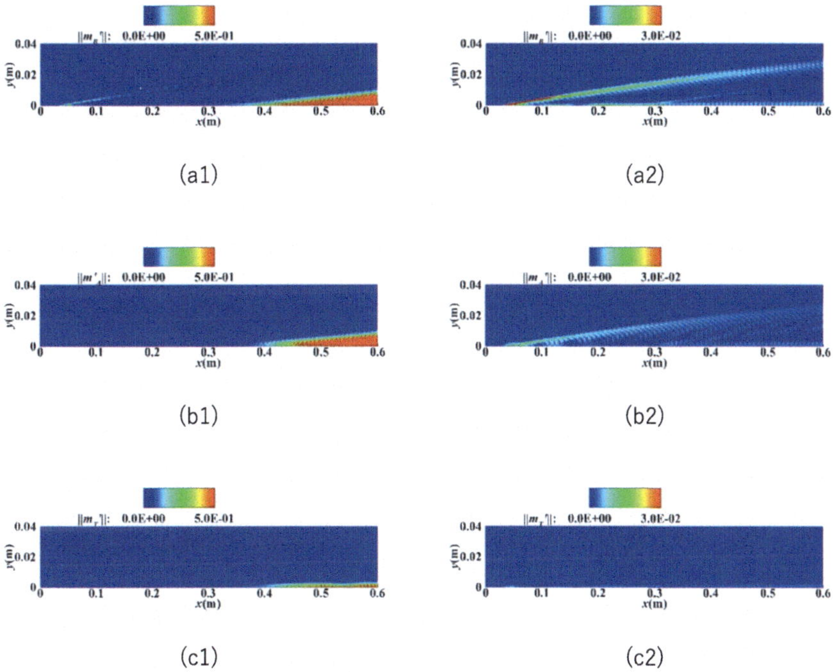

Fig. 5.3 Magnitudes of **a1**, **a2** m'_B, **b1**, **b2** m'_A, and **c1**, **c2** m'_T in the case of the solid wall (left column) and the porous wall (right column). (unit: kg/(m^2·s)

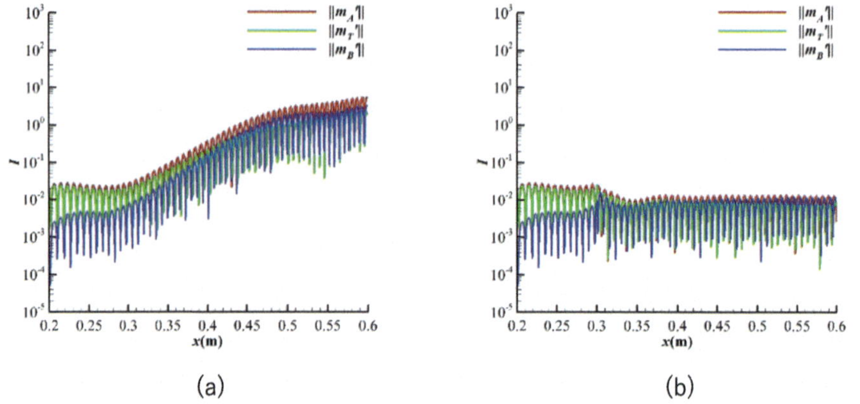

Fig. 5.4 Magnitudes of m'_B, m'_A, and m'_T at y = 0 m in the case of the solid wall (left column) and the porous wall (right column). (unit: kg/(m^2·s)

This observation indicates that the "sound radiation" is not sufficiently precise to describe the unstable supersonic mode. Both acoustic and vortical waves exist in the "radiation" region.

Figure 5.4 shows the distribution of $\boldsymbol{m'}_B$, $\boldsymbol{m'}_A$, and $\boldsymbol{m'}_T$ along the x axis at $y =$ 0 m in the solid wall case and the porous wall case. On the acoustic metasurface, the amplitude of each component is reduced by an order of magnitude at $x = 0.4$ m (Fig. 5.4a and b). The effect of the coating end (x = 0.3 m), associated with the juncture between the solid and porous parts, induces vortical and slight acoustic components along the expansion wave. The vortical component continues to be the dominant one in terms of exhibiting the highest amplitude among all FT components (Fig. 5.3a2). Moreover, the regions of "sound radiation" in the acoustic and vortical components disappears (Fig. 5.3b2 and c2).

5.2.2.5 Mean Energy Fluxes

In this section, the mean energy fluxes carried by different components of the MPT is discussed to provide insights into the development of perturbations in the BL. For convenience of discussion, the x gradient of the streamwise component of the mean energy flux $\frac{\partial}{\partial x}\left(\overline{m'_{Bx}H'}\right)$, $\frac{\partial}{\partial x}\left(\overline{m'_{Ax}H'}\right)$, and $\frac{\partial}{\partial x}\left(\overline{m'_{Tx}H'}\right)$ are denoted by $\boldsymbol{J}_{Bx}$, $\boldsymbol{J}_{Ax}$, and $\boldsymbol{J}_{Tx}$, respectively. Similarly, the y gradient of the normal component of the mean energy flux, $\frac{\partial}{\partial y}\left(\overline{m'_{By}H'}\right)$, $\frac{\partial}{\partial y}\left(\overline{m'_{Ay}H'}\right)$, and $\frac{\partial}{\partial y}\left(\overline{m'_{Ty}H'}\right)$ are denoted by $\boldsymbol{J}_{By}$, $\boldsymbol{J}_{Ay}$, and $\boldsymbol{J}_{Ty}$, respectively. The vortical source $-\overline{\boldsymbol{m'}_B \cdot \boldsymbol{\alpha'}}$, acoustic source $-\overline{\boldsymbol{m'}_A \cdot \boldsymbol{\alpha'}}$, and thermal source $-\overline{\boldsymbol{m'}_T \cdot \boldsymbol{\alpha'}}$ are designated as $\boldsymbol{P}_B$, $\boldsymbol{P}_A$, and $\boldsymbol{P}_T$, respectively. The sources due to thermo-acoustic coupling $-\frac{1}{\gamma R}\overline{S'\frac{\partial p'}{\partial t}}$ and $\frac{\gamma-1}{\gamma R}\overline{p'\frac{\partial S'}{\partial t}}$ in Eqs. (5.45), and (5.53) are respectively designated as $\boldsymbol{P}_{diff1}$ and $\boldsymbol{P}_{diff2}$. The energy exchange from component

a to b, $\overline{\bar{\rho}\boldsymbol{u}'_a \cdot \frac{\partial \boldsymbol{u}'_b}{\partial t}}$, is denoted by $\boldsymbol{Ex_{ab}}$. The sums of each of the right-hand-side source terms in Eqs. (5.37), (5.45), and (5.53) are $\boldsymbol{N_B}$, $\boldsymbol{N_A}$, and $\boldsymbol{N_T}$, respectively. Thus, the energy budgets of Eqs. (5.37), (5.45), and (5.53) can be expressed as

$$\boldsymbol{J}_{Bx} + \boldsymbol{J}_{By} = N_B = \boldsymbol{P_B} + \boldsymbol{Ex_{AB}} + \boldsymbol{Ex_{TB}} \tag{5.55}$$

$$\boldsymbol{J}_{Ax} + \boldsymbol{J}_{Ay} = N_A = \boldsymbol{P_A} + \boldsymbol{P_{diff1}} + \boldsymbol{Ex_{BA}} + \boldsymbol{Ex_{TA}} \tag{5.56}$$

$$\boldsymbol{J}_{Tx} + \boldsymbol{J}_{Ty} = N_T = \boldsymbol{P_T} + \boldsymbol{P_{diff2}} + \boldsymbol{Ex_{BT}} + \boldsymbol{Ex_{AT}} \tag{5.57}$$

$\boldsymbol{J_{Ax}}$, $\boldsymbol{J_{Bx}}$, $\boldsymbol{J_{Tx}}$, $\boldsymbol{J_{Ay}}$, $\boldsymbol{J_{By}}$, and $\boldsymbol{J_{Ty}}$ on the solid wall and the acoustic metasurface at $x = 0.48$ m are illustrated in Fig. 5.5. The magnitudes of $\boldsymbol{J_{Ax}}$, $\boldsymbol{J_{Bx}}$, and $\boldsymbol{J_{Tx}}$ are much smaller than those of $\boldsymbol{J_{Ay}}$, $\boldsymbol{J_{By}}$, and $\boldsymbol{J_{Ty}}$ in the BL. This indicates that the source terms on the right-hand side of Eqs. (5.55)–(5.57) influences the normal fluctuation energy transport in the first place, possibly because the y gradient of mean flow is much stronger than the x gradient in the BL flow field. In the case of the solid wall, $\boldsymbol{J_{Ay}}$ reaches its positive peak at around $y_c = 0.001$ m (the critical layer where $\overline{M}(y_c) = \frac{\bar{u}(y_c)-c}{\bar{a}(y_c)} = 0$) and then becomes negative above this peak. This indicates the acoustic source terms strengthen acoustic fluxes in the y-direction around the critical layer and then weaken them in the negative region. $\boldsymbol{J_{By}}$ has a negative peak around the critical layer while $\boldsymbol{J_{Ty}}$ is positive between two sonic lines. In the case of the porous wall, the acoustic flux $\boldsymbol{J_{Ay}}$ behaves similarly to that in the case of the solid wall. $\boldsymbol{J_{By}}$ first reaches a negative peak and then changes to a positive peak above the critical layer, and $\boldsymbol{J_{Ty}}$ shows a contrary trend to $\boldsymbol{J_{By}}$, but with a slightly smaller magnitude. In the detailed analysis of the source terms provided in the next section, the reversals in $\boldsymbol{J_{By}}$ and $\boldsymbol{J_{Ty}}$ are found to be related to the terms of energy exchange.

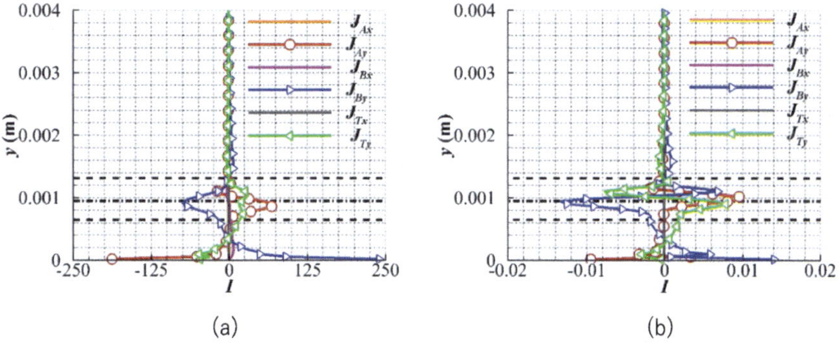

Fig. 5.5 Gradients of mean energy fluxes at $x = 0.48$ m in the cases of the solid wall (**a**) and the acoustic metasurface (**b**). The sonic line and critical layer are marked using a dashed line and dash-dotted line, respectively (unit: $kg/(m \cdot s^3)$)

Figure 5.6 shows the acoustic, vortical, and thermal flux lines in the range of $x =$ 0.45–0.5 m, which indicates the directions of the mean transport of the fluctuation energy according to its respective components in the flow field. The background shows contours of $\boldsymbol{N_A}$, $\boldsymbol{N_B}$, and $\boldsymbol{N_T}$. As shown in Fig. 5.6a1, acoustic fluxes originate around the critical layer owing to the acoustic source terms and diffused. Above this layer, the fluctuation energy is transported toward the bulk flow by $\boldsymbol{m_{A'}}$, and leading to the "spontaneous sound radiation." In Fig. 5.6b1, vortical source terms produce energy for the vortical fluxes in the vicinity of the surface of the plate. Then, vortical fluxes originating from the plate's surface sink near the critical layer. The fluctuation energy is transported from the bulk flow into the BL by $\boldsymbol{m_{B'}}$. Compared with $\boldsymbol{N_A}$ and $\boldsymbol{N_B}$, the intensity of the thermal source terms is so weak that the thermal component is not prominent in the region of "sound radiation" (Figs. 5.6c1 and 5.3c1). In the case of the acoustic metasurface (Fig. 5.6a2–c2), the sum of the source terms, i.e., $\boldsymbol{N_A}$, $\boldsymbol{N_B}$, and $\boldsymbol{N_T}$, is about four orders of magnitude smaller than that in the case of the solid wall. As shown in Fig. 5.6a2, $\boldsymbol{N_A}$ around the critical layer is so weak that the upward acoustic flux lines finally turn upward. The fluctuation energy is barely transported out of the BL by $\boldsymbol{m_{A'}}$. As a result, the "spontaneous radiation phenomenon" is absent in the case of the acoustic metasurface.

5.2.2.6 Source Mechanism

Figure 5.7 shows the contour of the acoustic source terms and Fig. 5.8 shows the distribution of the acoustic source terms along the y axis at $x = 0.48$ m. The left column represents the case of the solid wall and the right column that of the acoustic metasurface. In the former case, $\boldsymbol{P_A}$ is a primary energy producer near the critical layer and extracts fluctuation energy from the mean flow. Then, $\boldsymbol{P_A}$ dissipates this fluctuation energy during the upward transport of the fluctuation energy due to $\boldsymbol{m'_A}$ (Fig. 5.7a1). $\boldsymbol{P_{diff1}}$ always dissipates the fluctuation energy near the critical layer (Fig. 5.7b1). The fluctuation energy is transferred from the vortical part to the acoustic part near the critical layer (Fig. 5.7c1), while acoustic fluctuation energy is transferred into thermal energy Fig. 5.7d1). The amplitude of $\boldsymbol{Ex_{BA}}$ is larger than that of $\boldsymbol{Ex_{TA}}$. Therefore, $\boldsymbol{Ex_{BA}}$ compensates for the energy loss due to $\boldsymbol{Ex_{TA}}$ and became the other primary energy producer besides $\boldsymbol{P_A}$ near the critical layer. In this regard, the outward acoustic fluxes and "sound radiation" phenomenon are directly related to $\boldsymbol{P_A}$ and $\boldsymbol{Ex_{BA}}$. In the case of the porous wall, the energy exchange terms $\boldsymbol{Ex_{BA}}$ and $\boldsymbol{Ex_{TA}}$ have nearly the same magnitude (Fig. 5.7c2 and d2). Due to the balance between $\boldsymbol{Ex_{BA}}$ and $\boldsymbol{Ex_{TA}}$, $\boldsymbol{P_A}$ is the unique primary energy producer in this case (Fig. 5.7a2). However, $\boldsymbol{P_A}$ is significantly suppressed compared with that in the case of the solid wall, and this weakens the outward transport of the fluctuation energy by $\boldsymbol{m_{A'}}$.

Figure 5.9 shows the contour of vortical source terms and Fig. 5.10 shows the distribution of the vortical source terms along the y axis at $x = 0.48$ m. In the case of the solid wall, $\boldsymbol{P_B}$ is the energy producer in the vicinity of the surface of the plate (Fig. 5.9a1). Near the critical layer, vortical energy is transformed into acoustic energy (Fig. 5.9b1), which leads to a significant sink in the vortical fluxes (Fig. 5.6b1).

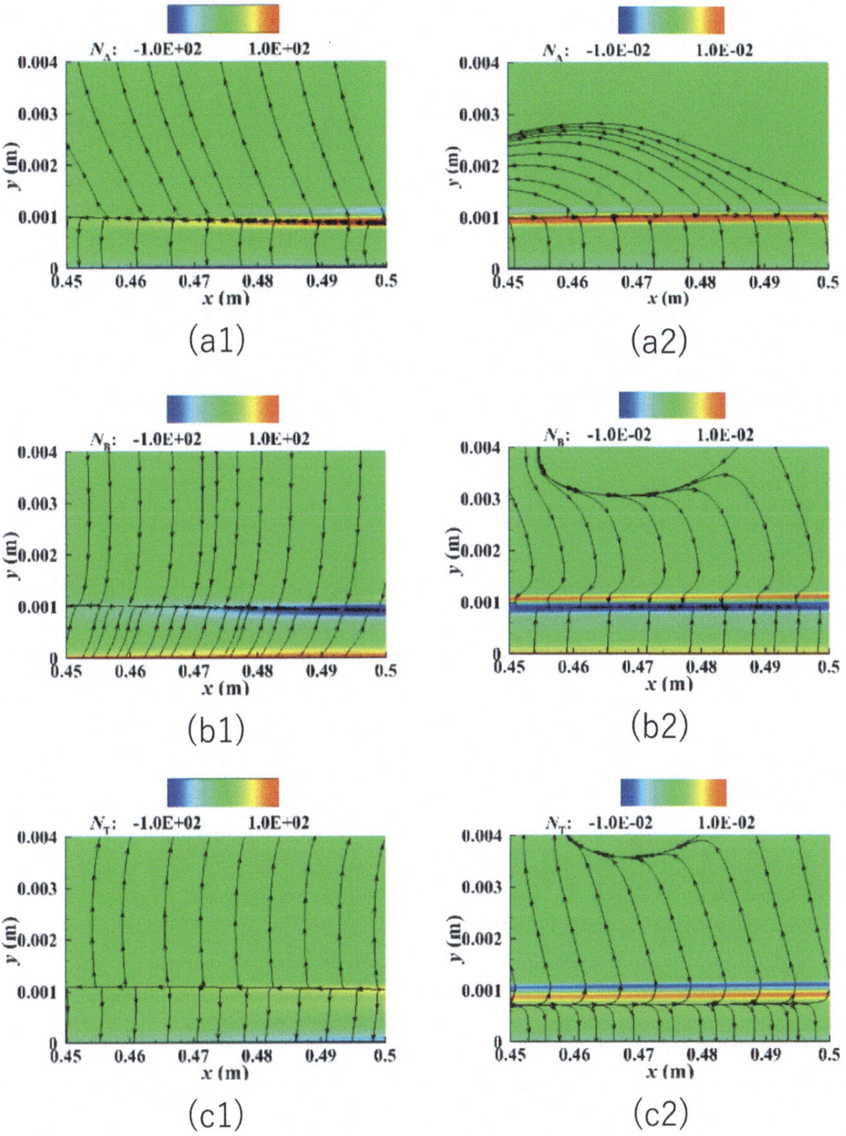

Fig. 5.6 **a1**, **a2** acoustic, **b1**, **b2** vortical, and **c1**, **c2** thermal fluxes in the cases of the solid wall (upper row) and the acoustic metasurface (lower row; unit: kg/(m·s^3))

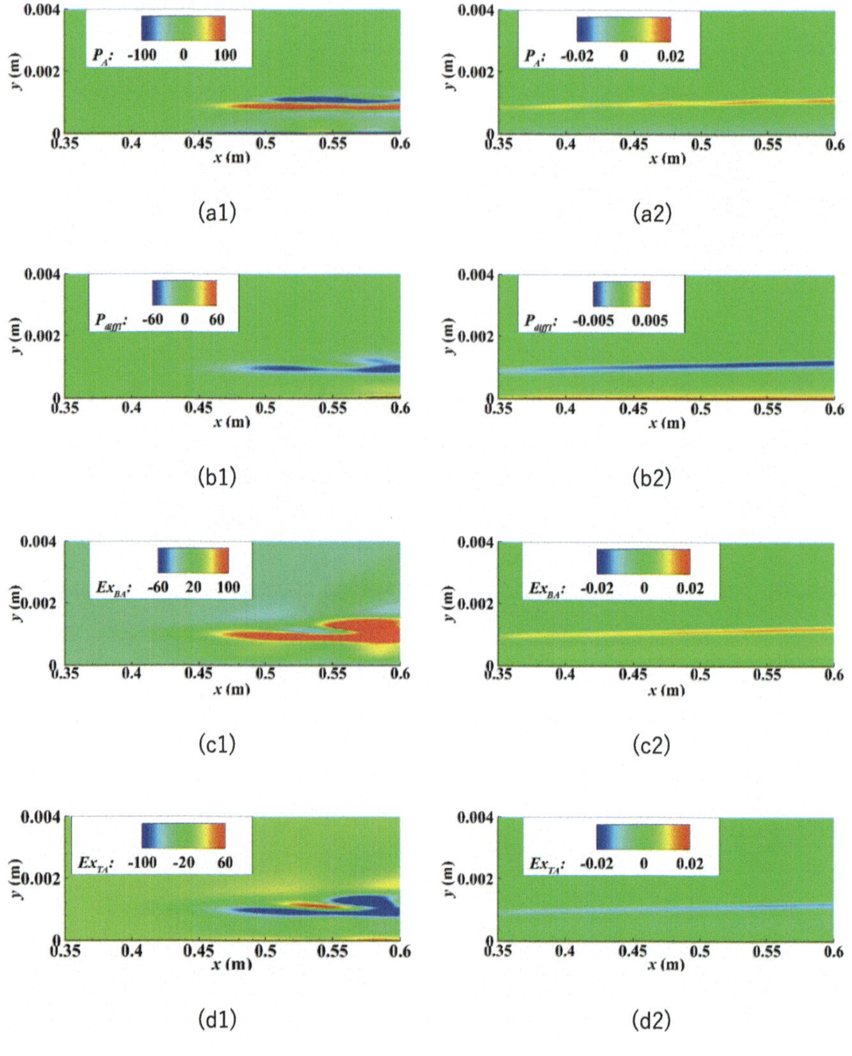

Fig. 5.7 Left column: acoustic source terms for the solid wall; right column: acoustic source terms for the acoustic metasurface (unit: kg/(m·s^3))

In the case of the porous wall, $\boldsymbol{P_B}$ is suppressed at the plate's surface and thus less vortical energy is transported to the critical layer. Furthermore, $\boldsymbol{Ex_{TB}}$ has a larger amplitude than $\boldsymbol{Ex_{AB}}$, which is in contrast to the case of the solid wall, and leads to reversals in $\boldsymbol{J_{By}}$ and $\boldsymbol{J_{Ty}}$ (Fig. 5.5b).

Figure 5.11 shows the contour of the thermal source terms and Fig. 5.12 shows the distribution of the thermal source terms along the y axis at $x = 0.48$ m. In the case of the solid wall, $\boldsymbol{Ex_{AT}}$ is the primary energy producer of thermal fluxes, which is

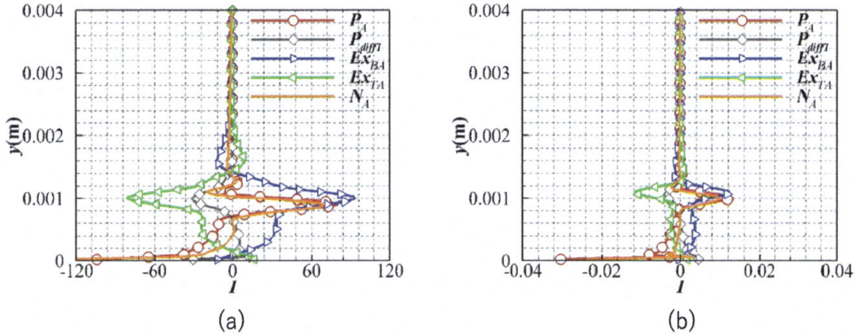

Fig. 5.8 Acoustic source terms at x = 0.48 m in the cases of the solid wall (**a**) and the acoustic metasurface (**b**) (unit: kg/(m·s^3))

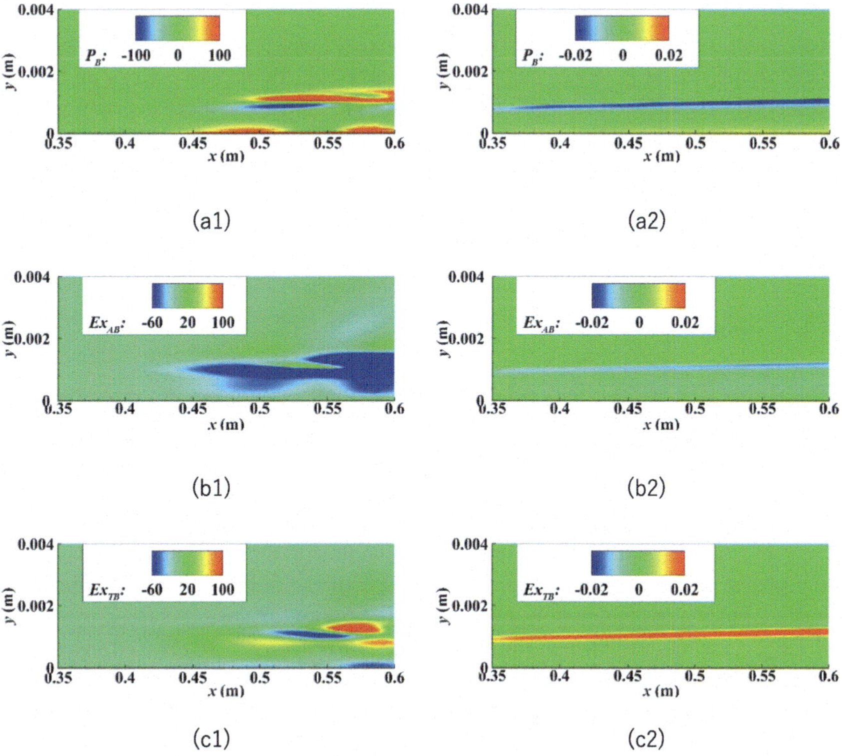

Fig. 5.9 Left column: vortical source terms for the solid wall; right column: vortical source terms for the acoustic metasurface (unit: kg/(m·s^3))

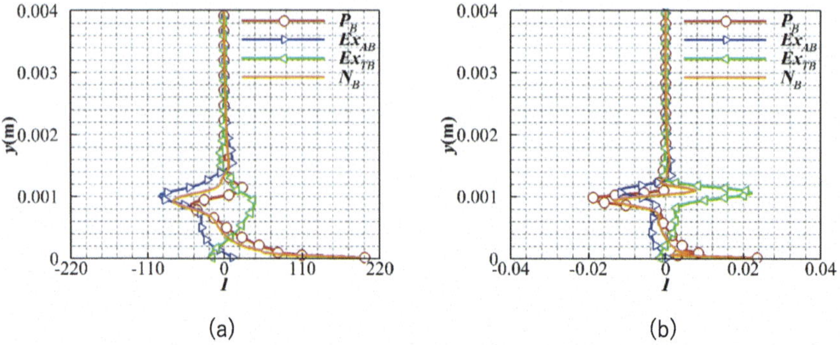

Fig. 5.10 Vortical source terms at $x = 0.48$ m in the cases of the solid wall (**a**) and the acoustic metasurface (**b**) (unit: kg/(m·s^3))

transformed from acoustic energy near the critical layer (Fig. 5.11c1). Nearly all of the energy produced by $\boldsymbol{Ex_{AT}}$ and $\boldsymbol{P_T}$ is consumed by the negative $\boldsymbol{Ex_{BT}}$ and $\boldsymbol{P_{diff2}}$, leading to a marginal value of $\boldsymbol{N_T}$ (Fig. 5.6c1). Therefore, the thermal component is not prominent in the region of "sound radiation." In the case of the porous wall, $\boldsymbol{P_T}$ is the primary energy producer, and is approximately balanced by the negative $\boldsymbol{Ex_{BT}}$ (Fig. 5.11d2). The magnitudes of the source terms on the acoustic metasurface are much smaller than those over the solid wall.

5.2.2.7 Summary

In this subsection, the vortical, acoustic, and thermal components are extracted from the results of DNS for hypersonic BLs on a solid wall and a acoustic metasurface by using Doak's MPT decomposition. Three independent energy budget equations for each component are adopted to clarify the contributions of the source terms to each energy flux term. The result shows that the normal transport of the fluctuation energy by the acoustic component is responsible for the mechanisms of "sound radiation" in the supersonic mode. In the case of the solid wall, the vortical and acoustic components coexist in the region of radiation of the supersonic mode. $\boldsymbol{P_A}$ and $\boldsymbol{Ex_{BA}}$ near the critical layer produce energy for acoustic fluxes to transport the fluctuation energy to the bulk flow, which is responsible for the "sound radiation." $\boldsymbol{P_B}$ produced energy for vortical fluxes in the vicinity of the plate surface. However, vortical fluxes sink near the critical layer and fluctuation energy is transformed from the vortical component to the acoustic component. $\boldsymbol{Ex_{AT}}$ and $\boldsymbol{P_T}$ produce energy for thermal fluxes near the critical layer and then fluctuation energy is consumed by $\boldsymbol{Ex_{BT}}$ and $\boldsymbol{P_{diff2}}$. In the case of the acoustic metasurface, $\boldsymbol{P_B}$ in the vicinity of the plate's surface is suppressed, and less vortical energy is transported to the critical layer. Therefore, less vortical energy is transformed into acoustic energy and $\boldsymbol{P_A}$ becomes the unique energy producer for acoustic fluxes. The energy produced by $\boldsymbol{P_A}$

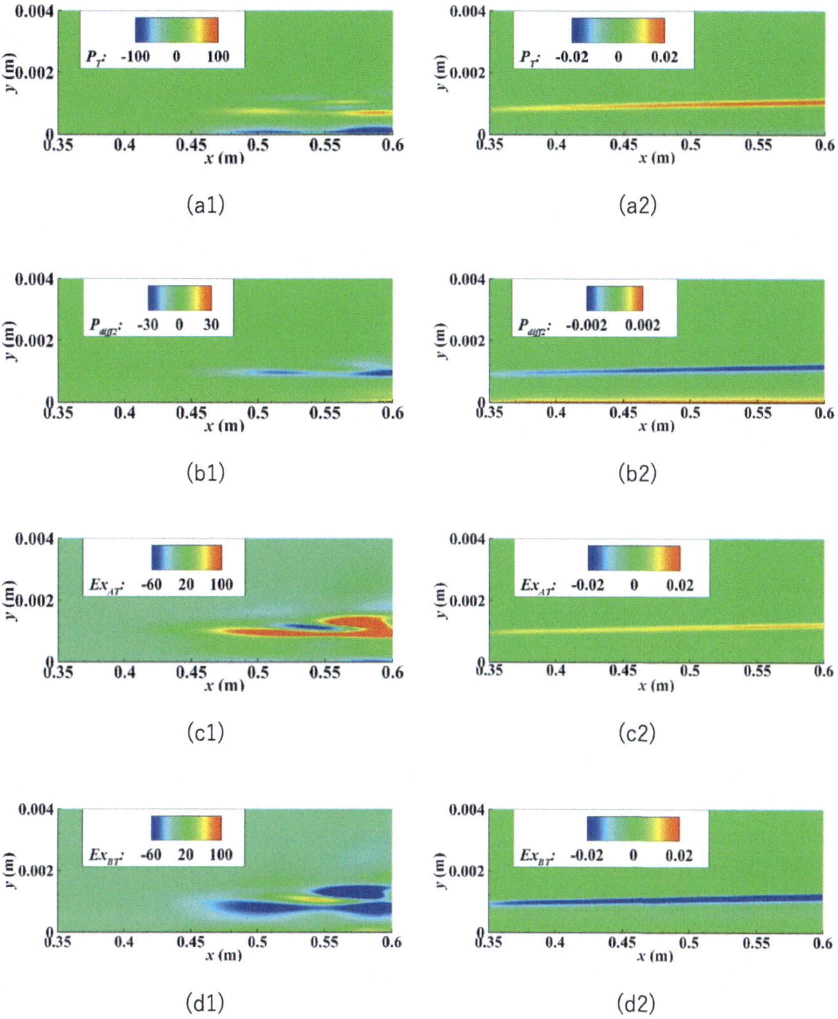

Fig. 5.11 Left column: thermal source terms for the solid wall; right column: thermal source terms for the acoustic metasurface (unit: kg/(m·s^3))

cannot compensate for energy loss during the outward transport of the fluctuation energy. Eventually, there is no "spontaneous radiation phenomenon" on the acoustic metasurface.

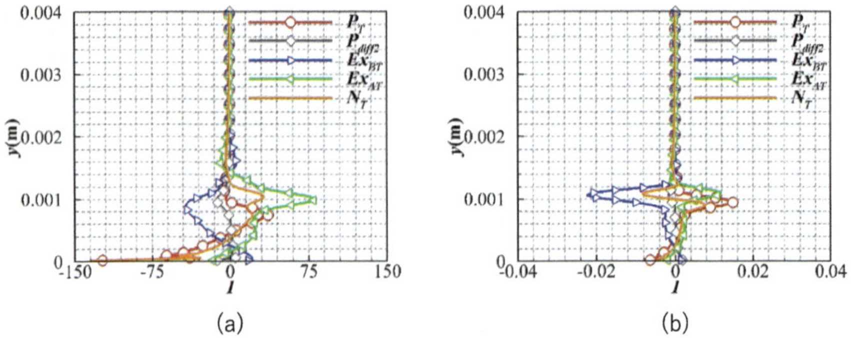

Fig. 5.12 Thermal source terms at $x = 0.48$ m in the cases of the solid wall (**a**) and the acoustic metasurface (**b**) (unit: kg/(m·s^3))

5.2.3 MPT Analysis of Modal Growth Rate

5.2.3.1 Motivation

Considering energy fluxes and source terms, the MPT approach provides a convincing physical interpretation of the sound radiation mechanism of the supersonic mode. Another important issue is the study of the high-speed boundary layer instability is the physical mechanism of the modal growth of Mack's second mode. Numerous existed LST analyses indicated that the synchronization of the S mode and the F mode causes unstable Mack's second mode. The LST results are intriguing. However, the origin of the fluctuation energy that drives the modal growth of Mack's second mode is still mysterious. Therefore, the MPT approach is applied to the LST result of Mack's second mode for the physical explanation of the modal growth of Mack's second mode. The energy analysis in the framework of the MPT approach provides a supplementary to the eigen analysis in the LST.

5.2.3.2 Growth Rate in MPT

The growth rate analysis was performed on the second mode obtained from 2-D spatial LST because the most unstable second mode is two-dimension.

In the 2D spatial LST, any two fluctuation quantities (real value) can be expressed as follows:

$$\begin{cases} f' = \hat{f} e^{-\alpha_i x + i(\alpha_r x - \omega t)} + \hat{f}^{+} e^{-\alpha_i x - i(\alpha_r x - \omega t)} \\ g' = \hat{g} e^{-\alpha_i x + i(\alpha_r x - \omega t)} + \hat{g}^{+} e^{-\alpha_i x - i(\alpha_r x - \omega t)} \end{cases} \tag{5.58}$$

where $\hat{f}$ and $\hat{g}$ are the complex-valued eigenfunctions of y in the wall normal direction, the superscript $^{+}$ indicates the corresponding conjugate. The product of these two fluctuation quantities is as follows:

$$f'g' = \hat{f}\hat{g}e^{-2\alpha_i x + i2(\alpha_r x - \omega t)} + \hat{f}\hat{g}^{+}e^{-2\alpha_i x} + \hat{f}^{+}\hat{g}e^{-2\alpha_i x} + \hat{f}^{+}\hat{g}^{+}e^{-2\alpha_i x - i2(\alpha_r x - \omega t)} \tag{5.59}$$

Due to

$$\int_0^{\frac{2\pi}{\omega}} \hat{f}\hat{g}e^{-2\alpha_i x + i2(\alpha_r x - \omega t)}dt = 0, \quad \int_0^{\frac{2\pi}{\omega}} \hat{f}^{+}\hat{g}^{+}e^{-2\alpha_i x - i2(\alpha_r x - \omega t)}dt = 0 \tag{5.60}$$

The time average of Eq. (5.60) is as follows:

$$\overline{f'g'} = \frac{\omega}{2\pi}\int_0^{\frac{2\pi}{\omega}} f'g'dt = \left(\hat{f}\hat{g}^{+} + \hat{f}^{+}\hat{g}\right)e^{-2\alpha_i x} \tag{5.61}$$

Take partial differential of Eq. (5.61) with x yields the following:

$$\frac{\partial\overline{f'g'}}{\partial x} = -2\alpha_i\left(\hat{f}\hat{g}^{+} + \hat{f}^{+}\hat{g}\right)e^{-2\alpha_i x} = -2\alpha_i\overline{f'g'} \tag{5.62}$$

Thus, we have

$$\frac{\partial\overline{(\rho u)'H'}}{\partial x} = -2\alpha_i\overline{(\rho u)'H'} = 2\sigma\overline{(\rho u)'H'} \tag{5.63}$$

Considering instability modes in a two-dimensional (2D) spatial LST, Eq. (5.30) can be rewritten as follows:

$$2\sigma\overline{(\rho u)'H'} + \frac{\partial\overline{(\rho v)'H'}}{\partial y} = P_B + P_A + P_T + P_{ta} \tag{5.64}$$

where u is the streamwise velocity, v is the surface normal velocity, and $\sigma = -\alpha_i$ is the growth rate in the 2D spatial LST. α_i is the imaginary part of the complex wavenumber α in the 2D spatial LST. There are two terms, $2\sigma\overline{(\rho u)'H'}$ and $\frac{\partial\overline{(\rho v)'H'}}{\partial y}$ on the left-hand side of Eq. (5.64). Consequently, it is difficult to clarify the effect of different source terms on the growth rate σ. To eliminate the y derivative term, Eq. (5.64) is integrated from $y = 0$ to $y \to +\infty$ with the following boundary condition:

$$\begin{cases} y = 0 : (\rho v)' = 0 \\ y \to +\infty : (\rho v)' = 0 \end{cases} \Rightarrow \int_0^\infty \frac{\partial \overline{(\rho v)' H'}}{\partial y} dy = \overline{(\rho v)' H'}\Big|_{y\to\infty} - \overline{(\rho v)' H'}\Big|_{y=0} = 0 \tag{5.65}$$

Finally, we have,

$$2\sigma \int_0^{+\infty} \overline{(\rho u)' H'} dy = \int_0^{+\infty} P_B dy + \int_0^{+\infty} P_A dy + \int_0^{+\infty} P_T dy + \int_0^{+\infty} P_{ta} dy \tag{5.66}$$

Furthermore, Eq. (5.66) is rewritten as follows:

$$\sigma = \frac{\int_0^{+\infty} P_B dy + \int_0^{+\infty} P_A dy + \int_0^{+\infty} P_T dy + \int_0^{+\infty} P_{ta} dy}{2Q_x} = \sigma_B + \sigma_A + \sigma_T + \sigma_{ta} \tag{5.67}$$

Here, is denoted as $\int_0^{+\infty} \overline{(\rho u)' H'} dy = Q_x$. The contribution of each source term to the growth rate σ is defined as follows:

$$\sigma_B = \frac{\int_0^{+\infty} P_B dy}{2Q_x}, \sigma_A = \frac{\int_0^{+\infty} P_A dy}{2Q_x}, \sigma_T = \frac{\int_0^{+\infty} P_T dy}{2Q_x}, \sigma_{ta} = \frac{\int_0^{+\infty} P_{ta} dy}{2Q_x} \tag{5.68}$$

5.2.3.3 Growth Rate Analysis of the Mode S

Two-dimensional spatial LST analyses are performed to calculate the eigenmode for the canonical $M = 6.0$ flat plate boundary layer case, and then MPT components and source terms are extracted from the LST mode and analyzed in detail. Firstly, the $M = 6.0$ flat plate boundary layer over the adiabatic wall with $Re = 1000$, 5000, and 9000 are studied to reveal the viscous effect on the modal growth of the second mode. The free stream temperature $T_\infty^* = 300$ K. In this case, the recovery temperature is $T_r \approx 7.02\ T_\infty^*$. Afterward, LST analyses of three cases with wall temperatures $T_w = 0.7\ T_r$, $1.0\ T_r$ (adiabatic wall), and $1.3\ T_r$ at Re = 5000 are performed to take the wall temperature effect into consideration. Self-similar solutions of the steady state of the Mach 6.0 flat plate boundary layer with $T_w = 0.7\ T_r$, $1.0\ T_r$, and $1.3\ T_r$ are plotted in Fig. 5.13. Here, Y is the non-dimensional wall normal coordinate by the local boundary layer thickness L_{ref}. It is evident that wall temperature significantly affects the boundary layer thickness. The boundary layer is thicker with higher wall temperature.

Figure 5.14 shows 2-D spatial results of the Mach 6.0 flat plate boundary layer over the adiabatic wall with Re = 1000, 5000, and 9000. Nondimensional phase speed

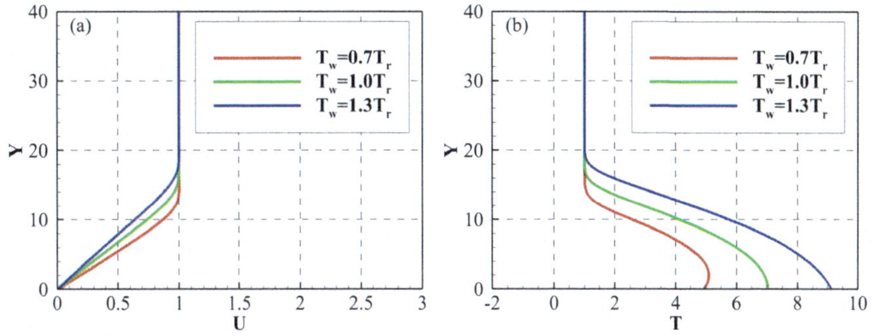

Fig. 5.13 Distributions of **a** streamwise velocity and **b** temperature in the Mach 6.0 flat plate boundary layer with different wall temperatures

$C_{ph} = \frac{\omega}{\alpha_r}$ of two discrete modes are plotted in Fig. 5.14a. Three horizontal dashed doted lines indicate the phase speed of fast acoustic waves (upper, $C_{ph} = 1 + \frac{1}{M}$), entropy/vorticity waves (middle, $C_{ph} = 1$), and slow acoustic waves (lower, $C_{ph} = 1 - \frac{1}{M}$) in the free stream. Two discrete modes exist in the 2D spatial LST results. The fast mode (mode F) has the same phase speed as fast acoustic waves at the leading edge (non-dimensional circular frequency $\omega = \frac{\omega^*}{u^*_\infty}\sqrt{\frac{x^*\mu^*_\infty}{\rho^*_\infty u^*_\infty}} = 0$ in Fig. 5.14), while the slow mode (mode S) has the same phase speed as slow acoustic waves here[17]. The mode F synchronizes with the Mode S at $\omega \approx 0.156$, 0.157, and 0.158 with Re = 1000, 5000, and 9000, respectively. Non-dimensional growth rate σ of the modes F and S are plotted Fig. 5.14b. The mode S is unstable and has the largest growth rate at $\omega \approx 0.159$, 0.166, and 0.166 with Re = 1000, 5000, and 9000, respectively. The mode S evolves into the second mode in the vicinity of the synchronization point [17], where the mode S has the same phase speed as the Mode *F*, in the considered case. Downstream of the synchronization point, the mode S is more unstable with a larger *Re*.

Figure 5.15 shows 2-D spatial results of the Mach 6.0 flat plate boundary layer with $T_w = 0.7\ T_r$, 1.0 T_r, and 1.3 T_r at Re = 5000. The mode F synchronizes with the mode S at $\omega \approx 0.157$ with different T_w. The cold wall destabilizes the second mode (the mode *S* around the synchronization point) significantly, while the hot wall stabilizes the second mode. The most unstable second mode locates at $\omega \approx 0.171$, 0.166, and 0.162, respectively.

Because the second mode in present cases is related to the unstable mode *S* where the synchronization occurs, the mode S obtained from LST is analyzed by MPT in the following sections. For convenience of discussion, the notation $q_x = \overline{(\rho u)' H'}$ is adopted. It is noted that $Q_x = \int_0^\infty q_x dy$ appears in Eqs. (5.67) and (5.68) as denominators, which makes an essential difference in the contribution of MPT sources to the amplification of the second mode. Figure 5.16 shows $\frac{q_x}{|Q_x|}$ of the mode S at $\omega = 0.1$(lower frequency than the second mode), 0.15(in the region of the second mode), and 0.2(higher frequency than the second mode) with different wall

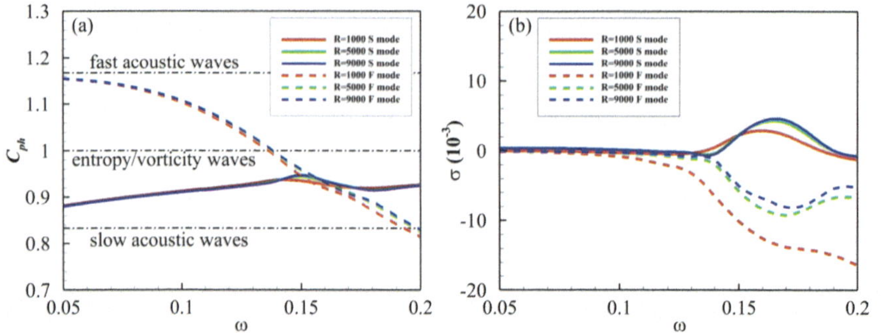

Fig. 5.14 **a** Phase speed $C_{ph} = \frac{\omega}{\alpha_r}$ and **b** growth rate $\sigma = -\alpha_i$ of the mode S and the mode F in the Mach 6.0 boundary layer over the adiabatic wall with different Re. ω is the non-dimensional circular frequency

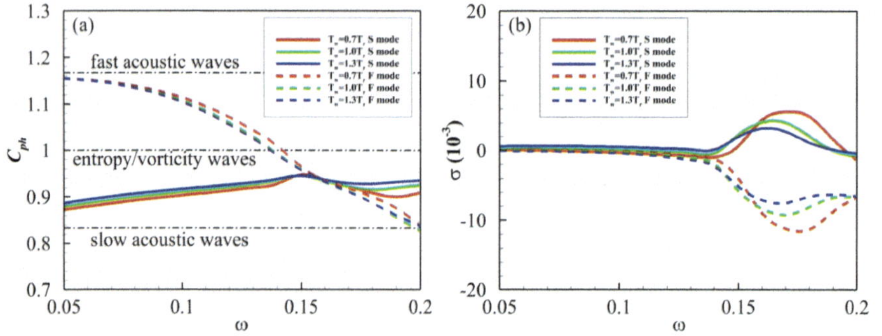

Fig. 5.15 **a** Phase speed $C_{ph} = \frac{\omega}{\alpha_r}$ and **b** growth rate $\sigma = -\alpha_i$ of the mode S and the mode F in the Mach 6.0 boundary layer with wall temperature $T_w = 0.7\ T_r$, $1.0\ T_r$, and $1.3\ T_r$ at Re = 5000

temperature and$R$. In all cases, the magnitude of $\frac{q_x}{|Q_x|}$ is relatively small in most regions of the boundary layer near the wall, and becomes negative with a large magnitude near the boundary layer edge. Consequently, Q_x is negative for the mode S. Based on Eq. (5.67), the results indicate that negative MPT sources in conjunction with a negative Q_x eventually destabilize the mode S.

Figure 5.17 shows the distribution of MPT sources of the mode S normalized by $2Q_x$. There exists a negative correlation between $\frac{P_B}{2Q_x}$ and $\frac{P_T}{2Q_x}$. This observation is consistent with the results of Unnikrishnan and Gaitonde [7], and they attribute this phenomenon to the vortical-entropy coupling effect. MPT sources are significant near the wall and around the critical layer ($u = C_{ph}$). Here, the critical layer is the location where singularity occurs in the inviscid stability equations and the solution is still large in the complete viscous theory. Another important observation is that $\frac{P_{ta}}{2Q_x}$ is positive everywhere in all cases, which indicates that the thermal-acoustic source P_{ta} always destabilizes the mode S.

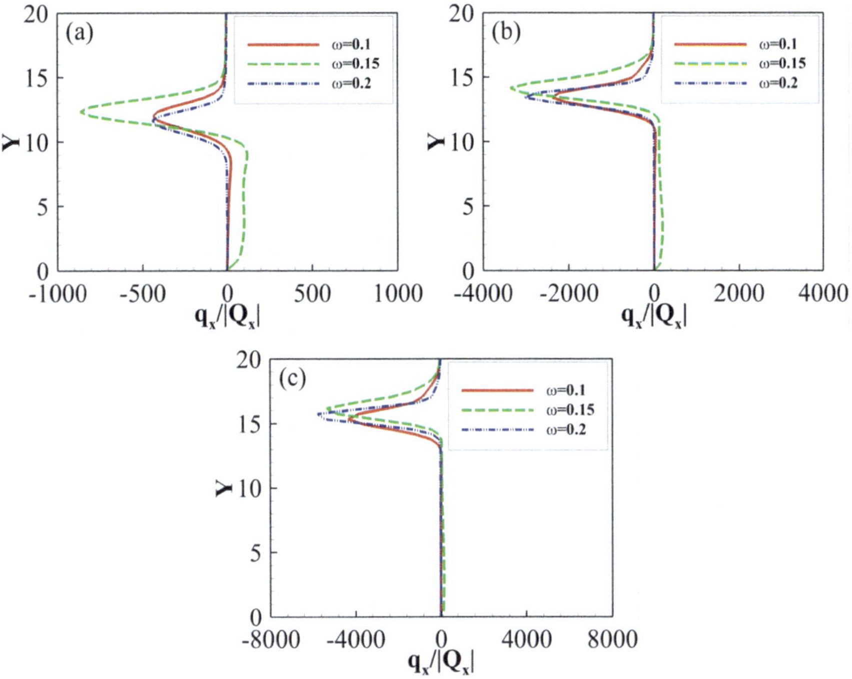

Fig. 5.16 $\frac{q_x}{|Q_x|}$ of the mode S at $\omega = 0.1, 0.15$, and 0.2 with **a** Re = 1000 and $T_w = 0.7\ T_r$; **b** Re = 5000 and $T_w = 1.0\ T_r$; **c** Re = 9000 and $T_w = 1.3\ T_r$

The contribution of MPT sources to the growth rate σ of the mode S over the adiabatic wall with different Re are plotted in Fig. 5.18. The vertical dashed-dotted line indicates the synchronization point (SP) of the mode F and mode S. In all cases with different Re, σ_{ta} and σ_B are positive, while σ_A and σ_T are negative. The thermal-acoustic source P_{ta} and the vortical source P_B destabilize the mode S while the acoustic source P_A and the thermal source P_T stabilize the mode S. For all three cases with different Re, similar tendencies are shown which may indicate a universal law in this state. More importantly, the term σ_{ta}, yielded by the thermal-acoustic source, is the most pronounced positive contribution among all in the vicinity of the SP. It has been shown that the synchronization between the fast and slow modes signify the unstable second mode. Consequently, Fig. 5.18 provides direct evidence that the thermal-acoustic source is highly important in the evolvement of the second mode instability. The contribution of MPT sources to the growth rate σ of the mode S at Re = 5000 with different wall temperatures are also plotted in Fig. 5.19. The results are similar to those in Fig. 5.18. A notable difference is that the magnitude of σ_A, σ_B, σ_T and σ_{ta} is larger with lower wall temperature. Meanwhile, the role of the peak σ_{ta} is not affected by the wall cooling effect, which makes the conclusion more general and convincing.

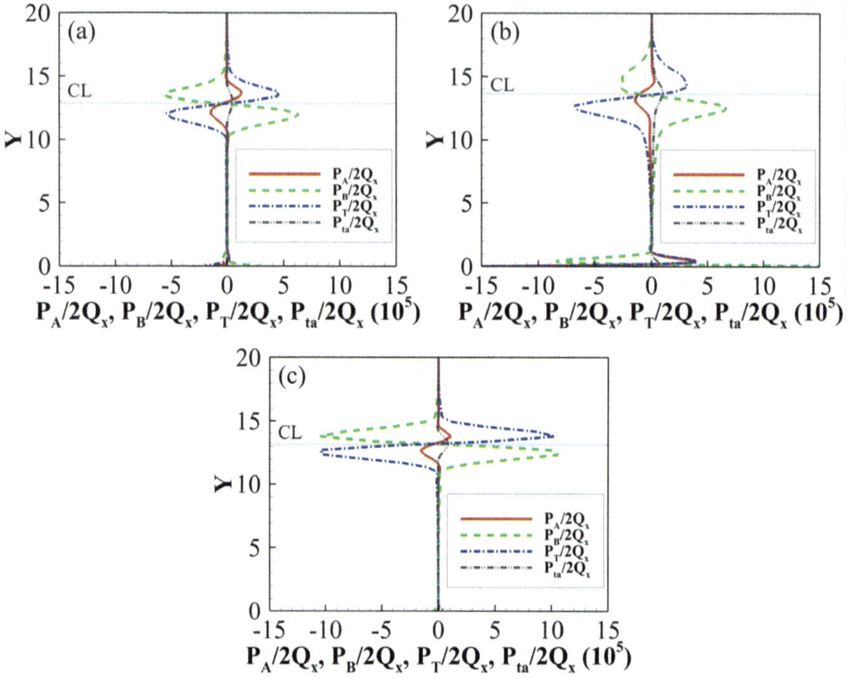

Fig. 5.17 Distribution of MPT sources of the mode S normalized by $2Q_x$ at Re = 5000 and $T_w = 1.0\,T_r$ with **a** $\omega = 0.1$; **b** $\omega = 0.15$; **c** $\omega = 0.2$. The location of the critical layer (CL) is marked with a horizontal dotted line

The negative correlation between σ_B and σ_T is attributed to the negative correlation between vortical source P_B and thermal source P_T due to the vortical-entropy coupling effect (see Fig. 5.17). The acoustic source P_A is not correlated to the thermal-acoustic source P_{ta} in Fig. 5.17, while their contribution to the growth rate σ_A and σ_{ta} is negatively correlated in Figs. 5.18 and 5.19. In author's previous MPT work on the supersonic mode, the independent energy budget equation for each MPT component is developed in the following form:

$$\begin{cases} \nabla \cdot \overline{H'\boldsymbol{m}'_B} = P_B + Ex_{AB} + Ex_{TB} \\ \nabla \cdot \overline{H'\boldsymbol{m}'_A} = P_A + Ex_{BA} + Ex_{TA} + \dfrac{P_{ta}}{\gamma} \\ \nabla \cdot \overline{H'\boldsymbol{m}'_T} = P_T + Ex_{BT} + Ex_{AT} + \dfrac{(\gamma - 1)P_{ta}}{\gamma} \end{cases} \tag{5.69}$$

On the right-hand side of Eqs. (5.69), terms in the form Ex_{ab} are energy exchange terms from a component and b component to the acoustic component ('A', 'B', and 'T' represents the acoustic, vortical, and thermal component, respectively). The thermal-acoustic source P_{ta} is split into two parts $\frac{P_{ta}}{\gamma}$ and $\frac{(\gamma-1)P_{ta}}{\gamma}$, which affect

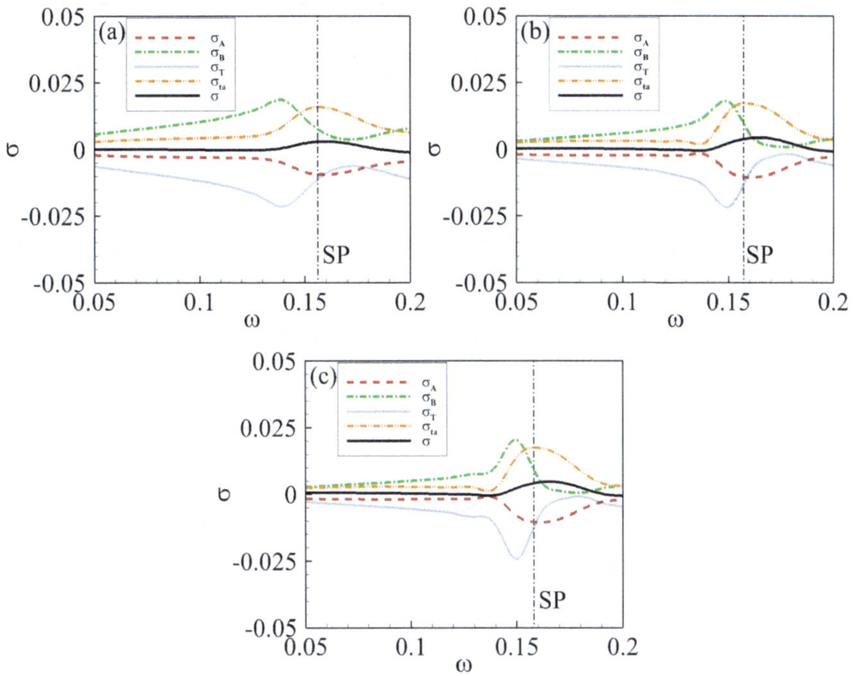

Fig. 5.18 Contribution of MPT sources on the growth rate σ of the mode S over the adiabatic wall with **a** Re = 1000, **b** Re = 5000, and **c** Re = 9000. The location of the synchronization point (SP) is marked with a vertical dashed-dotted line

the acoustic energy flux $\overline{H'\boldsymbol{m}'_A}$ and the thermal energy flux $\overline{H'\boldsymbol{m}'_T}$ respectively. Especially, acoustic source P_A is found correlated to $\frac{P_{ta}}{\gamma}$ in the acoustic energy budget equation in Eq. (5.69). As shown in Figs. 5.20 and 5.21, $\frac{\sigma_{ta}}{\gamma}$ is perfectly balanced with σ_A except for the second mode. However, the difference between σ_A and $-\frac{\sigma_{ta}}{\gamma}$ for the second mode is small compared to the magnitude of σ_{ta}. Therefore, the net effect of σ_A and σ_{ta} can be approximated as $\frac{(\gamma-1)\sigma_{ta}}{\gamma}$. Because the net effect of correlated σ_B and σ_T always stabilize the mode S, it is concluded that the rise of the second mode is attributed to the thermal part of the thermal-acoustic source, $\frac{(\gamma-1)P_{ta}}{\gamma}$.

5.2.3.4 Thermal-Acoustic Source

By analyzing the contribution of MPT sources to the growth rate σ of the mode S, the thermal part of the thermal-acoustic source $\frac{(\gamma-1)P_{ta}}{\gamma}$ is revealed to be the cause of the modal growth of the second mode. In this section, the thermal-acoustic source P_{ta} is further analyzed based on the linearized energy conservation equation. Particularly, the viscous and inviscid effects are discussed in detail.

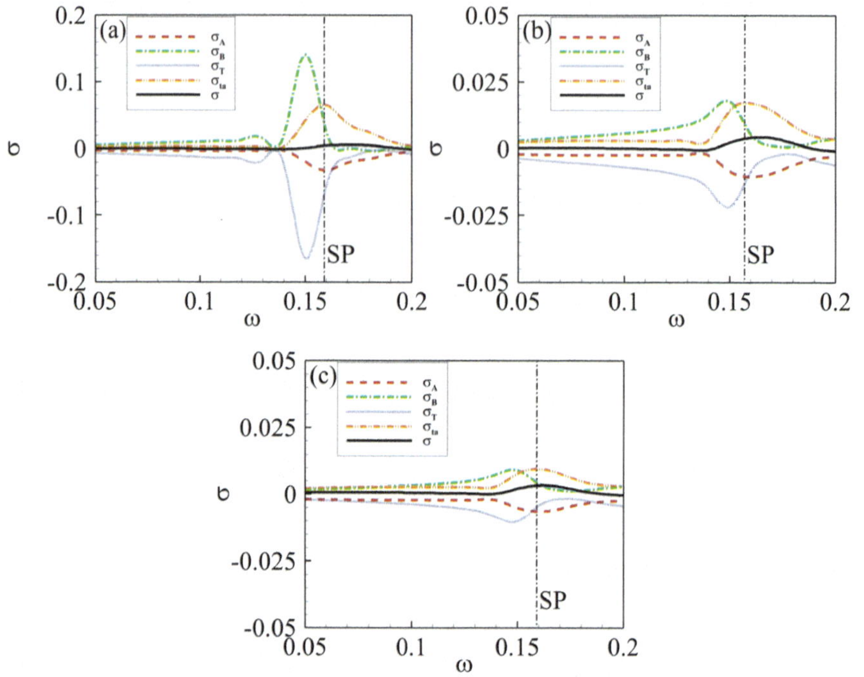

Fig. 5.19 Contribution of MPT sources on the growth rate σ of the mode S at Re = 5000 with **a** $T_w = 0.7\,T_r$, **b** $T_w = 1.0\,T_r$, and **c** $T_w = 1.3\,T_r$. The location of the synchronization point (SP) is marked with a vertical dashed-dotted line

The linearization of the energy conservation equation for a 2D parallel flow field with unique mean pressure gives the following:

$$\rho T\left(\frac{\partial S'}{\partial t} + u\frac{\partial S'}{\partial x} + v'\frac{\partial S}{\partial y}\right) = \Phi' \tag{5.70}$$

Here, $\Phi = \nabla \cdot (k\nabla T) + \boldsymbol{\tau} \cdot (\nabla \boldsymbol{u})$ includes both the thermal conduction effect and viscous dissipation effect. k is the coefficient of thermal conductivity. Equation (5.70) can be rewritten as follows:

$$\frac{\partial S'}{\partial t} + u\frac{\partial S'}{\partial x} = -v'\frac{\partial S}{\partial y} + \frac{\Phi'}{\rho T} \tag{5.71}$$

The left-hand side of Eq. (5.71) is the rate of change of the fluctuation entropy observed in the reference frame with the mean streamwise velocity u. The first term on the right-hand side of Eq. (5.71), $-v'\frac{\partial S}{\partial y}$, is the entropy exchange between neighboring layers with different mean entropy due to normal fluctuation velocity, while the second term $\frac{\Phi'}{\rho T}$ is related to irreversible processes.

Considering disturbances with the harmonic waveform, we have the following:

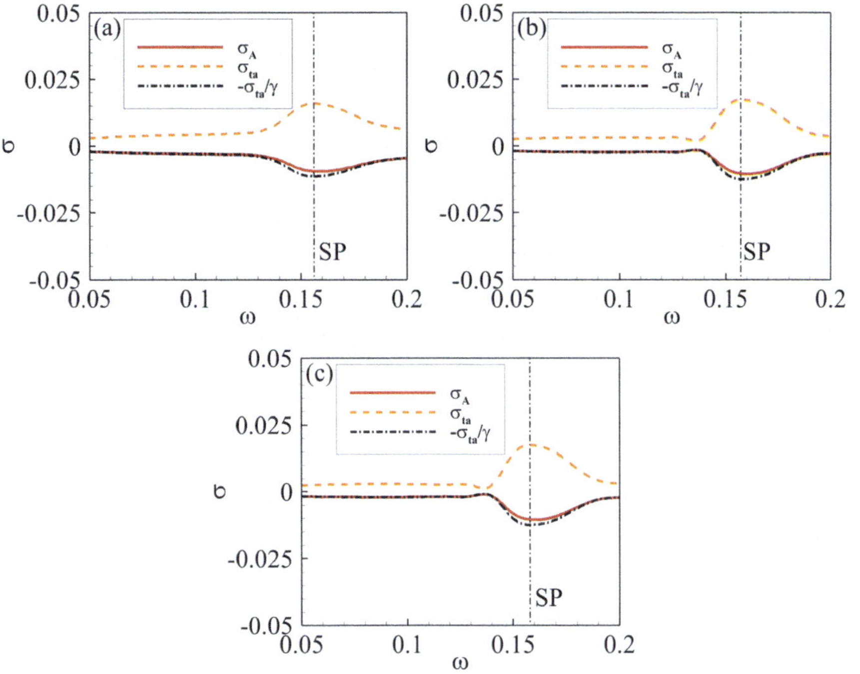

Fig. 5.20 Balance between σ_A and $\frac{\sigma_{ta}}{\gamma}$ over the adiabatic wall with **a** Re = 1000, **b** Re = 5000, and **c** Re = 9000

$$\hat{S} = -\frac{\hat{v}}{i(\alpha u - \omega)}\frac{\partial S}{\partial y} + \frac{1}{\rho T}\frac{\hat{\Phi}}{i(\alpha u - \omega)} \tag{5.72}$$

Equation (5.72) is valid for nonneutral instability waves. For a neutral instability wave, $(\alpha u - \omega) = 0$ and $\frac{\Phi'}{\rho T} = v'\frac{\partial S}{\partial y}$ at the critical layer. However, we are concerned about the physical mechanism of the unstable second mode. The neutral instability wave case is not taken into consideration in the present study.

Furthermore, two new variables η and Ψ' are introduced as follows:

$$\frac{\partial \eta}{\partial t} + u\frac{\partial \eta}{\partial x} = v', \quad \frac{\partial \Psi'}{\partial t} + u\frac{\partial \Psi'}{\partial x} = \Phi' \tag{5.73}$$

η is the disturbance displacement, and Ψ' is the energy loss per volume due to the viscous effect observed in the reference frame with the mean streamwise velocity u.

The harmonic waveform of η and Ψ' are as follows:

$$\hat{\eta} = \frac{\hat{v}}{i(\alpha u - \omega)}, \quad \hat{\Psi} = \frac{\hat{\Phi}}{i(\alpha u - \omega)} \tag{5.74}$$

Then, Eq. (5.72) can be expressed as follows:

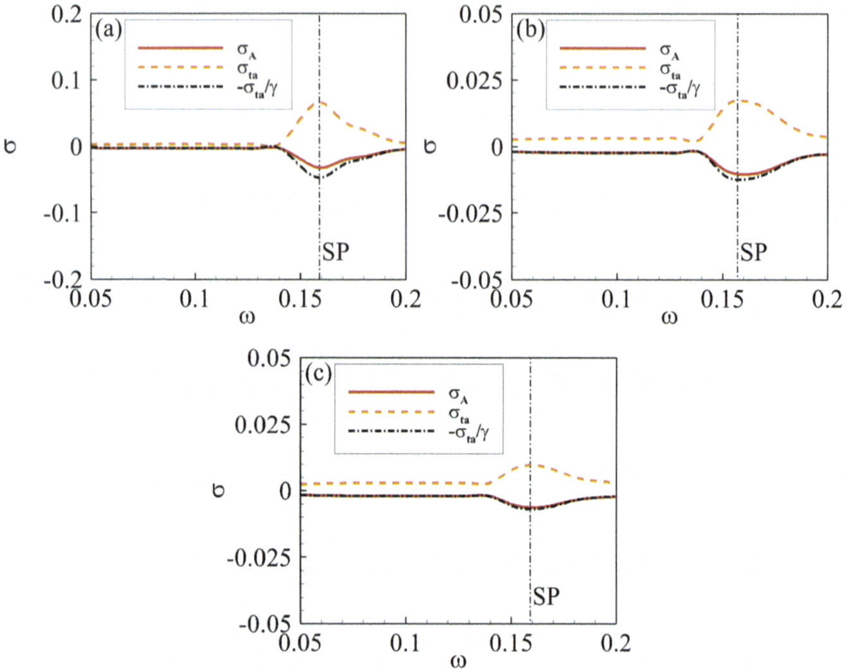

Fig. 5.21 Balance between σ_A and $\frac{\sigma_{ta}}{\gamma}$ at Re = 5000 with **a** $T_w = 0.7\ T_r$, **b** $T_w = 1.0\ T_r$, and **c** $T_w = 1.3\ T_r$

$$\hat{S} = -\hat{\eta}\frac{\partial S}{\partial y} + \frac{\hat{\Psi}}{\rho T} \tag{5.75}$$

Finally, we have

$$S' = -\eta\frac{\partial S}{\partial y} + \frac{\Psi'}{\rho T} \tag{5.76}$$

The first part is fluctuation entropy caused by disturbance displacement between neighboring layers with different mean entropy. The second part is fluctuation entropy due to the dissipative effect (viscosity and thermal conductivity). Thus, the fluctuation entropy S' is considered to be composed of the "non-dissipative part" $S'_{nd} = -\eta\frac{\partial S}{\partial y}$ and the "dissipative part" $S'_d = \frac{\Psi'}{\rho T}$.

Furthermore, the thermal-acoustic source P_{ta} is split into two parts:

$$P_{ta} = \overline{\frac{p'}{R}\frac{\partial S'}{\partial t}} = -\overline{\frac{p'}{R}\frac{\partial \eta}{\partial t}\frac{\partial S}{\partial y}} + \overline{\frac{p'}{R}\frac{\partial \Psi'}{\partial t}} \tag{5.77}$$

Again, the first part $-\overline{\frac{p'}{R}\frac{\partial \eta}{\partial t}\frac{\partial S}{\partial y}}$ is due to the non-dissipative effect denoted as P_{nd} and the second part $\overline{\frac{p'}{R}\frac{\partial \Psi'}{\partial t}}$ is due to the dissipative effect denoted as P_d.

Figure 5.22 shows the P_{ta}, P_{nd}, and P_d (normalized by $2Q_x$) of the mode S at $\omega = 0.1, 0.15$, and 0.2 with $\mathrm{Re} = 5000$, $T_w = 1.0\,T_r$. Figure 5.23 shows the P_{ta}, P_{nd}, and P_d (normalized by $2Q_x$) of the mode S at $\omega = 0.15$ with different T_w and Re. In all cases, P_{ta} has two positive peaks near the wall and the critical layer, which indicates that the thermal-acoustic source destabilizes the mode S effectively in these two regions. At the critical layer where singularity occurs in the inviscid case, both the magnitude of P_{nd} and P_d become particularly large. At $\omega = 0.15$ (the second mode), P_{ta} always overlaps P_d near the wall. The viscous effect dominates the thermal-acoustic source near the wall for the second mode, while P_{nd} is negligible there. P_{nd} dominates the positive P_{ta} around the critical layer, especially for high Re cases. The results indicate that P_{nd} dominates and destabilizes the second mode around the critical layer, which is consistent with the inviscid nature of the second mode. At $\omega = 0.1$ and 0.2, P_{ta} becomes negligible near the wall. P_d dominates P_{ta} in the vicinity of the critical layer except for a very small region where the inviscid singularity effect is significant. The results demonstrate that (1) the viscous and inviscid effects dominate the thermal-acoustic term P_{ta} for the second mode near the wall and the critical layer respectively and (2) the viscous effects dominate the thermal-acoustic term P_{ta} near the critical layer (except for a small region around the critical layer where the singularity effect arises) for mode S not related to the second mode.

5.2.3.5 Summary

Theoretical analyses based on momentum potential theory were conducted to clarify the dominant source resulting in the amplification of the second mode in the Mach 6.0 boundary layer flow field. By integrating Doak's energy budget equation along the normal direction, the contribution of the MPT source term to the growth rate of the unstable mode S is identified. The vortical part σ_B is negatively correlated to the entropic part σ_T due to the vortical-entropy coupling effect, while the thermal-acoustic part σ_{ta} is negatively correlated to the acoustic part σ_A. The net effect of σ_B and σ_T is always negative, which stabilize the second mode. The net effect of σ_{ta} and σ_A can be approximated as the thermal part of σ_{ta}, $\frac{(\gamma-1)}{\gamma}\sigma_{ta}$, which is the cause of the amplification of the second mode. The pronounced positive peak of the thermal-acoustic source P_{ta}, appearing in the vicinity of the synchronization point, provides direct evidence that P_{ta} is the most likely to be responsible for the evolvement of the second-mode instability. The thermal-acoustic source P_{ta} is further split into the non-dissipative part and the dissipative part. The dissipative (viscous) thermal-acoustic source term destabilizes the second mode near the wall. In the vicinity of the critical layer, both dissipative (viscous) and non-dissipative (inviscid) terms can be responsible for the resulting positive thermal-acoustic contribution, depending on the wall temperature and Reynolds number. These results correct the misconception in

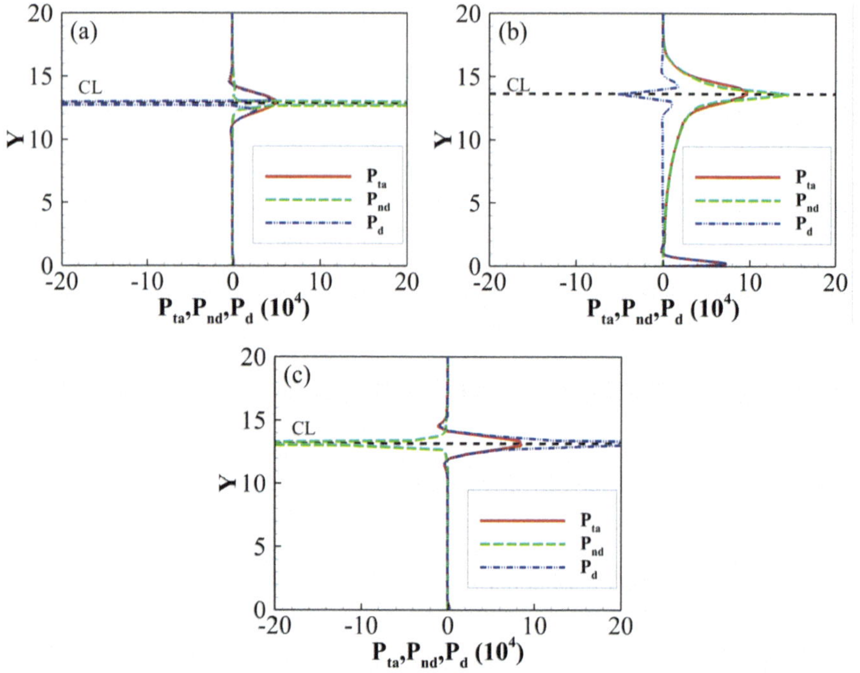

Fig. 5.22 P_{ta}, P_{nd}, and P_d normalized by $2Q_x$ at **a** $\omega = 0.1$; **b** $\omega = 0.15$; **c** $\omega = 0.2$ with Re = 5000, $T_w = 1.0\, T_r$. The location of the critical layer (CL) is marked with a horizontal dashed line

previous MPT studies of the second mode and offer a new perspective to understand the modal growth of the second mode.

5.2.4 MPT Analysis of Various Acoustic Metasurfaces

5.2.4.1 Motivation

The present subsection is to use Doak's MPT to provide a physical interpretation of the differences in the stabilization mechanisms of the absorptive, impedance-near-zero and reflection-controlled metasurfaces. After numerically simulating the flow field, the fluctuating momentum density fields on the rigid surface and the acoustic metasurfaces are decomposed into vortical, acoustic and thermal components. The contributions of source terms to the growth rate of the Mack second mode are clarified, and the source mechanisms of the different surfaces are compared. In addition, two types of macroslit surfaces are introduced and their stabilization mechanism are also discussed under Doak's MPT framework.

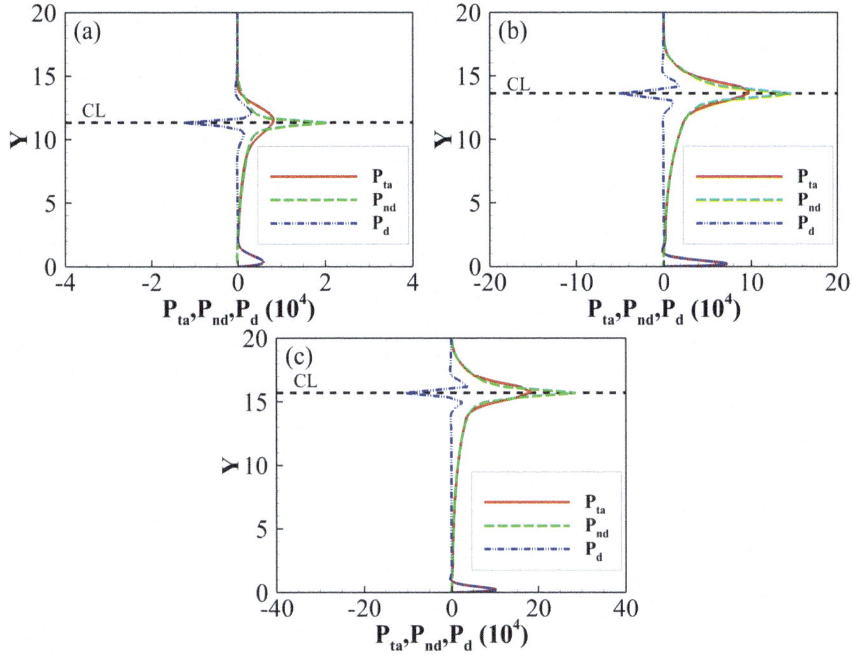

Fig. 5.23 P_{ta}, P_{nd}, and P_d normalized by $2Q_x$ at $\omega = 0.15$ with **a** Re $= 1000$, $T_w = 0.7\ T_r$; **b** Re $= 5000$, $T_w = 1.0\ T_r$; **c** Re $= 9000$, $T_w = 1.3\ T_r$. The location of the critical layer (CL) is marked with a horizontal dashed line

5.2.4.2 Momentum Potential Theory

For the DNS of an acoustic metasurface with a regular periodic microstructure, integrating the fluctuation energy equation in a periodic unit cell (Fig. 5.24):

$$\int_{y_w}^{+\infty}\int_{x}^{x0}\left(\frac{\partial\overline{(\rho u)'H'}}{\partial x}+\frac{\partial\overline{(\rho v)'H'}}{\partial y}\right)dxdy = \oiint P_A dxdy + \oiint P_B dxdy + \oiint P_T dxdy + \oiint P_{ta} dxdy \tag{5.78}$$

where s is the period of the microstructure, and $y_w = 0$ for the rigid part, and $y_w = -h$ indicating the bottom of slits. Since $(\rho v)\prime = 0$ for $y = y_w$ and $y \to \infty$, the MPT energy budget equation in two-dimensional integral form within a period is derived as:

$$\left.\int_{y_w}^{+\infty}\overline{(\rho u)'H'}dy\right|_{x_0}^{x_0+s} = \oiint P_A dxdy + \oiint P_B dxdy + \oiint P_T dxdy + \oiint P_{ta} dxdy \tag{5.79}$$

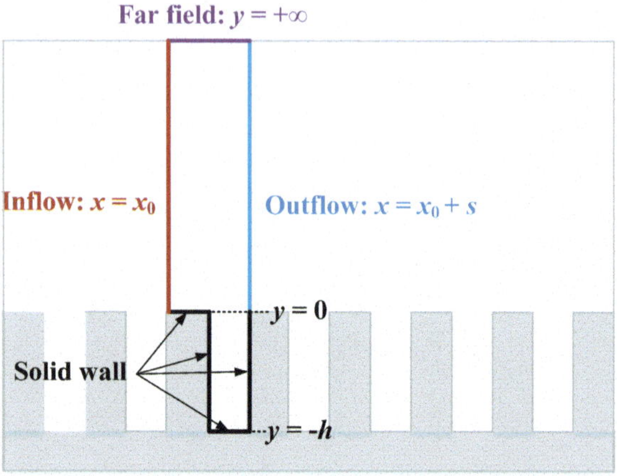

Fig. 5.24 Schematic of the integration area of a periodic unit cell

Since the mean flow field over the acoustic metasurface is periodic, the growth rate σ_m satisfies

$$\int_{y_w}^{+\infty} \overline{(\rho u)' H'} dy \Bigg|_{x_0+s} = e^{2\sigma_m s} \int_{y_w}^{+\infty} \overline{(\rho u)' H'} dy \Bigg|_{x_0} \tag{5.80}$$

As a result,

$$\sigma_m = \frac{1}{2s} ln \left(\frac{\int_{y_w}^{+\infty} \overline{(\rho u)' H'} dy \Big|_{x_0+s}}{\int_{y_w}^{+\infty} \overline{(\rho u)' H'} dy \Big|_{x_0}} \right) \tag{5.81}$$

From Eq. (5.81) we can derive the contribution of each source term to σ_m:

$$\sigma_{mj} = \frac{1}{2s} ln \left(\frac{\oint P_j dx dy}{\int_{y_w}^{+\infty} \overline{(\rho u)' H'} dy \Big|_{x_0}} + 1 \right). \tag{5.82}$$

5.2.4.3 Acoustic Characteristics and Stabilization Effects

Before considering the hypersonic flow field, we perform full-wave finite element simulations to compare the acoustic characteristics of the different metasurfaces in quiescent air. For completeness, the contours of all kinds of surfaces are presented below, although Fig. 5.25a and d have been reported in aforementioned subsections. The operating frequency is $f = 138.74$ kHz, and the angle of the incident wave is assumed to be $\theta_i = 0°$, *i.e.*, normal to the surface. The incident acoustic waves have the same amplitude in all cases, and by which the scattered acoustic pressure is normalized. The acoustic pressure on the optimized absorptive metasurface (Fig. 5.25b) is minimal, two orders of magnitude smaller than that on the other metasurfaces. The acoustic pressure amplitudes on the other two metasurfaces are similar, and slightly lower than that on the rigid surface because of insufficient viscous dissipation. For the impedance-near-zero metasurface, a minimum pressure amplitude can be observed at the surface due to the out-of-phase behavior between the incident and reflected waves. The pressure field pattern of the reflection-controlled metasurface (Fig. 5.25d) is quite different from that of the other metasurfaces, and the reflected waves propagate upstream parallel to the surface because of the linearly-changed acoustic phase of the microslits. These wavefront manipulation characteristics were all verified to successfully restrain the Mack second mode.

The instantaneous fluctuating pressure contours for all cases are summarized in Fig. 5.26. The typical "two-cell" structure is observed on the rigid surface in Fig. 5.26a. The fluctuations are considerably amplified downstream ($x > 0.1$ m), indicating the dominance of the Mack second mode. However, on the acoustic metasurfaces, the typical "two-cell" patterns are changed in different ways. For the absorptive metasurface (Fig. 5.26b), the upper mode cells tend to fuse with the lower mode cells and the trailing legs sink into the microslits. For the impedance-near-zero metasurface (Fig. 5.26c), a minimum value appears at the surface as a result of the phase opposition of the instability waves, and the mode structure is divided in the slit and above the surface. For the reflection-controlled metasurface, the mode structures are divided into several segments in the normal direction as shown in Fig. 5.26d. The

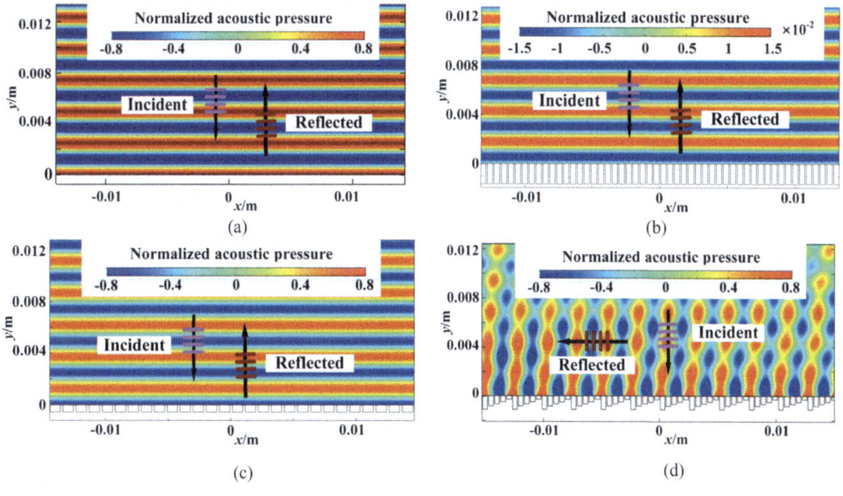

Fig. 5.25 Normalized scattered pressure fields for various acoustic metasurfaces in the absence of hypersonic fluid flow: **a** rigid surface, **b** absorptive metasurface, **c** impedance-near-zero metasurface, and **d** reflection-controlled metasurface

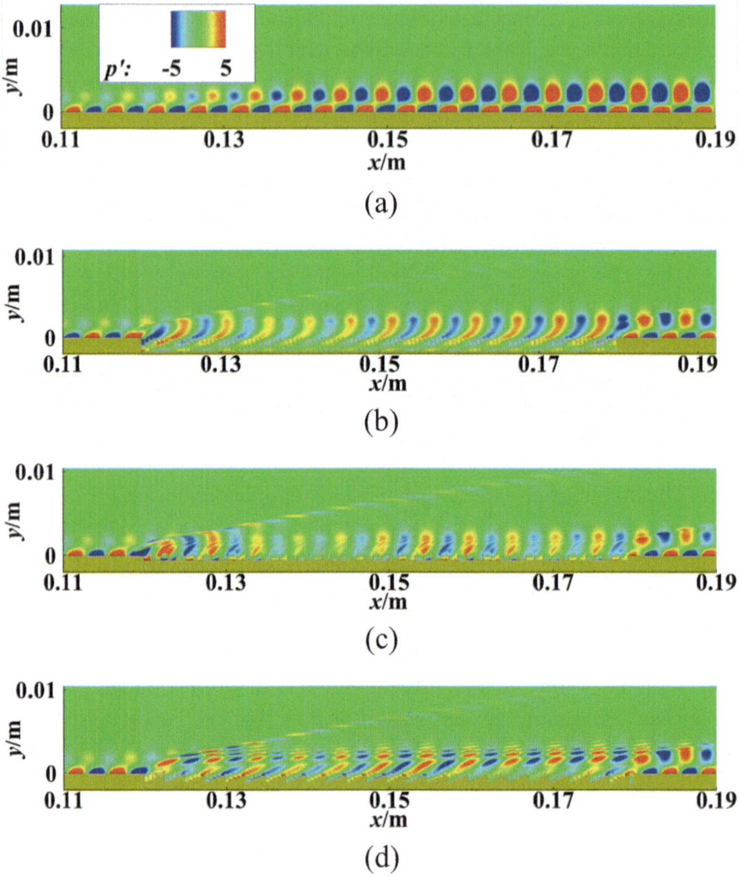

Fig. 5.26 Fluctuating pressure contours for **a** rigid surface, **b** absorptive metasurface, **c** impedance-near-zero metasurface, and **d** reflection-controlled metasurface (unit: Pa)

amplitudes of the fluctuating pressure at the wall are shown in Fig. 5.27 Although the mode structures differ in various ways, for all three metasurfaces, the intensities of the Mack second mode are obviously weakened, and the maximum amplitudes of the fluctuating pressures at the wall are significantly reduced.

5.2.4.4 Interpretation of Stabilization Mechanisms

(1) MPT decomposition

The different components are extracted by applying Doak's MPT approach to the instantaneous fluctuation fields. Snapshots of the magnitudes of the acoustic component ($\|\boldsymbol{m}_{A'}\|$), vortical component ($\|\boldsymbol{m}_{B'}\|$), and thermal component ($\|\boldsymbol{m}_{T'}\|$) of the

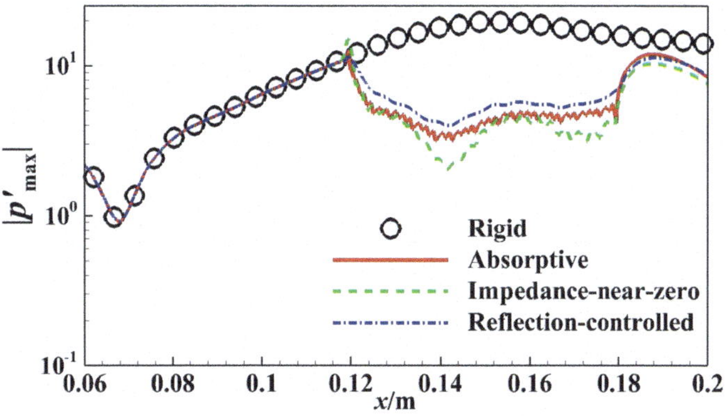

Fig. 5.27 Comparisons of the maximum fluctuating pressure amplitude along the wall for various surfaces (unit: Pa)

rigid surface and absorptive metasurface are depicted in the left column and the right column of Fig. 5.28, respectively. In the MPT framework, the Mack second mode has the largest magnitudes of vortical perturbation and the smallest magnitudes of acoustic perturbation. Its thermal perturbation is confined within the BL. Above the absorptive metasurface, all components are suppressed to varying degrees. The expansion wave associated with the juncture between the rigid surfaces and the metasurfaces ($x = 0.12$ m) induces radiations of the vortical and acoustic components. The component contours of the impedance-near-zero and reflection-controlled metasurfaces are nearly identical to those of the absorptive metasurface, and will not be repeated here. To clarify the differences of stabilization mechanism of various metasurfaces, the source contributions to growth rate of disturbances and the corresponding source terms are calculated and analyzed in the following sections.

(2) Source contributions to the growth rate of the Mack second mode

Figure 5.29 presents the streamwise distributions of σ and its components contributed by each source term based on the DNS results. Similar to the LST results in subsection 5.5, σ_{rta} and σ_{rB} are positive, while σ_{rA} and σ_{rT} are negative. σ_{rta} is the largest component in the region of Mack second mode amplification. σ_{rta} and σ_{rA} are negatively correlated, while σ_{rB} and σ_{rT} are negatively correlated. The sum of the growth rates contributed by all the source terms, σ_r, remains positive in the region of the Mack second mode amplification, indicating that the Mack second mode is unstable on rigid surface (Fig. 5.29a). Figure 5.29b, c and d show the growth rate distributions on the three different acoustic metasurfaces. The overall growth rate of each metasurface σ_m is decreased to nearly zero, implying that the metasurfaces greatly stabilize the Mack second mode. σ_{mta} is still the dominant component of the disturbance growth rates contributed by the source terms.

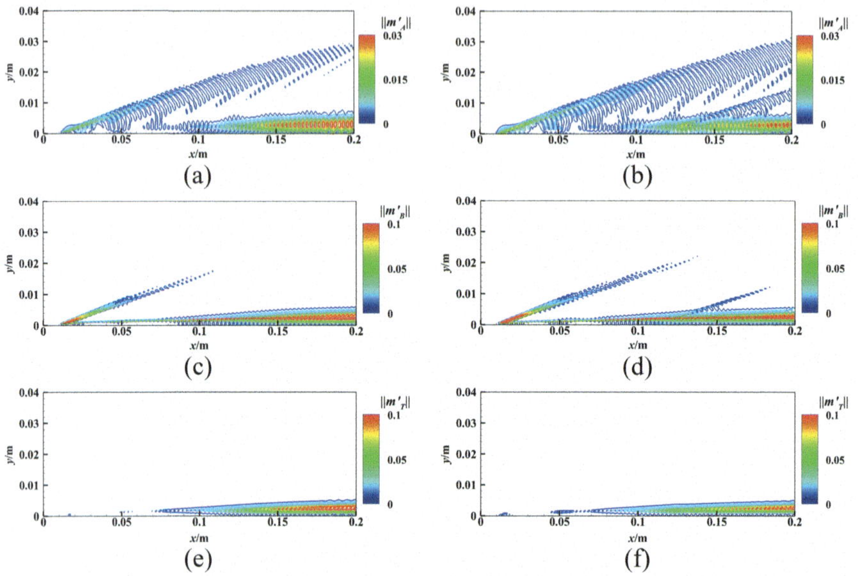

Fig. 5.28 Magnitudes of **a**, **b** m'_A, **c**, **d** m'_B, and **e**, **f** m'_T for the rigid surface (left column) and the absorptive metasurface (right column) (unit: $\mathrm{kg}/(\mathrm{m}^2 \bullet \mathrm{s})$)

(3) Source mechanisms

In this section, the distributions of the source terms are discussed to further explore the stabilization mechanism of different acoustic metasurfaces. Figure 5.8 shows the contours of the acoustic source terms P_A on various surfaces. It is worth noting that a negative source term leads to a positive growth rate and causes instability in the HBL. In most regions, the acoustic source terms are close to zero except near the surface and the critical layer. Above the rigid surface, the acoustic source is negative near the surface, thus destabilizing the Mack second mode. On the other hand, for most of the region near the critical layer, the acoustic source is positive, indicating that it stabilizes the Mack second mode in the HBL. According to Fig. 5.29a, σ_{rA} is negative in the entire flow area, implying that the positive acoustic source near the critical layer is dominant. For the three types of acoustic metasurfaces (Fig. 5.30b, c and d), the negative source near the surface is attenuated by the microslits, whereas the acoustic source near the critical layer remain positive. It can also be observed that the negative source near the critical layer is enhanced, indicating that the acoustic metasurfaces introduce instability near the critical layer. Since σ_{mA} is generally larger than σ_{rA} in Fig. 5.29, it can be concluded that acoustic metasurfaces contribute to the instability by enhancing the negative acoustic source near the critical layer.

Figure 5.31 shows the distributions of the vortical and thermal source terms (P_B and P_T) in the left column and right column, respectively. The wall-normal distributions of the thermal source terms are nearly negatively correlated with the corresponding vortical source terms on both rigid surface and acoustic metasurfaces.

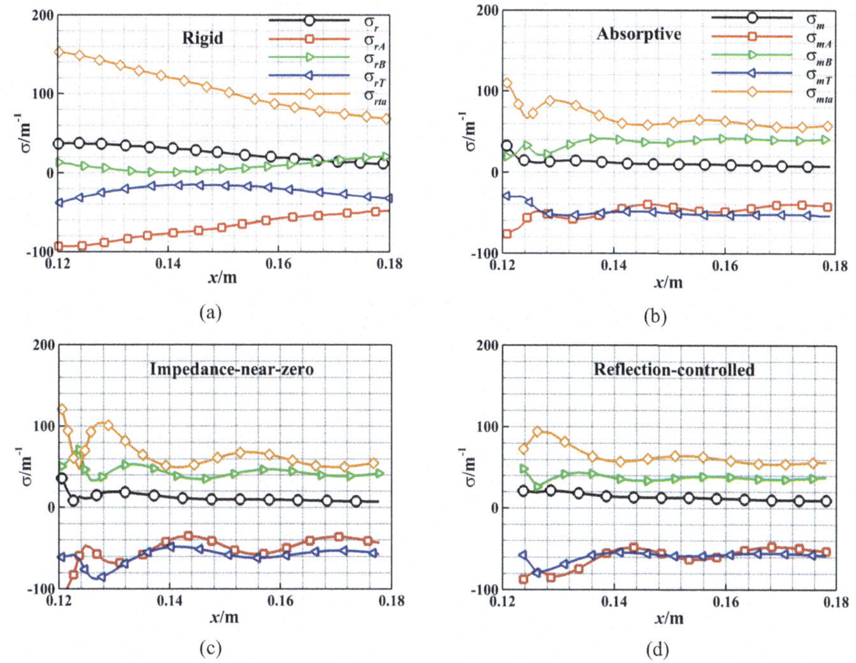

Fig. 5.29 Streamwise distributions of the growth rate on various surfaces based on the DNS results: **a** rigid surface, **b** absorptive metasurface, **c** impedance-near-zero metasurface, and **d** reflection-controlled metasurface

The reason for this may be the occurrence of the vortical–entropy coupling effect between $\boldsymbol{m}_{B}{}'$ and $\boldsymbol{m}_{T}{}'$, as suggested by Crocco's theorem. Similarly, in most regions, the vortical and thermal source terms are nearly zero, except near the surface and the critical layer. For the rigid surface (Fig. 5.31a), the vortical source term P_B is positive near the critical layer and the wall, which stabilizes the Mack second mode. Between the critical layer and the wall, P_B is negative and therefore destabilizes the Mack second mode. Since σ_{rB} is positive in Fig. 5.29a, the negative vortical source is most significant. For all the acoustic metasurfaces, the positive region of P_B at the near wall is considerably reduced due to the microslits on the wall, and the negative zone between the wall and the critical layer is weakened. As shown in Fig. 5.29, σ_{mB} is larger than σ_{rB}, indicating that the acoustic metasurfaces mainly contribute to instability by attenuating the positive source at the near wall. The distributions of the thermal source terms are opposite to each other.

Through the quantitative analyses of the source contributions to the growth rate of the Mack second mode (Fig. 5.29), it has been demonstrated that the thermal-acoustic source term P_{ta} is critical in the amplification of the Mack second mode. Thus, the effect of microslits on the thermal-acoustic source term is the key to explain the stabilization mechanism of different acoustic metasurfaces. As shown in Fig. 5.32a, the dominant negative contours cause σ_{rta} to be the largest component, and thus σ_{rta}

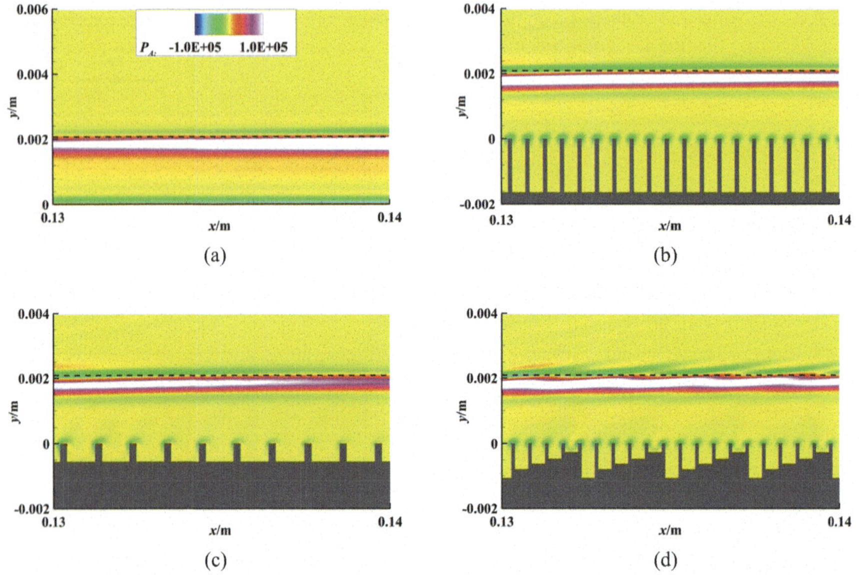

Fig. 5.30 Acoustic source terms for the: **a** rigid surface, **b** absorptive metasurface, **c** impedance-near-zero metasurface, and **d** reflection-controlled metasurface. The critical layer is indicated by a dashed line (unit:$\mathrm{kg}/\left(\mathrm{m} \bullet \mathrm{s}^3\right)$)

plays a leading role in the amplification of the Mack second mode. The changes in the source term distribution caused by the various acoustic metasurfaces are quite different. In Fig. 5.32b, the absorptive metasurface completely dissipates the negative source near the wall, and the negative source term near the critical layer is also attenuated slightly. The attenuation of the negative P_{ta} term leads to a reduction in σ_{ta}, thus stabilizing the Mack second mode. Above the impedance-near-zero acoustic metasurface (Fig. 5.32c), the negative source near the wall is also completely dissipated. However, a positive region exists below the critical layer. For the reflection-controlled metasurface Fig. 5.32d), the positive source intertwines along the critical layer and the period is consistent with the distribution period of the slit groups. This indicates that the impedance-near-zero and reflection-controlled metasurfaces affect not only the source terms near the wall but also the source terms near the critical layer.

5.2.4.5 Macroslits Stabilization Mechanisms

All of the aforementioned acoustic metasurfaces are corrugated with subwavelength microslits to achieve particular acoustic characteristics. When the slit spacing increases to the same order of magnitude as the disturbance wavelength, the scattered waves at the surface become strong, and that will weaken the acoustic ability in the suppression of Mack second mode. In this subsection, the possible stabilization mechanisms of macroslit surfaces will be discussed in the MPT framework. Two

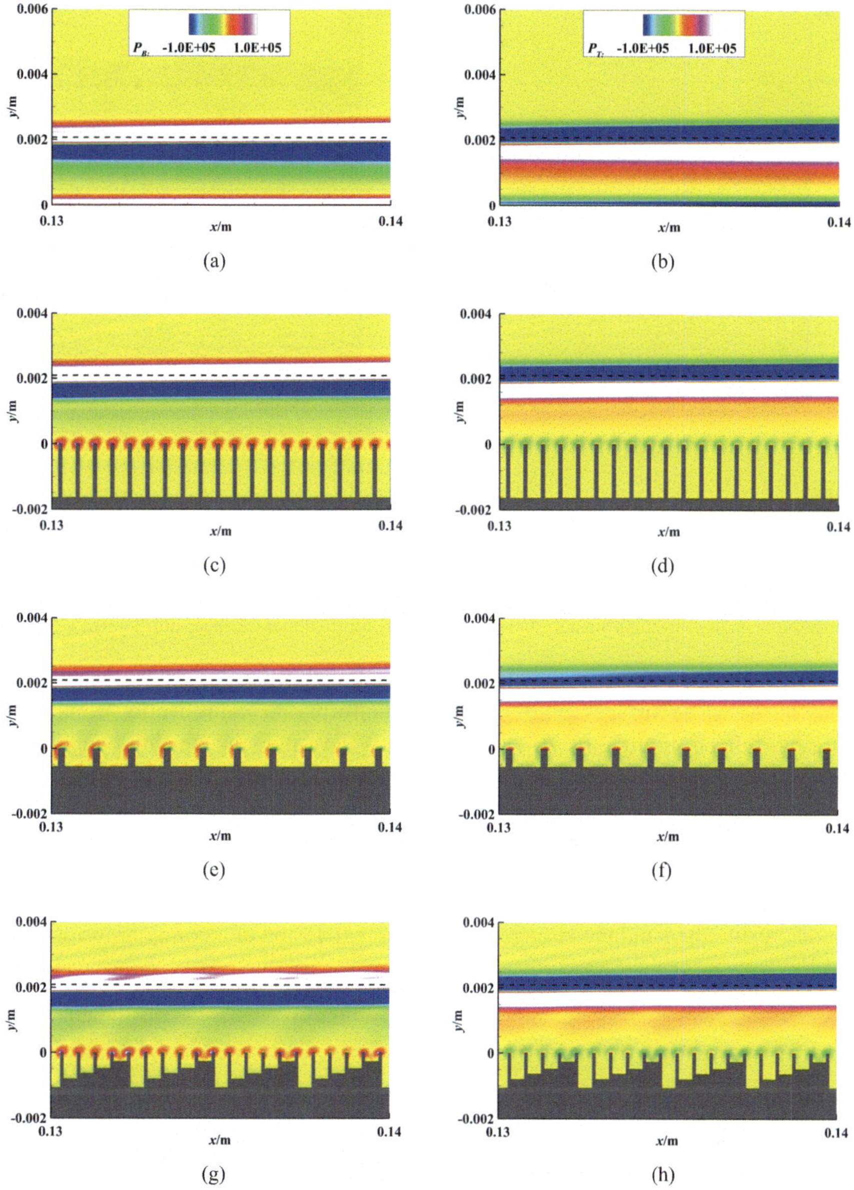

Fig. 5.31 Vortical source terms (left column) and thermal source terms (right column) for: **a**, **b** rigid surface, **c**, **d** absorptive, **e**, **f** impedance-near-zero, and **g**, **h** reflection-controlled metasurfaces (unit:$\text{kg}/\left(\text{m} \bullet \text{s}^3\right)$)

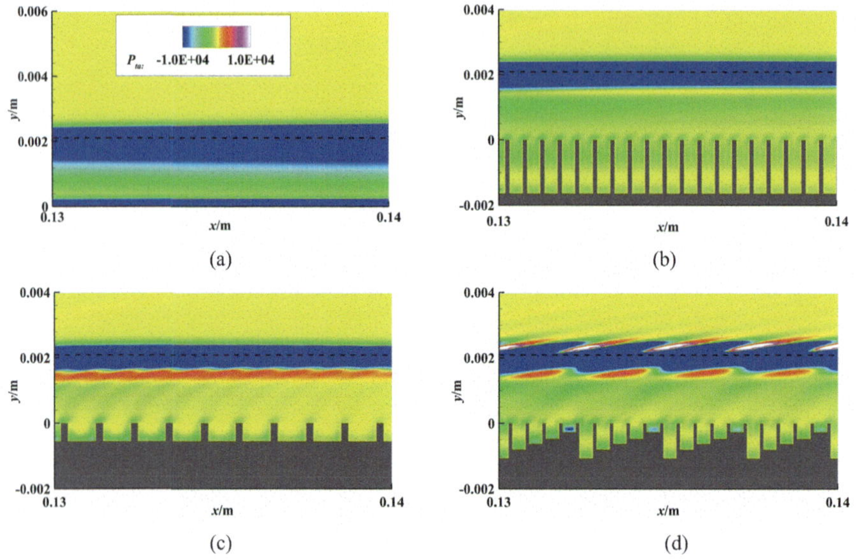

Fig. 5.32 Thermal-acoustic source terms for: **a** rigid surface, **b** absorptive metasurface, **c** impedance-near-zero metasurface, and **d** reflection-controlled metasurface (unit:$\mathrm{kg}/\left(\mathrm{m} \bullet \mathrm{s}^3\right)$)

kinds of macroslit surfaces are designed: the uniform one and reflection-controlled one with reflection angle of 70°. Both surfaces have the slits spacing equaling to the wavelength of disturbance of frequency 138.74 kHz, and the same porosity of $\phi = 0.76$ as the absorptive acoustic metasurface. The uniform one has the depth $h = 1.34 \times 10^{-3}$ m, while the reflection-controlled one is designed with two slits with different depths in every group ($h_1 = 6.7 \times 10^{-2}$ m and $h_2 = 1.34 \times 10^{-3}$ m). From the steady calculations shown in Fig. 5.33, compared to that above the metasurfaces, the macroslit edges induce stronger alternating expansion and compression waves, and the streamlines shows comparatively intensive oscillations near the macroslit surfaces.

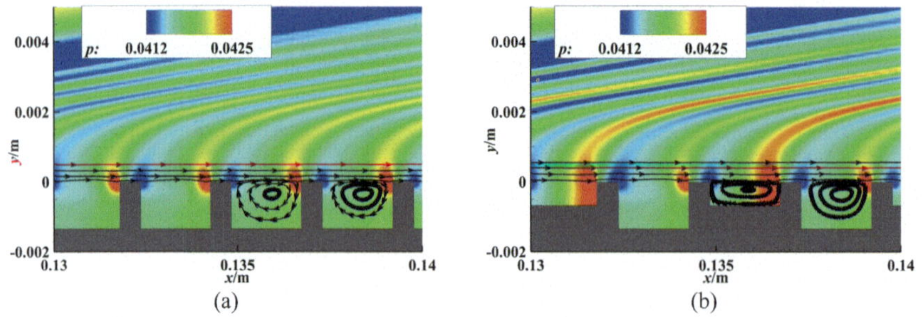

Fig. 5.33 Local pressure contour around the macroslits for: **a** uniform macroslit surface, **b** reflection-controlled macroslit surface

Figure 5.34 shows the instantaneous pressure flowfield from the unsteady calculations. Similar to the contours above the impedance-near-zero and reflection-controlled metasurfaces depicted in Fig. 5.26c and d, the typical "two-cell" structure of Mack second mode is destroyed on both macroslit surfaces. However, the intensities of fluctuating pressure do not diminish as obviously as those above the metasurfaces. From the comparisons of the maximum fluctuating pressure distributions at the surfaces (Fig. 5.35), the macroslit surfaces could also suppress the Mack second mode to some extent, and the reflection-controlled surface shows a better stabilization performance than the uniform one.

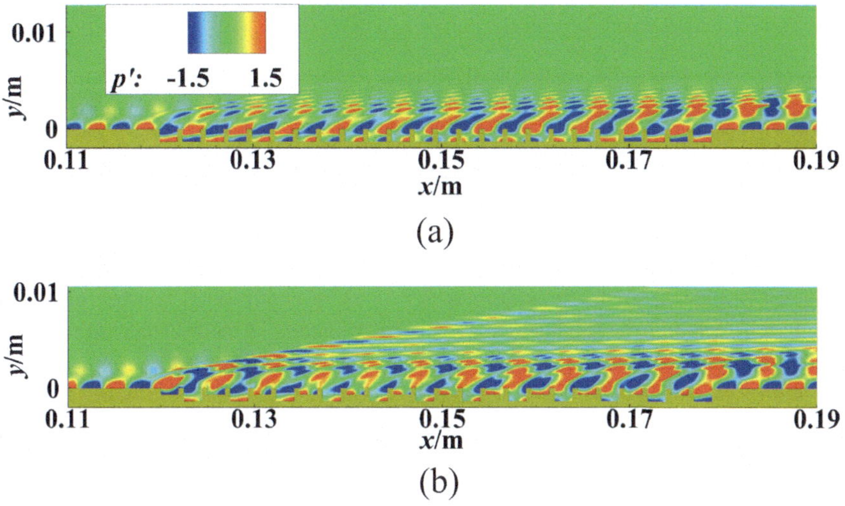

Fig. 5.34 Fluctuating pressure contours for: **a** uniform macroslit surface, **b** reflection-controlled macroslit surface

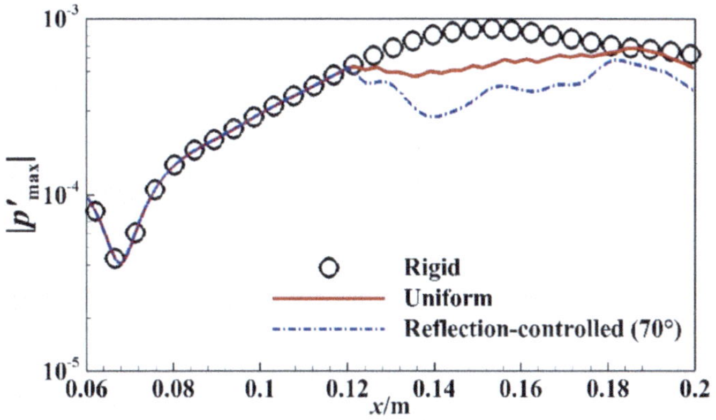

Fig. 5.35 Streamwise variation of upper envelope of maximum fluctuating pressure

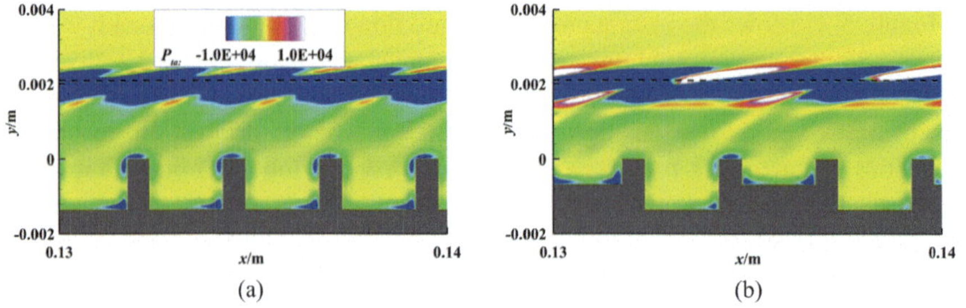

Fig. 5.36 Thermal-acoustic source term for: **a** uniform macroslit surface, **b** reflection-controlled macroslit surface. The critical layer is marked using a dashed line

Figure 5.36 shows the spatial distributions of the thermal-acoustic source term P_{ta} that acts as the dominant factor to excite the unstable Mack second mode. Same as those above the metasurfaces (Fig. 5.32), the negative P_{ta} near the surfaces is attenuated in the macroslits. The intensive alternating expansion and compression waves induced at the macroslit edges seem affect P_{ta} around the critical layer by producing positive source (Fig. 5.36a). Compared to the uniform macroslit surface, the reflection-controlled one induces a larger positive P_{ta} that wrapping along the critical layer, which causes a more prominent stabilization performance (Fig. 5.36b). Including the source mechanisms of acoustic metasurfaces, the recirculation zones inside the slits tend to attenuate the near-surface negative P_{ta} regardless of the slit widths, while the particular acoustic manipulations mainly affect the stabilization performance by diminishing negative P_{ta} (the absorptive one) or increasing positive P_{ta} (the impedance-near-zero, reflection-controlled ones, and the reflection-controlled macroslit surface) around the critical layer.

5.2.4.6 Summary

The stabilization mechanisms of various acoustic metasurfaces, including the macroslit surfaces, on the Mack second mode in the HBL were clarified within the MPT framework. The relationships between the commonly defined streamwise growth rates of the unstable modes in the HBL and the source terms in the MPT energy budget equation were established based on the DNS results of acoustic metasurface.

The vortical, acoustic and thermal components were extracted from the DNS results with MPT decomposition. The vortical component contributes most prominently to the perturbation, while the acoustic component has the smallest contribution. The thermal perturbation is confined in the BL. All the components are suppressed on the acoustic metasurfaces. On both rigid surface and acoustic metasurfaces, all the source terms are nearly zero in most flow regions, and the positive/negative source terms that affect the disturbance growth rate are mainly located near the wall and the critical layer. For the various acoustic metasurfaces, the distributions of the vortical, acoustic and thermal source terms are similar, with the main differences being in the thermal-acoustic source term P_{ta}. The narrow microslits in the absorptive metasurface largely restrain the negative P_{ta} term near the surface and slightly attenuate it near the critical layer. The other two types of metasurfaces mainly strengthen the positive P_{ta} term near the critical layer. The impedance-near-zero metasurface produces a positive contour under the critical layer, while the positive P_{ta} term intertwines along the critical layer above the reflection-controlled metasurface.

Two types of macroslit surfaces with slit spacing equal to disturbance wavelength were introduced to verify the stabilization performances and clarify the inherent mechanisms within the MPT framework. The macroslits induce intensive alternating expansion and compression waves, and slightly cause the BL thinner. The second mode disturbances are suppressed although the scattered waves at the uniform macroslit surface are too strong to represent particular acoustic characteristics. When deflecting the reflection wave to 70°, the stabilization effect becomes more prominent. By comparing the distributions of P_{ta}, the recirculation zones inside the slits were thought responsible for the attenuation of the near-surface negative P_{ta} regardless of the slit widths, while the acoustic manipulation produces positive source around the critical layer that largely diminishes the growth rate of Mack second model.

In addition, it suggests a further physical interpretation of the thermal-acoustic source term because of its importance in the amplification of the Mack second mode and also its different behaviors in the metasurface stabilizations. A receptivity study will be carried to find the way how thermal-acoustic source term excites the Mack second mode. Also, the reasons why there exist negative correlations between different source terms should be clarified in a future study.

5.3 Thermal-Acoustic Coupling Theory

5.3.1 Motivation

Unlike Tollmien-Schlichting waves, second-mode instability is inviscid and believed to be acoustic instability trapped between the wall and the sonic line during propagation. Kuehl analyzed the acoustic impedance along the wall normal for second-mode waves and found that the region trapping second-mode waves forms an acoustic impedance well, resulting in the resonation of second-mode waves. He also proposed a thermoacoustic interpretation of second-mode instability that the energy to sustain the resonant standing waves comes from thermoacoustic Reynolds stress and thermodynamic work. Unnikrishnan and Gaitonde decomposed the fluid-thermodynamic components, which consist of acoustic, vortical and thermal components, of fluctuations using MPT and found that though the vortical component is the largest in magnitude, the thermal component is the primary source for the growth of perturbations dominated by the second mode. However, the detailed physical mechanisms of how the second-mode disturbances are sustained and amplified still need to be explored. The authors theoretically investigate the growth mechanisms of second-mode instability. From the shape of eigenfunctions of the second mode, we learn that the temperature perturbation is remarkably strong in the adjacent region of the critical layer, which inspires us to consider the effect of internal energy change on second-mode instability. Furthermore, we are motivated by the mechanism of the Rijke tube in generating acoustic waves to take into account the relative phases of energy sources. Hence, we demonstrate the magnitude and phase of energy terms in linear disturbance energy equations. We also utilize acoustic metasurfaces to show the relation between the stabilization of the second mode and the variations in the energy sources.

5.3.2 Simplified Linear Stability Equations

The eigenvector of an eigenmode within supersonic/hypersonic boundary layers can provide the shape information of a disturbance component, as well as its relative weight. Substituting the eigenvector and eigenvalue obtained from LST calculations to the linear stability equation and then eliminating the low-weight terms, we can simplify the linear stability equations.

Here we employ the mode at $\omega = 0.152$ with $\alpha = 0.164 - 0.004i$ and $\beta = 0$, which is the most unstable second mode under the current boundary layer over the smooth solid wall, as an instance to simplify linear stability equation. The components of the eigenvector are depicted in Fig. 5.37. Each term in linear stability equation is evaluated, and then the explicit form for 2D boundary layer flow ($W = 0$) can be recasted as followings:

$$\tfrac{1}{T}(i\alpha U - i\omega)\hat{u} = -\tfrac{dU}{dy}\tfrac{\hat{v}}{T} - i\alpha\hat{p} + \tfrac{1}{R}\tfrac{d}{dy}\left(\mu\tfrac{d\hat{u}}{dy} + \tfrac{dU}{dy}\tfrac{d\mu}{dT}\hat{T}\right) + res. \tag{5.83}$$

$$\tfrac{1}{T}(i\alpha U - i\omega)\hat{v} = -\tfrac{d\hat{p}}{dy} + \tfrac{4}{3R}\tfrac{d}{dy}\left(\mu\tfrac{d\hat{v}}{dy}\right) + res. \tag{5.84}$$

$$\tfrac{1}{T}(i\alpha U - i\omega)\hat{w} = -i\beta\hat{p} + \tfrac{1}{R}\tfrac{d}{dy}\left(\mu\tfrac{d\hat{w}}{dy}\right) + res. \tag{5.85}$$

$$\begin{aligned}\frac{1}{T}(i\alpha U - i\omega)\hat{T} = &-\frac{dT}{dy}\frac{\hat{v}}{T} - (\gamma - 1)\left(i\alpha\hat{u} + \frac{d\hat{v}}{dy} + i\beta\hat{w}\right)\\ &+ s\frac{\gamma}{RPr}\frac{d^2(\mu\hat{T})}{dy^2} + res.\end{aligned} \tag{5.86}$$

where *res.* denotes the residual terms. The first three Eqs. (5.83)–(5.85) are the linearized disturbance momentum equations. Equation (5.86) is a rewrite of the linearized disturbance energy equation based on the internal energy rather than enthalpy in order to reveal the internal energy variation straightforwardly.

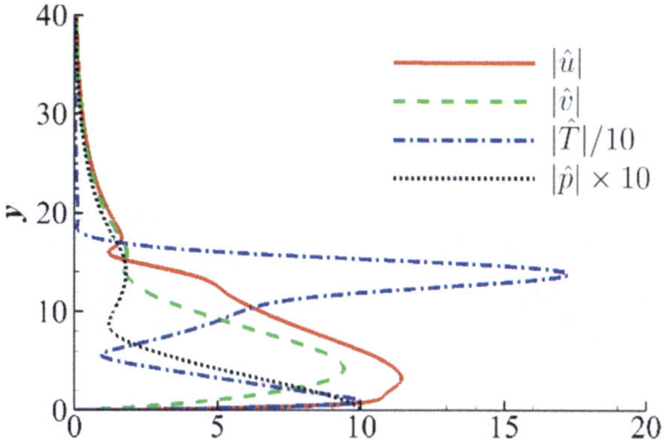

Fig. 5.37 Disturbance Components of the most unstable mode of the smooth-solid-wall case with $\omega = 0.152$, $\alpha = 0.164 - 0.004i$, and $\beta = 0$. All the components are normalized by the wall-pressure-disturbance magnitude $|\hat{p}_w|$

Essentially, the residual terms in Eqs. (5.83)–(5.86) are negligible for both the first and second modes, including oblique waves. Here we choose two modes from the smooth-solid-wall case at $\omega = 0.063$ and $\omega = 0.152$ with a propagation phase angle of $\arctan\left(\frac{\beta}{\alpha_r}\right) = 30°$, which respectively represents the 3D first- and second-mode waves, for validation. The corresponding eigenvalues are $\alpha = 0.071 - 0.00067i$ and $\alpha = 0.163 - 0.0017i$, respectively. The magnitudes of the terms in Eqs. (5.83)–(5.86) are examined for the first and second modes and their profiles along y coordinate are illustrated in Figs. 5.38 and 5.39, respectively.

Figures 5.38 and 5.39 show, both for the first and second mode, the magnitudes of the residual terms of Eq. (5.83)–(5.86) are fairly small along y coordinate except the magnitude of the residual of Eq. (5.86), which deviates from zero near the wall due to the viscous dissipation. But this deviation is negligible in contrast to the terms $\frac{\gamma}{RPr}\frac{d^2(\mu\hat{T})}{dy^2}$ and $-(\gamma-1)\left(i\alpha\hat{u} + \frac{\partial\hat{v}}{\partial y} + i\beta\hat{w}\right)$ near the wall, as shown in Fig. 5.39d. Therefore, Eq. (5.83)–(5.86) can be reduced to

$$\frac{1}{T}(i\alpha U - i\omega)\hat{u} = -\frac{dU}{dy}\frac{\hat{v}}{T} - i\alpha\hat{p} + \frac{1}{R}\frac{d}{dy}\left(\mu\frac{d\hat{u}}{dy} + \frac{dU}{dy}\frac{d\mu}{dT}\hat{T}\right) \tag{5.87}$$

$$\frac{1}{T}(i\alpha U - i\omega)\hat{v} = -\frac{d\hat{p}}{dy} + \frac{4}{3R}\frac{d}{dy}\left(\mu\frac{d\hat{v}}{dy}\right) \tag{5.88}$$

$$\frac{1}{T}(i\alpha U - i\omega)\hat{w} = -i\beta\hat{p} + \frac{1}{R}\frac{d}{dy}\left(\mu\frac{d\hat{w}}{dy}\right) \tag{5.89}$$

$$\frac{1}{T}(i\alpha U - i\omega)\hat{T} = -\frac{dT}{dy}\frac{\hat{v}}{T} - (\gamma - 1)\left(i\alpha\hat{u} + \frac{d\hat{v}}{dy} + i\beta\hat{w}\right) + \frac{\gamma}{RPr}\frac{d^2(\mu\hat{T})}{dy^2} \tag{5.90}$$

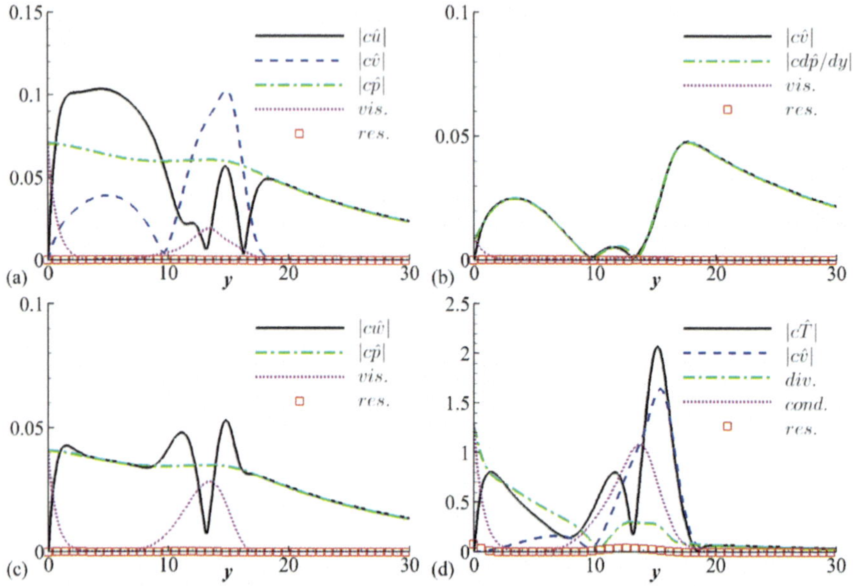

Fig. 5.38 Magnitude of the terms in Eq. (5.83)–(5.86) at $\omega = 0.063$ with $\arctan\left(\frac{\beta}{\alpha_i}\right) = 30^o$ and eigenvalue of $\alpha = 0.071 - 0.00067i$: **a** Eq. (5.83), **b** Eq. (5.84), **c** Eq. (5.85) and **d** Eq. (5.86). Here c denotes the coefficient of each components, $vis.$ the viscous term in the disturbance momentum equations, $res.$ the sum of the absolutes of the residual terms, $div.$ the term $-(\gamma - 1)\left(i\alpha\hat{u} + \frac{\partial\hat{v}}{\partial y} + i\beta\hat{w}\right)$ and $cond.$ the thermal conduction term $\frac{\gamma}{RPr}\frac{d^2\left(\mu\hat{T}\right)}{dy^2}$

The continuity equation is retained and rewritten herein, as shown below

$$\gamma Ma_e^2(i\alpha U - i\omega)\hat{p} - (i\alpha U - i\omega)\frac{\hat{T}}{T} + i\alpha\hat{u} + \frac{\partial\hat{v}}{\partial y} + i\beta\hat{w} - \frac{dT}{dy}\frac{\hat{v}}{T} = 0 \quad (5.91)$$

Equations (5.87)–(5.91) are the simplified linear stability equations. In the following sections, the disturbance energy Eq. (5.90) is frequently utilized. Therefore, it is necessary to interpret the meaning of each term of Eq. (5.90). For convenience, all the energy terms henceforth represent those in disturbance flow fields. On the left side of Eq. (5.90), $-i\omega\frac{\hat{T}}{T}$. denotes the time rate of change of internal energy, and $i\alpha U\frac{\hat{T}}{T}$ is the heat transport by the mean flow. These two terms are significant for the change of internal energy in a finite control volume and their sum $\frac{1}{T}(i\alpha U - i\omega)\hat{T}$ can be deemed as the total time rate of change of internal energy. On the right side of Eq. (5.90), the first term $-\frac{dT}{dy}\frac{\hat{v}}{T}$ is the heat transport by the wall-normal convection, the second term $-(\gamma - 1)\left(i\alpha\hat{u} + \frac{\partial\hat{v}}{\partial y} + i\beta\hat{w}\right)$ represents the rate of change of volume due to dilatation fluctuations, and the last term $\frac{1}{RPr}\frac{d^2}{dy^2}\left(\mu\hat{T}\right)$ denotes the thermal conduction. The viscous dissipation is neglected in Eq. (5.90) due to its rather modest contribution.

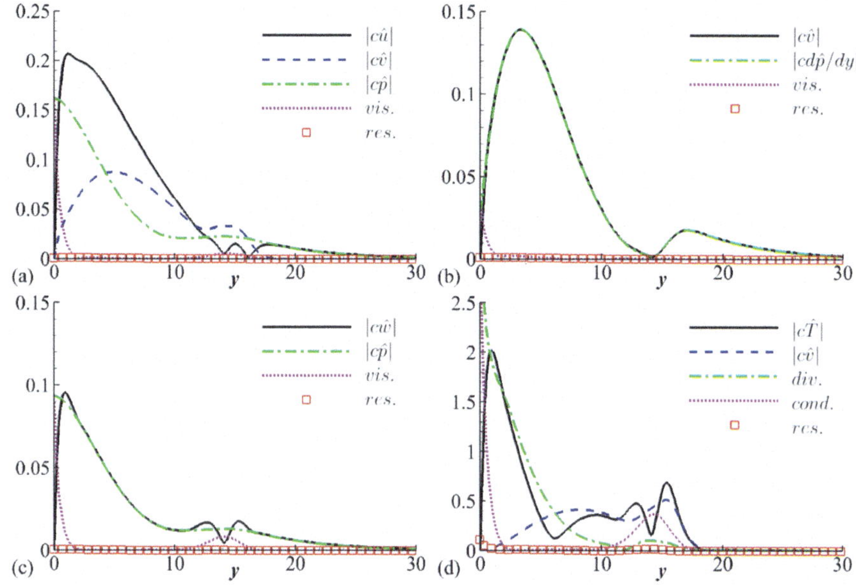

Fig. 5.39 Magnitude of the terms in Eqs. (5.83)–(5.86) aat $\omega = 0.152$ with $\arctan\left(\frac{\beta}{\alpha_i}\right) = 30^o$ and eigenvalue of$\alpha = 0.164 - 0.0017i$: **a** Eq. (5.83), **b** Eq. (5.84), **c** Eq. (5.85) and **d** Eq. (5.86). Here c denotes the coefficient of each components, *vis.* the viscous term in the disturbance momentum equations, *res.* the sum of the absolutes of the residual terms, *div.* the term $-(\gamma - 1)\left(i\alpha\hat{u} + \frac{\partial\hat{v}}{\partial y} + i\beta\hat{w}\right)$ and *cond.* the thermal conduction term $\frac{\gamma}{RPr}\frac{d^2(\mu\hat{T})}{dy^2}$

It should be noted that $-\frac{dT}{dy}\frac{\hat{v}}{T}$ is a source term that draws energy from the mean-flow field to the disturbance-flow field. In the current boundary layer, $\frac{dT}{dy} \leq 0$ (the equality holds at the wall and the upper boundary layer edge), therefore, for a control volume in the boundary layer, the flow moving upwards brings high-temperature fluid elements into it, while the flow moving downwards brings low-temperature fluid elements into it. In this work, the simplified linear stability equations are not used to solve the eigenproblem, but to analyze the energy exchange in the boundary layer. All the eigenmodes are obtained still by resolving linear stability equations.

5.3.3 Growth Mechanisms of the First and Second Modes

Unnikrishnan and Gaitonede [7] analyzed the energy interactions among the acoustic, vortical and thermal components and found that the thermal component plays a great role in the growth of acoustic waves. With respect to the most unstable mode in the current boundary layer, the internal energy fluctuations around the critical layer (the phase speed of disturbances is equal to local mean-flow speed) and near the wall

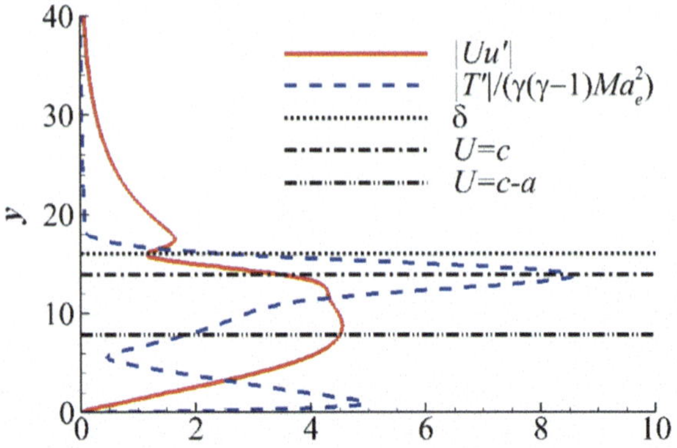

Fig. 5.40 Comparison of the kinetic ($|Uu'|$) and internal energy ($|T'|/[\gamma(\gamma-1)Ma_e^2]$) fluctuations of the mode at $\omega = 0.152$ with $\alpha = 0.164 - 0.004i$. Here δ denotes the boundary layer thickness, $U = c$ represents the critical layer and c is the phase speed, and $U = c - a$ is the sonic line and a is the local speed of sound

are comparable to the kinetic energy fluctuations sandwiched between the peaks of internal energy fluctuations, as shown in Fig. 5.40. Therefore, it can be speculated that the internal energy fluctuations may have an influence on the local instabilities.

It is well known that the Rijke tube can turn heat into sound [18]. According to Rayleigh's criterion, the excitation of acoustic waves by heat addition depends on the phase of the vibration at which heat transfer occurs. The statements are given below:

> If heat be communicated to, and abstracted from, a mass of air vibrating (for example) in a cylinder bounded by a piston, the effect produced will depend upon the phase of the vibration at which the transfer of heat takes place. If heat be given to the air at the moment of greatest condensation, or be taken from it at the moment of greatest rarefaction, the vibration is encouraged. On the other hand, if heat be given at the moment of greatest rarefaction, or abstracted at the moment of greatest condensation, the vibration is discouraged. [19]

This motivates us to consider the phase relations of the energy terms in the boundary layer instabilities by examining the magnitude and phase of the terms of the simplified disturbance energy Eq. (5.90).

5.3.3.1 Second-Mode Instability

We first investigate the mechanisms of the second mode instabilities in the boundary layer over the smooth solid wall. Here we consider 2D instabilities as 2D waves are more unstable than 3D waves in terms of the second mode. We choose the modes belonging to the second mode family at different frequencies $\omega = 0.13$, 0.152, and 0.17 for comparison. Their eigenvalues are $\alpha = 0.13884 - 0.00048i$, $0.16406 - 0.00401i$, and $0.18543 - 0.0019i$, respectively. The corresponding eigenvectors are substituted into the simplified disturbance energy Eq. (5.90). Figure 5.41 demonstrate the profiles of the magnitude and phase of the left-hand side terms of Eq. (5.90) and Fig. 5.42 illustrates those of the right-hand side terms of Eq. (5.90). With respect to each eigenvector, all the components are normalized by the pressure component at the wall.

The magnitudes of $-i\omega\frac{\hat{T}}{T}$ and $i\alpha U\frac{\hat{T}}{T}$ shown in the left plots of Fig. 5.41 are considerably large in the vicinity of the critical layer ($y \approx 14$ for the current three cases). As their magnitudes are comparable in this region, the magnitude of their sum $(i\alpha U - i\omega)\frac{\hat{T}}{T}$ is fairly small. It indicates the convection by the mean flow is dominant for the internal energy variation in the region around the critical layer. Since $(i\alpha U - i\omega)\frac{\hat{T}}{T} = i\alpha_r(U - c)\frac{\hat{T}}{T} - \alpha_i U\frac{\hat{T}}{T}$ and $\alpha_r \gg -\alpha_i$, the sign alteration of

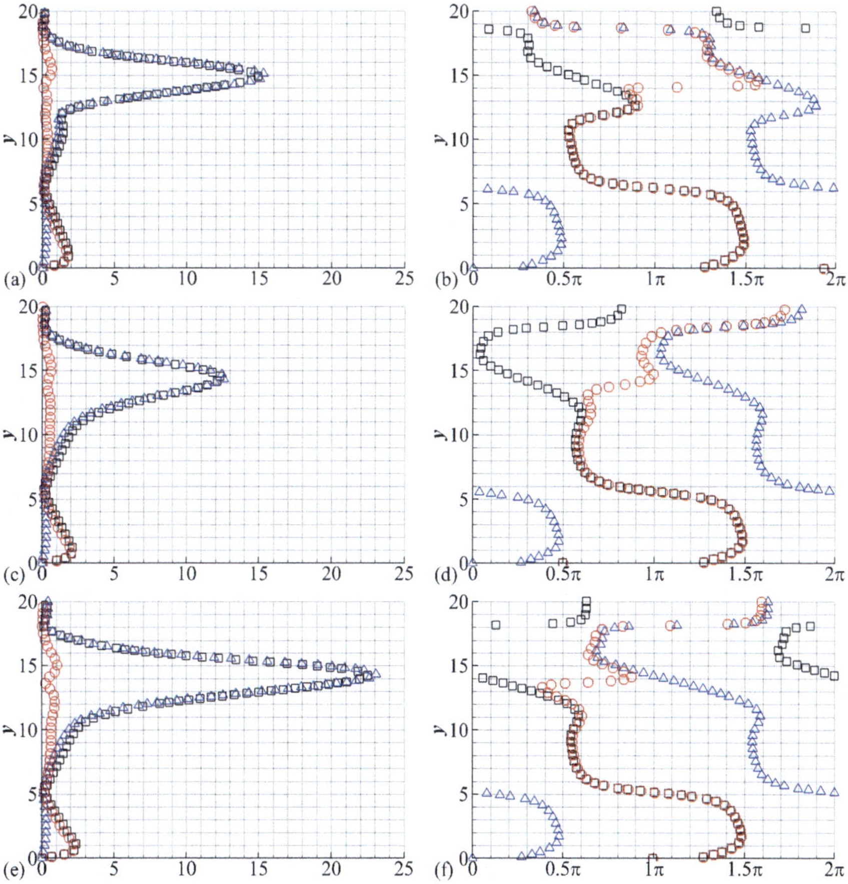

Fig. 5.41 Magnitude (left-hand plots) and phase (right-hand plots) of $-i\omega\frac{\hat{T}}{T}$ (□), $i\alpha U\frac{\hat{T}}{T}$ (△) and $(i\alpha U - i\omega)\frac{\hat{T}}{T}$ (○) under different ω: **a**, **b** $\omega = 0.13$, $\alpha = 0.13884 - 0.00048i$; **c**, **d** $\omega = 0.152$, $\alpha = 0.16406 - 0.00401i$; **e**, **f** $\omega = 0.17$, $\alpha = 0.18543 - 0.0019i$

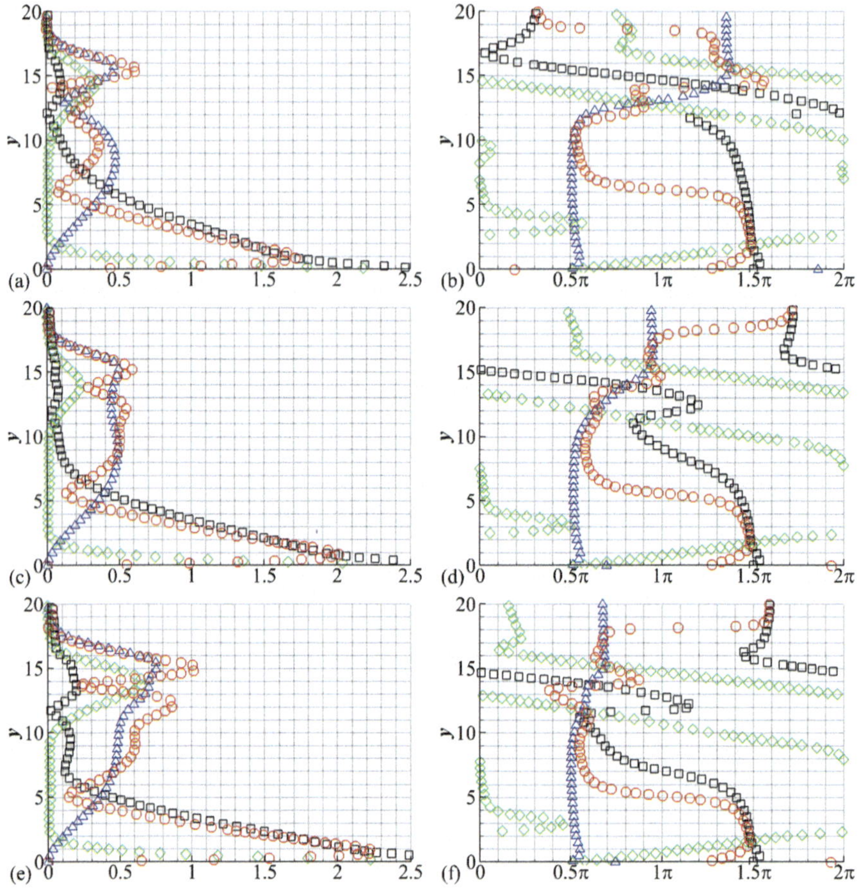

Fig. 5.42 Magnitude (left-hand plots) and phase (right-hand plots) of $-(\gamma-1)\left(i\alpha\hat{u}+\frac{\partial\hat{v}}{\partial y}\right)$(□), $-\frac{dT}{dy}\frac{\hat{v}}{T}$ (△), $\frac{\gamma}{RPr}\frac{d^2}{dy^2}(\mu\hat{T})$ (◇) and $(i\alpha U-i\omega)\frac{\hat{T}}{T}$ (○) under different ω: **a**, **b** $\omega=0.13, \alpha=0.13884-0.00048i$; **c**, **d** $\omega=0.152, \alpha=0.16406-0.004i$; **e**, **f** $\omega=0.17, \alpha=0.18543-0.0019i$

$U-c$ accounts for a phase shift of $(i\alpha U-i\omega)\frac{\hat{T}}{T}$ across the critical layer, as shown in the right plots of Fig. 5.41.

Below the critical layer, on the left side of Eq. (5.90), $(i\alpha U-i\omega)\frac{\hat{T}}{T}$ is in phase with $-i\omega\frac{\hat{T}}{T}$ and has a new phase shift, as depicted in the right plots of Fig. 5.41. This phase shift corresponds to a valley in the magnitude profile of $(i\alpha U-i\omega)\frac{\hat{T}}{T}$ shown in the left plots of Fig. 5.41. Below this valley, the thermal convection $i\alpha U\frac{\hat{T}}{T}$ is negligible and hence $(i\alpha U-i\omega)\frac{\hat{T}}{T}\sim -i\omega\frac{\hat{T}}{T}$. On the right side of Eq. (5.90), below the phase shift of $(i\alpha U-i\omega)\frac{\hat{T}}{T}$, $(i\alpha U-i\omega)\frac{\hat{T}}{T}$ basically has the same magnitude and phase with $-(\gamma-1)\left(i\alpha\hat{u}+\frac{\partial\hat{v}}{\partial y}\right)$, excluding the region very close to the wall. Therefore, the dilatation fluctuation accounts for the variation of internal energy. This conclusion agrees with the experimental findings by Zhu et al. [20] that dilatation fluctuations ruled by the second mode generate intense aerodynamic heating.

Against the wall, $\frac{\gamma}{RPr}\frac{d^2}{dy^2}(\mu\hat{T})$ has identical magnitude and counter phase with $-(\gamma-1)\left(i\alpha\hat{u}+\frac{\partial\hat{v}}{\partial y}\right)$. Indeed, the thermal conduction $\frac{\gamma}{RPr}\frac{d^2}{dy^2}(\mu\hat{T})$ is passively induced by the near-wall thermal fluctuations due to the fluctuating thermal expansion and contraction. In the following studies on the second-mode instabilities, the effect of $\frac{\gamma}{RPr}\frac{d^2}{dy^2}(\mu\hat{T})$ will not be considered.

Figure 5.42d shows, of the case $\omega = 0.152$ which has the largest growth rate among these three cases, $-\frac{dT}{dy}\frac{\hat{v}}{T}$ is nearly in phase with $(i\alpha U - i\omega)\frac{\hat{T}}{T}$ throughout the phase shift of $(i\alpha U - i\omega)\frac{\hat{T}}{T}$ crossing the critical layer. Moreover, Fig. 5.42c shows $-\frac{dT}{dy}\frac{\hat{v}}{T}$ is on the same order of magnitude with $(i\alpha U - i\omega)\frac{\hat{T}}{T}$ on both the two sides of the critical layer. As $-\alpha_i$ cannot be negligible at the critical layer, its value is significant for the phase of $(i\alpha U - i\omega)\frac{\hat{T}}{T}$ in the vicinity of the critical layer. The large value of $-\alpha_i$ in this case is associated with the contribution of the convective heat transfer $-\frac{dT}{dy}\frac{\hat{v}}{T}$ to the internal energy perturbations.

By contrast, of the cases $\omega = 0.13$ and $\omega = 0.17$ which have lower growth rates compared with the case $\omega = 0.152$, the phase of $(i\alpha U - i\omega)\frac{\hat{T}}{T}$ diverges from the phase of $-\frac{dT}{dy}\frac{\hat{v}}{T}$ manifestly across the critical layer, as demonstrated in Fig. 5.42b and f. As a result, the heat transport by the wall-normal convection does not synchronize with the total time rate of change of internal energy around the critical layer, which diminishes the contribution of the heat transport by the wall-normal convection to the local internal energy fluctuations.

Therefore, it could be concluded that the contribution of the convective heat transfer $-\frac{dT}{dy}\frac{\hat{v}}{T}$ to the internal energy fluctuations accounts for the amplification of the second-mode instabilities. When it is in phase with $(i\alpha U - i\omega)\frac{\hat{T}}{T}$ in the vicinity of the critical layer, the local internal fluctuations are promoted by the energy supply from the mean flow, which results in a large growth rate.

5.3.3.2 First-Mode Instability

In terms of the first mode, we continue to conduct phase analysis on the simplified disturbance energy equation. Likewise, we focus on 2D instabilities in the smooth-solid-wall case. The 3D instabilities will be analyzed in the following sections. We choose three non-dimensional angular frequencies: $\omega = 0.03$, 0.063 and 0.09. and the corresponding eigenvalues are $\alpha = 0.0348 - 0.0002i$, $0.07068 - 0.00039i$ and $0.09913-0.00028i$, respectively. These three eigenmodes all belong to the first mode family. The magnitude and phase profiles of the left-hand side terms of Eq. (5.90) are depicted in Fig. 5.43 and those of the right-hand side terms of Eq. (5.90) are depicted in Fig. 5.43, respectively.

Identical to the second mode, the magnitudes of $-i\omega\frac{\hat{T}}{T}$ and $i\alpha U\frac{\hat{T}}{T}$ are remarkably large across the critical layer ($y \approx 13$ for the current three cases), and only $-i\omega\frac{\hat{T}}{T}$ is

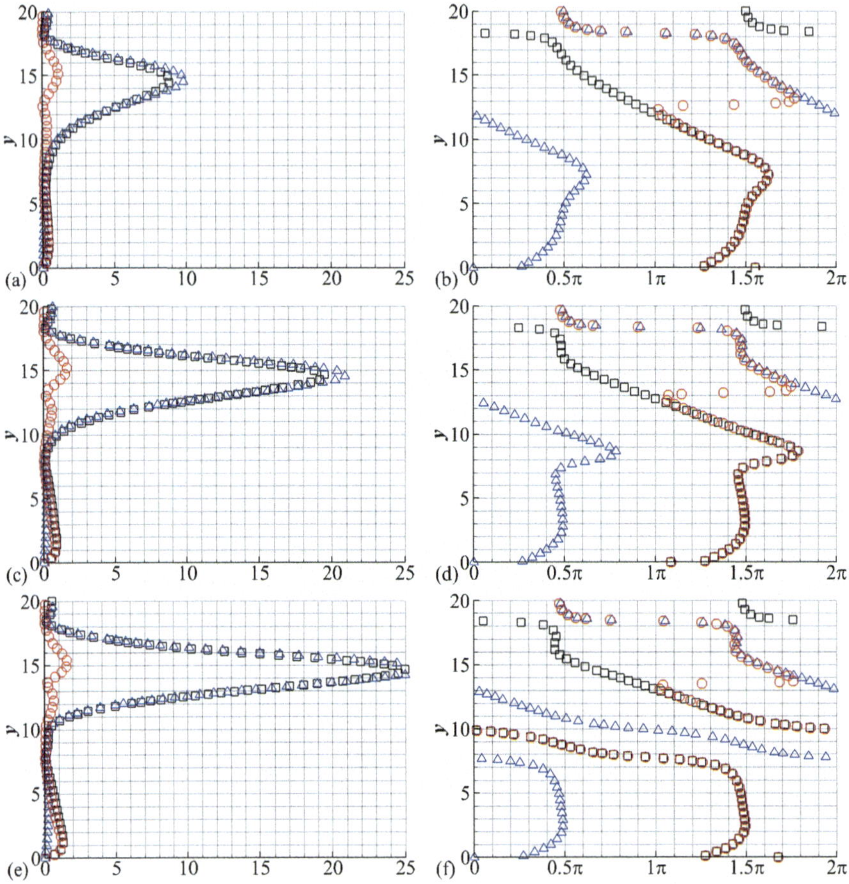

Fig. 5.43 Magnitude (left-hand plots) and phase (right-hand plots) of $-i\omega\frac{\hat{T}}{T}$ (□), $i\alpha U\frac{\hat{T}}{T}$ (△) and $(i\alpha U - i\omega)\frac{\hat{T}}{T}$ (○) under different ω: **a**, **b** $\omega = 0.03$, $\alpha = 0.0348 - 0.0002i$; **c**, **d** $\omega = 0.063$, $\alpha = 0.07068 - 0.00039i$; **e**, **f** $\omega = 0.09$, $\alpha = 0.09913 - 0.00028i$

significant near the wall, as shown in the left-hand plots of Fig. 5.43. As $-\alpha_i$ is fairly small in these three cases, $(i\alpha U - i\omega)\frac{\hat{T}}{T}$ is in phase with $i\alpha U\frac{\hat{T}}{T}$ above the critical layer and with $-i\omega\frac{\hat{T}}{T}$ below the critical layer, as shown in the right-hand plots of Fig. 5.43. Figure 5.43b, d and f demonstrate that across the critical layer the phase of $-\frac{dT}{dy}\frac{\hat{v}}{T}$ differs from that of $(i\alpha U - i\omega)\frac{\hat{T}}{T}$ evidently, which indicates the heat transport by the wall-normal convection slightly contributes to the internal energy fluctuations in the vicinity of the critical layer.

A little below the critical layer, the phase of $-(\gamma - 1)\left(i\alpha\hat{u} + \frac{\partial\hat{v}}{\partial y}\right)$ superposes on that of $(i\alpha U - i\omega)\frac{\hat{T}}{T}$ in a certain range along y coordinate, as shown in the right plots of Fig. 5.44. Moreover, the width of the phase superposition is roughly proportional to the growth rate $-\alpha_i$. We conjecture this phase superposition is associated with the amplification of the first mode wave. According to Rayleigh's criterion, the excitation of the acoustic wave emerges as the time rate of change of heat adding synchronizes with that of the condensation. It is perfectly held by the phase overlap of $-(\gamma-1)\left(i\alpha\hat{u}+\frac{\partial\hat{v}}{\partial y}\right)$ and $(i\alpha U-i\omega)\frac{\hat{T}}{T}$ herein. Therefore, the phase overlap of $(\gamma-1)\left(i\alpha\hat{u}+\frac{\partial\hat{v}}{\partial y}\right)$ and $(i\alpha U-i\omega)\frac{\hat{T}}{T}$ in a wide region below the critical layer facilitates the local fluctuations.

Against the wall, the thermal conduction $\frac{\gamma}{RPr}\frac{d^2}{dy^2}(\mu\hat{T})$ sustains the near-wall dilatation fluctuation, which is the same as the second mode instability. The thermal conduction $\frac{\gamma}{RPr}\frac{d^2}{dy^2}(\mu\hat{T})$ is passively generated by the non-uniform temperature

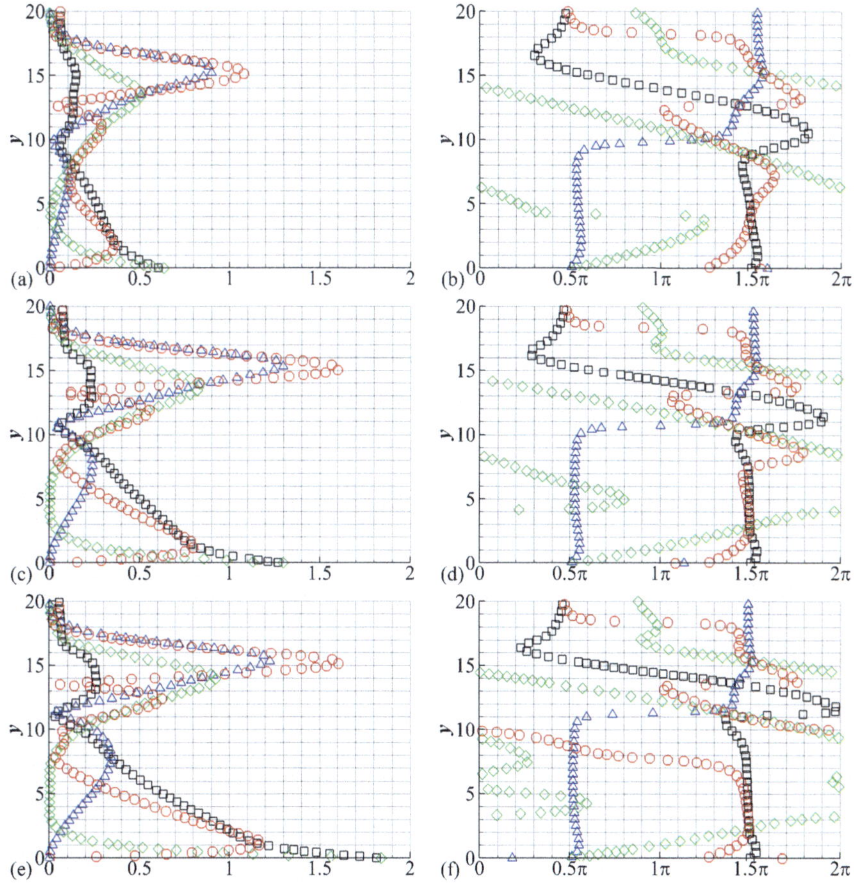

Fig. 5.44 Magnitude (left-hand plots) and phase (right-hand plots) of $-(\gamma-1)\left(i\alpha\hat{u}+\frac{\partial\hat{v}}{\partial y}\right)$(□), $-\frac{dT}{dy}\frac{\hat{v}}{T}$ (△), $\frac{\gamma}{RPr}\frac{d^2}{dy^2}(\mu\hat{T})$ (◇) and $(i\alpha U-i\omega)\frac{\hat{T}}{T}$ (○) under different ω: **a**, **b** $\omega=0.03$, $\alpha=0.0348-0.0002i$; **c**, **d** $\omega=0.063$, $\alpha=0.07068-0.00039i$; **e**, **f** $\omega=0.09$, $\alpha=0.09913-0.00028i$

fluctuations. In the following investigations on the first mode, the interactions in $(i\alpha U - i\omega)\frac{\hat{T}}{T}$, $-\frac{dT}{dy}\frac{\hat{v}}{T}$, and $-(\gamma - 1)\left(i\alpha\hat{u} + \frac{\partial\hat{v}}{\partial y}\right)$ will be further studied.

5.3.4 *Mechanisms of the Effects of Acoustic Metasurfaces on Unstable Modes*

This section is to investigate the mechanisms of the effects of porous walls on the first and second modes using phase analysis on the simplified linear stability equations. The mechanisms of the instabilities of the oblique first mode waves are also analyzed.

5.3.4.1 Stabilization and Destabilization of the Second Mode

According to the foregoing LST analyses, the stabilization effect of porous walls on the second mode is efficient when the admittance phase is close to π, and the destabilization effect of porous wall occurs as the admittance has a positive real part. Here the admittance of $A = -4$ and $A = 4$ are utilized to investigate the mechanisms of the stabilization and destabilization of the second mode. The non-dimensional angular frequency $\omega = 0.152$ is chosen. And the eigenvalues are $\alpha = 0.16406 - 0.004i$, $\alpha = 0.1646 - 0.00165i$ and $\alpha = 0.16251 - 0.01278i$ for the smooth solid wall $A = 0$, the porous wall with $A = -4$, the porous wall with $A = 4$, respectively. The profiles of the magnitude and phase angle of the terms $-(\gamma - 1)\left(i\alpha\hat{u} + \frac{\partial\hat{v}}{\partial y}\right)$, $-\frac{dT}{dy}\frac{\hat{v}}{T}$, and $(i\alpha U - i\omega)\frac{\hat{T}}{T}$ in the porous-wall cases and the smooth-solid-wall case are demonstrated in Fig. 5.45.

Firstly, we focus on the phase overlap between $-\frac{dT}{dy}\frac{\hat{v}}{T}$ and $(i\alpha U - i\omega)\frac{\hat{T}}{T}$ around the critical layer ($y \approx 14$). In the stabilization case ($A = -4$), the phase of $(i\alpha U - i\omega)\frac{\hat{T}}{T}$ crossing the critical layer shifts swiftly and it differs from the phase of $-\frac{dT}{dy}\frac{\hat{v}}{T}$ evidently. By contrast, in the destabilization case ($A = 4$), the phase of $(i\alpha U - i\omega)\frac{\hat{T}}{T}$ changes slowly across the critical layer and is completely superposed on by the phase of $-\frac{dT}{dy}\frac{\hat{v}}{T}$. Recalling the growth rates of these three cases, we again find the critical-layer phase overlap of $-\frac{dT}{dy}\frac{\hat{v}}{T}$ and $(i\alpha U - i\omega)\frac{\hat{T}}{T}$ is positively correlated with the growth rate $-\alpha_i$, which confirms the amplification of the second mode depends upon the contribution of the thermal convection $-\frac{dT}{dy}\frac{\hat{v}}{T}$ to the local internal energy fluctuations.

Below the critical layer, the phase relation between $-\frac{dT}{dy}\frac{\hat{v}}{T}$ and $-(\gamma - 1)\left(i\alpha\hat{u} + \frac{\partial\hat{v}}{\partial y}\right)$ is significantly affected by the porous walls. In the baseline case, $-\frac{dT}{dy}\frac{\hat{v}}{T}$ has a counter phase (i.e. phase difference of π) with $-(\gamma - 1)\left(i\alpha\hat{u} + \frac{\partial\hat{v}}{\partial y}\right)$ near the wall, as shown in Fig. 5.45b, which means the

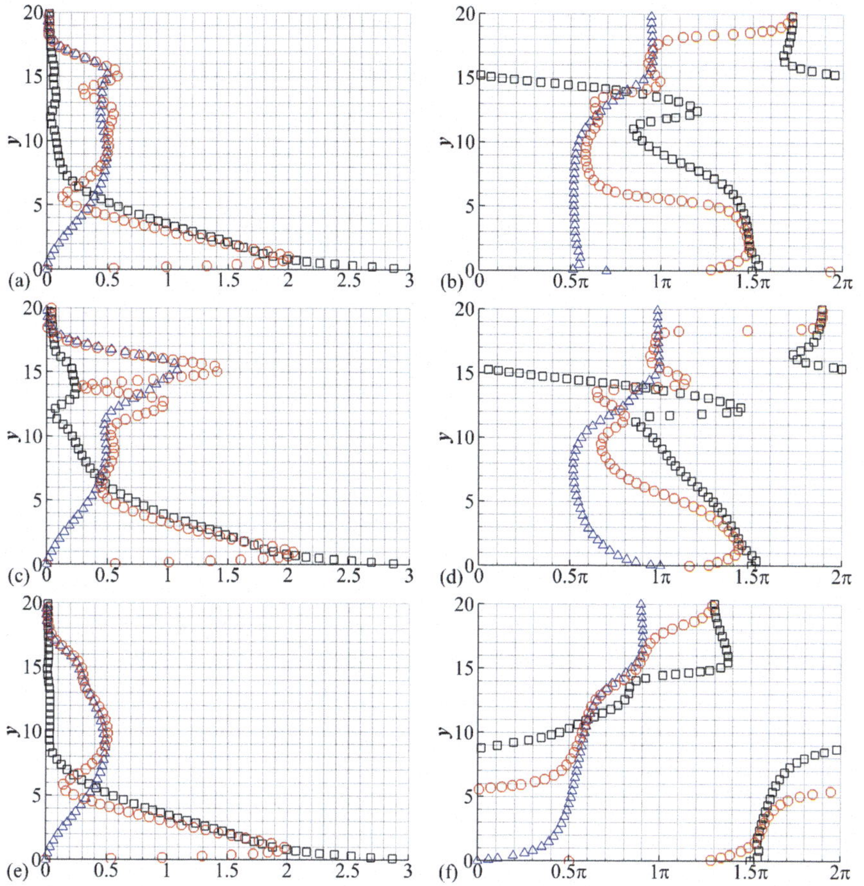

Fig. 5.45 Magnitude (left-hand plots) and phase (right-hand plots) of $-(\gamma-1)\left(i\alpha\hat{u}+\frac{\partial\hat{v}}{\partial y}\right)(\square)$, $-\frac{dT}{dy}\frac{\hat{v}}{T}$ ($\triangle$), and $(i\alpha U - i\omega)\frac{\hat{T}}{T}$ ($\circ$) under different walls at $\omega = 0.152$: **a**, **b** smooth solid wall, $\alpha = 0.16406 - 0.004i$; **c**, **d** porous wall with $A = -4$, $\alpha = 0.1646 - 0.00165i$; **e**, **f** porous wall with $A = 4$, $\alpha = 0.16251 - 0.01278i$

upward flow and downward flow correspond to the thermal expansion and contraction, respectively. For the disturbance defined by $\phi = \psi(y)e^{i(\alpha x+\beta z-\omega t)}$, the change of phase with time increasing is from right to left (i.e., from 2π to 0) in the right plots of Fig. 5.45. In the stabilization case, the phase of $-\frac{dT}{dy}\frac{\hat{v}}{T}$ shifts to the right at the wall, which indicates $\hat{v}$ is delayed by the porous wall. And this delay of $\hat{v}$ at the wall also causes a delay in the near-wall $\hat{v}$ due to the flow continuity. While in the destabilization case, the phase of $-\frac{dT}{dy}\frac{\hat{v}}{T}$ shifts to the left at the wall, which means an advance emerges in the phase of $\hat{v}$. Simlilarly, the phase of $\hat{v}$ in the near-wall region is driven in advance due to the flow continuity.

The near-wall change in the phase of $-\frac{dT}{dy}\frac{\hat{v}}{T}$ also results in variation in the magnitude of $-(\gamma-1)\left(i\alpha\hat{u}+\frac{\partial\hat{v}}{\partial y}\right)$ as these two terms are correlated with each other. As the disturbances around the critical layer interact with the near-wall perturbations mutually, the distortion in the near-wall perturbations results in the variation in the phase discrepancy between $-\frac{dT}{dy}\frac{\hat{v}}{T}$ and $(i\alpha U-i\omega)$ in the vicinity of the critical layer in the porous-wall cases.

5.3.4.2 Frequency Shift of the Second Mode

The LST results show the frequency band of the second mode instability shifts when the employed porous wall has an admittance phase close to 0.5π or 1.5π. Here we utilize the porous walls with the admittances of $A=4\exp(0.5\pi i)$ and $A=4\exp(1.5\pi i)$ to investigate the mechanisms of the frequency shift of the second mode. The non-dimensional angular frequency $\omega=0.152$ is chosen again. The eigenvalues for the smooth solid wall, the porous wall with $A=4\exp(0.5\pi i)$ and porous wall with $A=4\exp(1.5\pi i)$ are $\alpha=0.16406-0.00401i$, $\alpha=0.16689-0.002i$ and $\alpha=0.1621-0.00008i$, respectively. The profiles of the magnitude and phase of the terms $-(\gamma-1)\left(i\alpha\hat{u}+\frac{\partial\hat{v}}{\partial y}\right)$, $-\frac{dT}{dy}\frac{\hat{v}}{T}$ and $(i\alpha U-i\omega)\frac{\hat{T}}{T}$ for different walls are depicted in Fig. 5.46.

From the right plots of Fig. 5.46, we can see that around the critical layer ($y\approx 14$), that the high phase overlap of $-\frac{dT}{dy}\frac{\hat{v}}{T}$ and $(i\alpha U-i\omega)\frac{\hat{T}}{T}$ corresponds to a large growth rate still holds herein. It can also be seen when the admittance phase $\theta=0.5\pi$, the phase profiles of the terms $-(\gamma-1)\left(i\alpha\hat{u}+\frac{\partial\hat{v}}{\partial y}\right)$, $-\frac{dT}{dy}\frac{\hat{v}}{T}$ and $(i\alpha U-i\omega)\frac{\hat{T}}{T}$ approach those of the smooth-solid-wall case at large ω shown in Fig. 5.46f. On the other hand, when $\theta=1.5\pi$, the phase profiles of these three terms are similar to that of the smooth-solid-wall case at small ω shown in Fig. 5.46b. If we perform the analogy to a wide frequency spectrum, then the frequency shift is manifested.

The right-hand plots in Fig. 5.46 show that the phase shift of $-\frac{dT}{dy}\frac{\hat{v}}{T}$ crossing the critical layer in the porous wall cases is plainly distinct from the baseline. Apparently, this phase shift accounts for the critical-layer phase discrepancy between $-\frac{dT}{dy}\frac{\hat{v}}{T}$ and $(i\alpha U-i\omega)\frac{\hat{T}}{T}$. Moveover, phase shift of $-\frac{dT}{dy}\frac{\hat{v}}{T}$ crossing the critical layer is closely associated with the variation in the near-wall perturbations induced by porous walls.

Figure 5.47 compares the magnitude and phase of the wall-normal fluctuating velocity $\hat{v}$ among these three cases. Pertaining to the porous wall with the admittance of $A=4\exp(0.5\pi i)$, the phase of $\hat{v}$ at the wall is identical to that near the wall and meanwhile the magnitude of $\hat{v}$ is non-zero at the wall, hence the porous wall facilitates the magnitude of $\hat{v}$ near the wall. While for the case $A=4\exp(1.5\pi i)$, the disturbance velocity $\hat{v}$ at the wall with non-zero magnitude has a phase contrary to that against the wall, which leads to a decrease in the magnitude of $\hat{v}$ near the wall.

If we assume the phases of $\hat{v}$ beyond the boundary layer edge are uniform among these three cases, the phase of $\hat{v}$ below the critical layer will shift to the right in the case of $A=4\exp(0.5\pi i)$ and to the left in the case of $A=4\exp(1.5\pi i)$, which

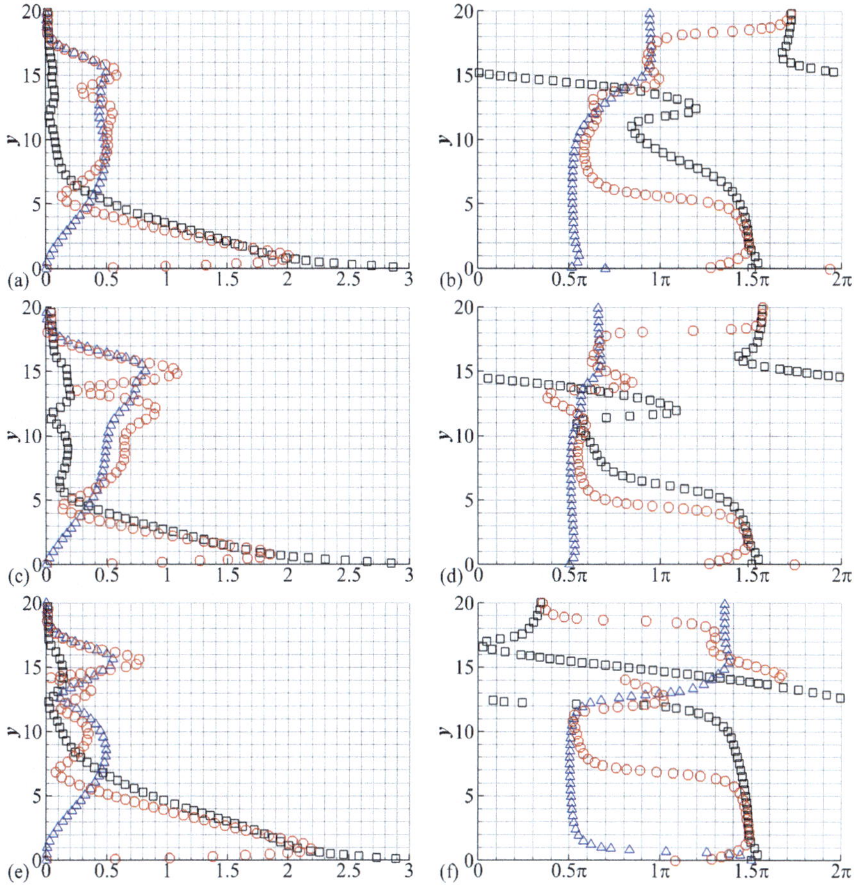

Fig. 5.46 Magnitude (left-hand plots) and phase (right-hand plots) of $-(\gamma - 1)\left(i\alpha\hat{u} + \frac{\partial\hat{v}}{\partial y}\right)$(□), $-\frac{dT}{dy}\frac{\hat{v}}{T}$ (△), and $(i\alpha U - i\omega)\frac{\hat{T}}{T}$ (○) of the smooth-solid-wall case and porous-wall cases at $\omega = 0.152$: **a**, **b** smooth solid wall, $\alpha = 0.16406 - 0.00401i$, **c**, **d** $A = 4\exp(0.5\pi i)$, $\alpha = 0.16689 - 0.002i$, **e**, **f** $A = 4\exp(1.5\pi i)$, $\alpha = 0.1621 - 0.00008i$

correspond a phase delay and phase advance of $\hat{v}$, respectively. Then the frequency shift is characterized by the phase shift of $\hat{v}$. In addition, the phase shift of $\hat{v}$ may be associated with the variation in its magnitude.

5.3.4.3 Stabilization and Destabilization of the First Mode

The stabilization and destabilization of the fist mode emerge when porous walls have admittance phases tending to 0.5π and π, respectively, according to the preceding calculations using LST. Here the porous walls with the admittance of

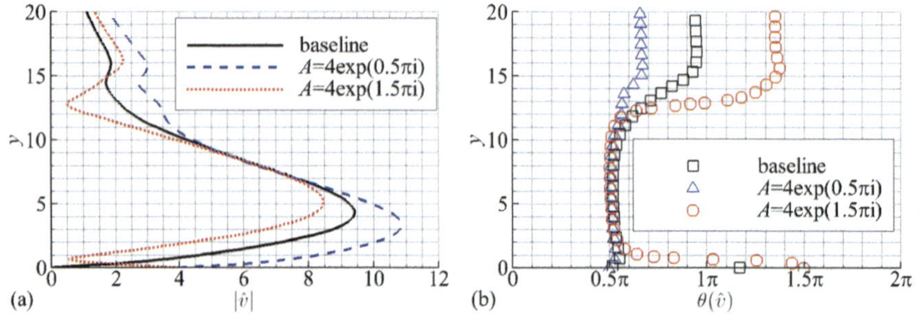

Fig. 5.47 Magnitude (left-hand plots) and phase (right-hand plots) of disturbance velocity $\hat{v}$ at $\omega = 0.152$ under different walls

$A = 4\exp(0.5\pi i)$ and $A = 4\exp(\pi i)$ are utilized to investigate the mechanisms of the stabilization and destabilization of the 2D first mode instabilities, respectively. The non-dimensional frequency $\omega = 0.063$ is chosen. The eigenvalues for the smooth solid wall, porous walls with the admittance of $= 4\exp(0.5\pi i)$ and admittance of $A = 4\exp(\pi i)$ are $\alpha = 0.07068 - 0.00039i$, $\alpha = 0.07025 - 0.00027i$ and $\alpha = 0.07062 - 0.00076i$, respectively. The magnitudes and phases of $(i\alpha U - i\omega)\frac{\hat{T}}{T}$, $-\frac{dT}{dy}\frac{\hat{v}}{T}$ and $-(\gamma - 1)\left(i\alpha\hat{u} + \frac{\partial\hat{v}}{\partial y}\right)$ of these three cases are illustrated in Fig. 5.48.

With respect to the case $A = 4\exp(0.5\pi i)$, the phase discrepancy between $-\frac{dT}{dy}\frac{\hat{v}}{T}$ and $(i\alpha U - i\omega)\frac{\hat{T}}{T}$ around the critical layer ($y \approx 13$) is analogous to that of the smooth-solid-wall case due to the small growth rate. While below the critical layer, the phase superposition of $-(\gamma - 1)\left(i\alpha\hat{u} + \frac{\partial\hat{v}}{\partial y}\right)$ and $(i\alpha U - i\omega)\frac{\hat{T}}{T}$ shrinks in y coordinate compared with the smooth-solid-wall case, as shown in Fig. 5.48b and d. Identical to Fig. 5.48b, d and f, the diminution of this phase superposition corresponds to a decrease in the growth rate. Therefore, it could be concluded that the amplification of the first mode in the current boundary layer depends upon the mutual interaction between the local dilatation fluctuations and internal energy fluctuations.

When the porous wall with the admittance of $A = 4\exp(\pi i)$ is employed, the phase of $-\frac{dT}{dy}\frac{\hat{v}}{T}$ is recast near the wall. Concurrently, the phase discrepancy between $-\frac{dT}{dy}\frac{\hat{v}}{T}$ and $(i\alpha U - i\omega)\frac{\hat{T}}{T}$ in the vicinity of the critical layer is diminished in contrast to the baseline, which promotes the contribution of thermal convection $-\frac{dT}{dy}\frac{\hat{v}}{T}$ to internal energy fluctuations. Therefore, the increase of the growth rate emerging in the case of the admittance phase close to π is associated with heat transport by the wall-normal convection around the critical layer.

5.3.4.4 Oblique First-Mode Wave

The LST results show that the oblique first-mode wave is more unstable than the 2D first-mode wave, which is not converted by porous walls. Here the non-dimensional

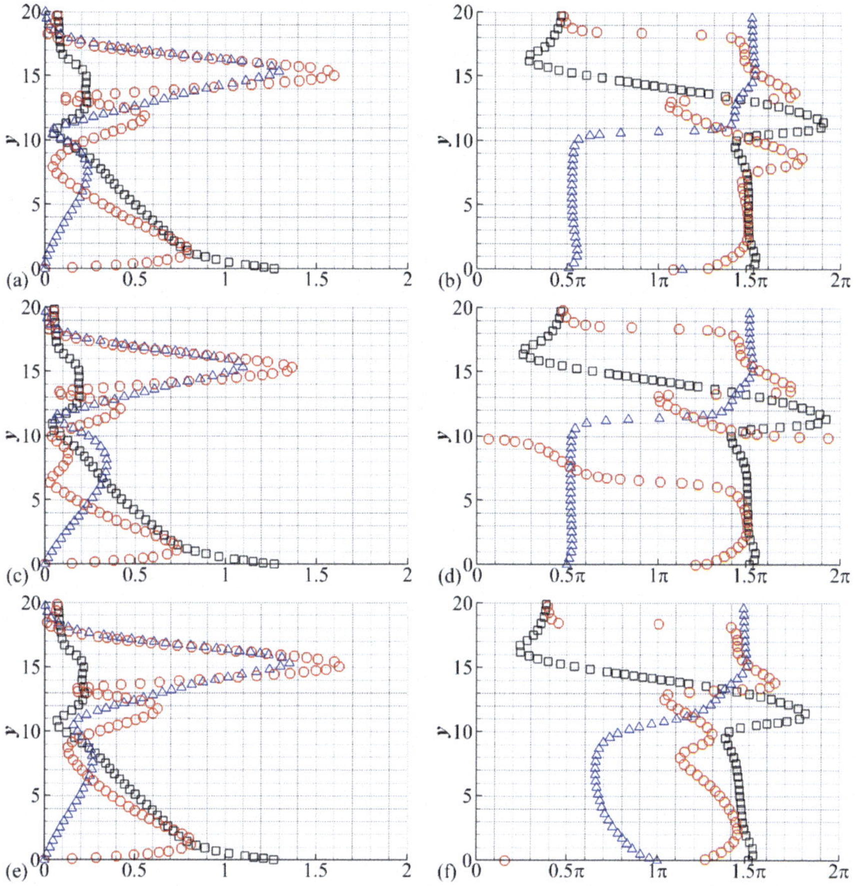

Fig. 5.48 Magnitude (left-hand plots) and phase (right-hand plots) of $-(\gamma-1)\left(i\alpha\hat{u}+\frac{\partial\hat{v}}{\partial y}\right)$(□), $-\frac{dT}{dy}\frac{\hat{v}}{T}$ (△), and $(i\alpha U-i\omega)\frac{\hat{T}}{T}$ (○) in the smooth-solid-wall case and porous-wall cases at $\omega=0.063$: **a**, **b** the smooth solid wall, $\alpha=0.07068-0.00039i$ **c**, **d** the porous wall with $A=4\exp(0.5\pi i)$, $\alpha=0.07025-0.00027i$, **e**, **f** the porous wall with $A=4\exp(\pi i)$, $\alpha=0.07062-0.00076i$

angular frequency $\omega=0.063$ is adopted again to compare the oblique-wave case with the 2D-wave case in the interactions of $-(\gamma-1)\left(i\alpha\hat{u}+\frac{\partial\hat{v}}{\partial y}+i\beta\hat{w}\right)$, $-\frac{dT}{dy}\frac{\hat{v}}{T}$ and $(i\alpha U-i\omega)\frac{\hat{T}}{T}$ to investigate the mechanisms of the growth of the oblique first-mode instability. The propagation angle of the oblique wave $\arctan\left(\frac{\beta}{\alpha_r}\right)=45^{\circ}$ is chosen herein. The eigenvalues of the oblique wave and 2D wave are $\alpha=0.07123-0.00103i$ and $\alpha=0.07068-0.00039i$, respectively. The magnitude and phase profiles of $-(\gamma-1)\left(i\alpha\hat{u}+\frac{\partial\hat{v}}{\partial y}+i\beta\hat{w}\right)$, $-\frac{dT}{dy}\frac{\hat{v}}{T}$ and $(i\alpha U-i\omega)\frac{\hat{T}}{T}$ of the two cases are illustrated in Fig. 5.49.

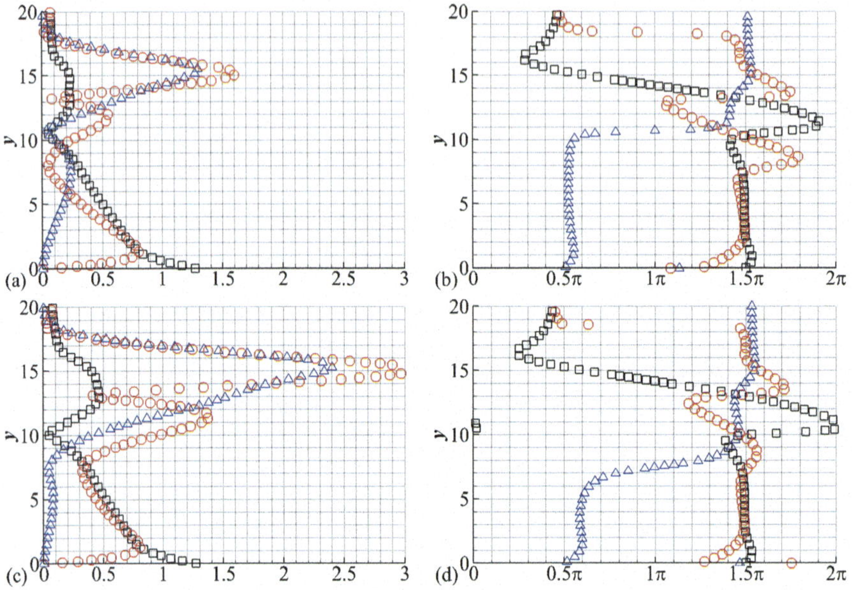

Fig. 5.49 Magnitude (left-hand plots) and phase (right-hand plots) of $-(\gamma-1)\left(i\alpha\hat{u}+\frac{\partial\hat{v}}{\partial y}+i\beta\hat{w}\right)$(□), $-\frac{dT}{dy}\frac{\hat{v}}{T}$ (△), and $(i\alpha U-i\omega)\frac{\hat{T}}{T}$ (○) in the 2D-wave and oblique-wave cases at $\omega=0.063$: **a**, **b** 2D wave, $\alpha=0.07068-0.00039i$ **c**, **d** oblique wave with propagation angle of $\arctan\left(\frac{\beta}{\alpha_r}\right)=45^{\circ}$, $\alpha=0.07123-0.00103i$

It is evident that around the critical layer the peak magnitude of $-\frac{dT}{dy}\frac{\hat{v}}{T}$ in the oblique-wave case is much larger than that in the 2D-wave case, as shown in Fig. 5.49a and c. By comparing Fig. 5.49b and d, it can be observed that in the vicinity of the critical layer, the phase discrepancy between $-\frac{dT}{dy}\frac{\hat{v}}{T}$ and $(i\alpha U-i\omega)\frac{\hat{T}}{T}$ of the oblique-wave case is relatively smaller than that of the 2D-wave case, meaning more contribution to the local internal energy fluctuations is provided by the thermal convection $-\frac{dT}{dy}\frac{\hat{v}}{T}$. On the other hand, the phase superposition of $(i\alpha U-i\omega)\frac{\hat{T}}{T}$ and $-(\gamma-1)\left(i\alpha\hat{u}+\frac{\partial\hat{v}}{\partial y}+i\beta\hat{w}\right)$ below the critical layer is wider in the oblique-wave case than the 2D-wave case, as shown in Fig. 5.49b and d. It reveals the mutual interactions between the internal energy fluctuation and dilatation fluctuation are promoted in the oblique-wave case.

Consequently, the large growth rate in the oblique-wave case may be associated with the phase discrepancy between $-\frac{dT}{dy}\frac{\hat{v}}{T}$ and $(i\alpha U-i\omega)\frac{\hat{T}}{T}$ in the vicinity of the critical layer and phase overlap between $(i\alpha U-i\omega)\frac{\hat{T}}{T}$ and $-(\gamma-1)\left(i\alpha\hat{u}+\frac{\partial\hat{v}}{\partial y}+i\beta\hat{w}\right)$ below the critical layer.

References

1. Reynolds O (1895) On the dynamical theory of incompressible viscous fluids and the determination of the criterion. Phil Trans R Soc A 186:123–164
2. Kuehl JJ (2018) Thermoacoustic interpretation of second-mode instability. AIAA J 56(9):3585–3592
3. Tian X, Wen C (2021) Growth mechanisms of second-mode instability in hypersonic boundary layers. J Fluid Mech 908
4. Rayleigh J, Lindsay RB (1945) The theory of sound. Dover Puclications
5. Kovásznauy L (1953) Turbulence in supersonic flow. J Aeronaut Sci 20:657–682. Art no 10. https://doi.org/10.2514/8.2793
6. Zhong X, Wang X (2012) Direct numerical simulation on the receptivity, instability, and transition of hypersonic boundary layers. Annu Rev Fluid Mech 44(1):527–561. https://doi.org/10.1146/annurev-fluid-120710-101208
7. Unnikrishnan S, Gaitonde DV (2019) Interactions between vortical, acoustic and thermal components during hypersonic transition. J Fluid Mech 868:611–647. https://doi.org/10.1017/jfm.2019.176
8. Doak PE (1989) Momentum potential theory of energy flux carried by momentum fluctuations. J Sound Vib 131(1):67–90. https://doi.org/10.1016/0022-460x(89)90824-9
9. Bitter N, Shepherd J (2015) Stability of highly cooled hypervelocity boundary layers. J Fluid Mech 778:586–620
10. Knisely CP, Zhong X (2018) Supersonic modes in hot-wall hypersonic boundary layers with thermochemical nonequilibrium effects. In: 2018 AIAA Aerospace Sciences Meeting, p 2085
11. Knisely CP, Zhong X (2018) Significant supersonic modes and the wall temperature effect in hypersonic boundary layers. AIAA J 57(4):1552–1566
12. Knisely CP, Zhong X (2019) Sound radiation by supersonic unstable modes in hypersonic blunt cone boundary layers. I. Linear stability theory. Phys Fluids 31(2):024103
13. Knisely CP, Zhong X (2019) Sound radiation by supersonic unstable modes in hypersonic blunt cone boundary layers. II. Direct numerical simulation. Phys Fluids 31(2):024104
14. Chuvakhov PV, Fedorov AV (2016) Spontaneous radiation of sound by instability of a highly cooled hypersonic boundary layer. J Fluid Mech 805:188–206
15. Mortensen CH (2018) Toward an understanding of supersonic modes in boundary-layer transition for hypersonic flow over blunt cones. J Fluid Mech 846:789–814
16. Unnikrishnan S, Gaitonde DV (2020) Instability characteristics of cooled hypersonic boundary layers. In: AIAA Scitech 2020 Forum, p 0588
17. Fedorov AV, Tumin A (2011) High-speed boundary-layer instability: old terminology and a new framework. AIAA J 49(8):1647–1657. https://doi.org/10.2514/1.j050835
18. Rijke PL (1859) Notice of a new method of causing a vibration of the air contained in a tube open at both ends. Lond Edinb Dublin Philos Mag J Sci 17(116):419–422
19. Strutt JW, Rayleigh B (1945) The theory of sound, vol 2. Dover
20. Zhu Y et al (2018) Aerodynamic heating in transitional hypersonic boundary layers: role of second-mode instability. Phys Fluids 30(1)

The manufacturer's authorised representative in the EU is Springer Nature Customer Service Centre GmbH, Europaplatz 3, 69115 Heidelberg, Germany. If you have any concerns regarding our products, please contact ProductSafety@springernature.com

Printed and bound by CPI Group (UK) Ltd, Croydon, CR0 4YY
01/10/2026
02225962-0002